A Level
Physics
for OCR

A

Year 1 and AS

Series Editor: Gurinder C

Graham Bone • Gurinder Chad rs

OXFORD
UNIVERSITY PRESS

OXFORD
UNIVERSITY PRESS

Great Clarendon Street, Oxford, OX2 6DP, United Kingdom

Oxford University Press is a department of the University of Oxford. It furthers the University's objective of excellence in research, scholarship, and education by publishing worldwide. Oxford is a registered trade mark of Oxford University Press in the UK and in certain other countries

British Library Cataloguing in Publication Data
Data available

978-0-19-835217-4

10 9 8

Paper used in the production of this book is a natural, recyclable product made from wood grown in sustainable forests. The manufacturing process conforms to the environmental regulations of the country of origin.

Printed and bound by CPI Group (UK) Ltd, Croydon, CR0 4YY

This resource is endorsed by OCR for use with specification H156 AS Level GCE Physics A and year 1 of H556 A Level GCE Physics A. In order to gain endorsement this resource has undergone an independent quality check. OCR has not paid for the production of this resource, nor does OCR receive any royalties from its sale. For more information about the endorsement process please visit the OCR website www.ocr.org.uk

AS/A Level course structure

This book has been written to support students studying for OCR AS Physics A and for students in their first year of studying for OCR A Level Physics A. It covers the AS modules from the specification, the content of which will also be examined at A Level. The modules covered are shown in the contents list, which also shows you the page numbers for the main topics within each module. There is also an index at the back to help you find what you are looking for. If you are studying for OCR AS Physics A, you will only need to know the content in the blue box.

AS exam

A level exam

Year 1 content

1. Development of practical skills in physics
2. Foundations in physics
3. Forces and motion
4. Electrons, waves, and photons

Year 2 content

5. Newtonian world and astrophysics
6. Particles and medical physics

A Level exams will cover content from Year 1 and Year 2 and will be at a higher demand. You will also carry out practical activities throughout your course.

Contents

How to use this book

Learning outcomes

→ At the beginning of each topic, there is a list of learning outcomes.

→ These are matched to the specification and allow you to monitor your progress.

→ A specification reference is also included.
Specification reference: 2.1.3

This book contains many different features. Each feature is designed to support and develop the skills you will need for your examinations, as well as foster and stimulate your interest in physics.

Terms that you will need to be able to define and understand are highlighted by **bold text**.

Application features

These features contain important and interesting applications of physics in order to emphasise how scientists and engineers have used their scientific knowledge and understanding to develop new applications and technologies. There are also practical application features, with the icon , to support further development of your practical skills.

1 All application features have a question to link to material covered with the concept from the specification.

Study Tips

Study tips contain prompts to help you with your understanding and revision.

Synoptic link

These highlight the key areas where topics relate to each other. As you go through your course, knowing how to link different areas of physics together becomes increasingly important. Many exam questions, particularly at A Level, will require you to bring together your knowledge from different areas.

Extension features

These features contain material that is beyond the specification. They are designed to stretch and provide you with a broader knowledge and understanding and lead the way into the types of thinking and areas you might study in further education. As such, neither the detail nor the depth of questioning will be required for the examinations. But this book is about more than getting through the examinations.

1 Extension features also contain questions that link the off-specification material back to your course.

Summary Questions

1 These are short questions at the end of each topic.

2 They test your understanding of the topic and allow you to apply the knowledge and skills you have acquired.

3 The questions are ramped in order of difficulty. Lower-demand questions have a paler background, with the higher-demand questions having a darker background. Try to attempt every question you can, to help you achieve your best in the exams.

Introduction at the beginning of each module summarises what you will cover.

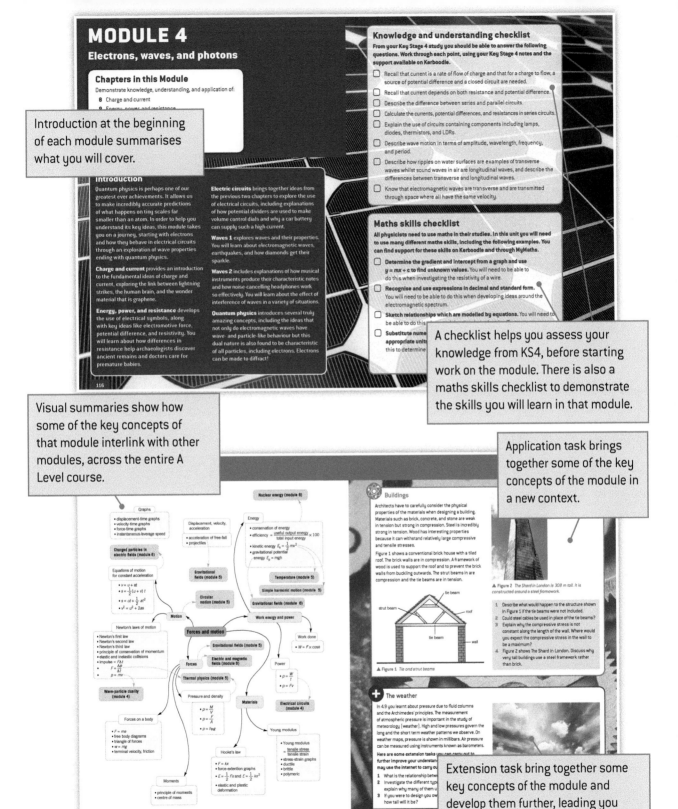

MODULE 4
Electrons, waves, and photons

Chapters in this Module
Demonstrate knowledge, understanding, and application of:

8 Charge and current
9 Energy, power, and resistance

Introduction

Quantum physics is perhaps one of our greatest ever achievements. It allows us to make incredibly accurate predictions of what happens on tiny scales far smaller than an atom. In order to help you understand its key ideas, this module takes you on a journey, starting with electrons and how they behave in electrical circuits through an exploration of wave properties ending with quantum physics.

Charge and current provides an introduction to the fundamental ideas of charge and current, exploring the link between lightning strikes, the human brain, and the wonder material that is graphene.

Energy, power, and resistance develops the use of electrical symbols, along with key ideas like electromotive force, potential difference, and resistivity. You will learn about how differences in resistance help archaeologists discover ancient remains and doctors care for premature babies.

Electric circuits brings together ideas from the previous two chapters to explore the use of electrical circuits, including explanations of how potential dividers are used to make volume control dials and why a car battery can supply such a high current.

Waves 1 explores waves and their properties. You will learn about electromagnetic waves, earthquakes, and how diamonds get their sparkle.

Waves 2 includes explanations of how musical instruments produce their characteristic notes and how noise-cancelling headphones work so effectively. You will learn about the effect of interference of waves in a variety of situations.

Quantum physics introduces several truly amazing concepts, including the ideas that not only do electromagnetic waves have wave- and particle-like behaviour but this dual nature is also found to be characteristic of all particles, including electrons. Electrons can be made to diffract!

116

Knowledge and understanding checklist
From your Key Stage 4 study you should be able to answer the following questions. Work through each point, using your Key Stage 4 notes and the support available on Kerboodle.

☐ Recall that current is a rate of flow of charge and that for a charge to flow, a source of potential difference and a closed circuit are needed.

☐ Recall that current depends on both resistance and potential difference.

☐ Describe the difference between series and parallel circuits.

☐ Calculate the currents, potential differences, and resistances in series circuits.

☐ Explain the use of circuits containing components including lamps, diodes, thermistors, and LDRs.

☐ Describe wave motion in terms of amplitude, wavelength, frequency, and period.

☐ Describe how ripples on water surfaces are examples of transverse waves whilst sound waves in air are longitudinal waves, and describe the differences between transverse and longitudinal waves.

☐ Know that electromagnetic waves are transverse and are transmitted through space where all have the same velocity.

Maths skills checklist
All physicists need to use maths in their studies. In this unit you will need to use many different maths skills, including the following examples. You can find support for these skills on Kerboodle and through MyMaths.

☐ **Determine the gradient and intercept from a graph and use** $y = mx + c$ **to find unknown values.** You will need to be able to do this when investigating the resistivity of a wire.

☐ **Recognise and use expressions in decimal and standard form.** You will need to be able to do this when developing ideas around the electromagnetic spectrum.

☐ **Sketch relationships which are modelled by equations.** You will need to be able to do this...

☐ **Substitute nume[rical]... appropriate unit[s]...** this to determine...

A checklist helps you assess your knowledge from KS4, before starting work on the module. There is also a maths skills checklist to demonstrate the skills you will learn in that module.

Visual summaries show how some of the key concepts of that module interlink with other modules, across the entire A Level course.

Application task brings together some of the key concepts of the module in a new context.

Buildings

Architects have to carefully consider the physical properties of the materials when designing a building. Materials such as brick, concrete, and stone are weak in tension but strong in compression. Steel is incredibly strong in tension. Wood has interesting properties because it can withstand relatively large compressive and tensile stresses.

Figure 1 shows a conventional brick house with a tiled roof. The brick walls are in compression. A framework of wood is used to support the roof and to prevent the brick walls from buckling outwards. The strut beams are in compression and the tie beams are in tension.

▲ Figure 2 *The Shard in London is 308 m tall. It is constructed around a steel framework.*

▲ Figure 1 *Tie and strut beams*

1. Describe what would happen to the structure shown in Figure 1 if the tie beams were not included.
2. Could steel cables be used in place of the tie beams?
3. Explain why the compressive stress is not constant along the length of the wall. Where would you expect the compressive stress in the wall to be a maximum?
4. Figure 2 shows The Shard in London. Discuss why very tall buildings use a steel framework rather than brick.

The weather

In 4.9 you learnt about pressure due to fluid columns and the Archimedes' principles. The measurement of atmospheric pressure is important in the study of meteorology (weather). High and low pressures govern the long and the short term weather patterns we observe. On weather maps, pressure is shown in millibars. Air pressure can be measured using instruments known as barometers.

Here are some extension tasks you can carry out to further improve your understan[ding]... may use the internet to carry ou[t]...

1. What is the relationship betw[een]...
2. Investigate the different typ[es]... explain why many of them u[se]...
3. If you were to design you ow[n]... how tall will it be?

Extension task bring together some key concepts of the module and develop them further, leading you towards greater understanding and further study.

114

Practice questions at the end of each chapter and the end of each module, including questions that cover practical and math skills.

b A spring has a force constant of $160\,\mathrm{Nm^{-1}}$. The energy stored in the spring is used to propel an object of mass 80 g. The spring is

▲ Figure 5

Use Figure 5 to answer to following questions.

(i) Explain how the graph shows that Hooke's law is obeyed. *(2 marks)*

(ii) Determine the force constant of the spring. *(2 marks)*

(iii) Determine the energy stored in the spring when its extension is 80 mm. *(3 marks)*

8 a Define *tensile stress* and *tensile strain*. *(1 mark)*

b Derive the units for stress in base units. *(3 marks)*

c A group of students are carrying out an experiment to determine the Young modulus of a metal wire. The wire has original length of 1.640 m and it is suspended vertically from a support. The wire is loaded in steps of 10.0 N up to 50.0 N and then unloaded. Table 1 shows the experimental results from the group.

▼ Table 1

force F/N	loading the wire extension x/mm	unloading the wire extension x/mm
0.0	0.00	0.00
10.0	0.50	0.49
20.0	1.01	1.00
30.0	1.49	1.50
40.0	2.00	1.99
50.0	2.52	2.52

(i) Use Table 1 to describe the behaviour of the wire when forces up to 50.0 N are applied to it. *(2 marks)*

(ii) Name the most likely instrument used to determine the extension of the wire. *(1 mark)*

▲ Figure 4

(i) Determine the total extension of the two springs. State any assumptions made. *(3 marks)*

(ii) Determine the force constant of the combination of these springs. *(2 marks)*

7 a State *Hooke's law*. *(1 mark)*

b Define the force constant of a spring. *(1 mark)*

c Describe how you can determine the force constant of an extendable spring in the laboratory. In your description pay particular attention to
• how the apparatus is used
• what measurements are taken
• how the data is analysed. *(4 marks)*

d Figure 5 shows the variation of force F with extension x for a spring.

(iii) The cross-sectional area of the wire is $3.2 \times 10^{-7}\,\mathrm{m^2}$. Use the value of the extension for the force of 50.0 N to calculate a value for the Young modulus of the metal. *(3 marks)*

(iv) Describe how the table of results can be used to plot a graph and hence determine a precise value for the Young modulus of the metal. *(3 marks)*

9 a Define the *Young modulus*. *(1 mark)*

b Figure 6 shows a violin.

▲ Figure 6

Two of the wires used on the violin, labelled A and G, are made of steel. The two wires are both 500 mm long between the pegs and support. The 500 mm length of wire labelled G has a mass of $2.0 \times 10^{-3}\,\mathrm{kg}$. The density of steel is $7.8 \times 10^3\,\mathrm{kgm^{-3}}$.

(i) Show that the cross-sectional area of wire G is $5.1 \times 10^{-7}\,\mathrm{m^2}$. *(2 marks)*

(ii) The wires are put under tension by turning the wooden pegs shown in Figure 5. The Young modulus of steel is $2.0 \times 10^{11}\,\mathrm{Pa}$. *(3 marks)*

(iii) Wire A has a diameter that is half that of wire G. Determine the tension required for wire A to produce an extension of $16 \times 10^{-4}\,\mathrm{m}$. *(1 mark)*

(iv) State the law that has been assumed in the calculations in (ii) and (iii). *(1 mark)*
June 2007 2821

10 Figure 7 shows the force F against extension x for a metal wire.

▲ Figure 7

a State the value of the force and extension at the elastic limit of the wire. *(1 mark)*

b Calculate the elastic potential energy for the wire when its extension is 0.25 mm. *(3 marks)*

c The Young modulus of the metal is $1.2 \times 10^{11}\,\mathrm{Pa}$ and the length of the wire is 1.82 m.

Use Figure 7 to determine the cross-sectional area of the wire. *(3 marks)*

11 Figure 8 shows an arrangement used by a student to investigate the energy stored in a compressible spring.

▲ Figure 8

a The spring is compressed by a distance x and then released. It climbs a vertical height h along the length of the rod. Show that h is directly proportional to x^2. *(3 marks)*

b Figure 9 shows a graph of h against x^2 for the spring.

▲ Figure 9

(i) Use Figure 9 to predict the height h when $x = 2.0\,\mathrm{cm}$. State any assumptions made. *(4 marks)*

(ii) The mass of the spring is 8.0 g.
Use Figure 9 to determine the force constant of the spring. *(3 marks)*

Paper 1 style questions

Practice questions at the end of the book, with multiple choice questions and synoptic style questions, also covering the practical and math skills.

SECTION A

Answer **all** the questions.

1 A student investigating an electrical experiment records the following measurements in the lab book.

by the spring?

A 0.020 J
B 0.080 J
C 0.140 J
D 1.00 J
(1 mark)

3 A wooden block is held under water and then released, as shown in Figure 1.

▲ Figure 1

The wooden block moves towards the surface of the water.

Which of the following statements is/are true about the block as soon as it is released?

1 The force experienced by the face B due to water is greater than the force experienced by the face A.
2 The upthrust on the block is equal to its weight.

3 The mass of the water displaced is equal to the weight of the block.

A 1, 2 and 3 are correct
B Only 1 and 2 are correct
C Only 2 and 3 are correct
D Only 1 is correct
(1 mark)

4 Figure 2 shows a stationary wave pattern formed in an air column.

▲ Figure 2

Which point **A**, **B**, **C**, or **D** has a phase difference of 180° with reference to **P**? *(1 mark)*

5 A ray of monochromatic light is incident at a boundary between two transparent materials. The refractive index of the materials is 1.30 and 1.50. The angle of refraction for the emergent ray is 60°.

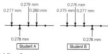
▲ Figure 3

What is the angle θ of incidence?

A 42°
B 49°
C 60°
D 88°
(1 mark)

6 Figure 4 shows the cross-section of a metal wire connected to a power supply. The charge carriers within the metal wire move from right to left.

▲ Figure 4

The section Q of the wire is thinner than section P.

Which statement is correct?

A The direction of the conventional current is from right to left.
B The section Q of the wire has fewer charge carriers per unit volume.
C The current in both sections is the same.
D The charge carriers are negative ions.
(1 mark)

7 A resistor **R** is connected in parallel with a resistor of resistance $10\,\Omega$. The total resistance of the combination is $6.0\,\Omega$. What is the resistance of resistor **R**?

A $0.067\,\Omega$
B $3.8\,\Omega$
C $4.0\,\Omega$
D $15\,\Omega$
(1 mark)

8 What is a reasonable estimate for the energy of a photon of visible light?

A $4 \times 10^{-19}\,\mathrm{J}$
B $4 \times 10^{-18}\,\mathrm{J}$
C $4 \times 10^{-16}\,\mathrm{J}$
D $4 \times 10^{-11}\,\mathrm{J}$
(1 mark)

9 Students A and B use micrometer screw gauges to measure the diameter of a copper wire in three different places along its length. The diameter of the wire according to the manufacturer is 0.278 mm. The results recorded by students A and B are shown in Figure 5.

Student A: 0.279 mm, 0.277 mm, 0.280 mm, 0.278 mm
Student B: 0.276 mm, 0.275 mm, 0.277 mm, 0.278 mm
▲ Figure 5

Which statement is correct about the measurements made by the student B compared with those of student A?

A The measurements are more accurate.
B The measurements are not as precise.
C The measurements are both more accurate and more precise.
D The measurements are not accurate but are more precise.
(1 mark)

10 The circuit in Figure 6 is constructed by a student in the laboratory.

▲ Figure 6

The e.m.f. of the cell is 1.5 V and it has an internal resistance of $3.0\,\Omega$. A resistor of resistance $2.0\,\Omega$ and a variable resistor R are connected in series to the terminals of the cell. The variable resistor is set to a resistance value of $7.0\,\Omega$.

What is the value of the ratio $\dfrac{\text{power dissipated in R}}{\text{power supplied by the cell}}$?

A 0.17
B 0.25
C 0.58
D 0.75
(1 mark)

SECTION B

Answer **all** the questions.

11 a Define *velocity*. *(1 mark)*

b The mass of an ostrich is 130 kg. It can run at a maximum speed of 70 kilometers per hour.

(i) Calculate the maximum kinetic energy of the ostrich when it is running. *(3 marks)*

(ii) Scientists have recently found fossils of a prehistoric bird known as Mononykus. Figure 7 shows what the Mononykus would have looked like.

Kerboodle

This book is supported by next generation Kerboodle, offering unrivalled digital support for independent study, differentiation, assessment, and the new practical endorsement.

If your school subscribes to Kerboodle, you will also find a wealth of additional resources to help you with your studies and with revision.

- Study guides
- Maths skills boosters and calculation worksheets
- On your marks activities to help you achieve your best
- Practicals and follow up activities to support the practical endorsement
- Interactive objective tests that give question-by-question feedback
- Animations and revision podcasts
- Self-assessment checklists

Revise with ease using the study guides to guide you through each chapter and direct you towards the resources you need.

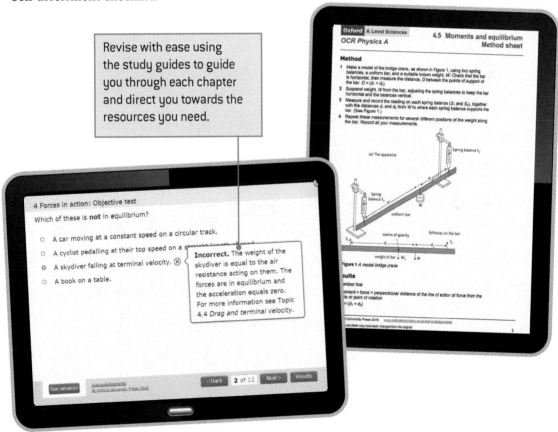

For teachers, Kerboodle also has plenty of further assessment resources, answers to the questions in the book, and a digital markbook along with full teacher support for practicals and the worksheets, which include suggestions on how to support and stretch students. All of the resources are pulled together into teacher guides that suggest a route through each chapter.

MODULE 1
Development of practical skills in physics

Physics is a practical subject and experimental work provides you with important practical skills, as well as enhancing your understanding of physical theory. You will be developing practical skills by carrying out practical and investigative work in the laboratory throughout both the AS and the A level Physics course. You will be assessed on your practical skills in two different ways:

- written examinations (AS and A level)
- practical endorsement (A level only)

Practical coverage throughout this book

Practical skills are a fundamental part of a complete education in science, and you are advised to keep a record of your practical work from the start of your A level course that you can later use as part of your practical endorsement. You can find more details of the practical endorsement from your teacher or from the specification.

In this book and its supporting materials practical skills are covered in a number of ways. By studying Application boxes and

Exam-style questions in this student book, and by using the Practical activities and Skills sheets in Kerboodle you will have many opportunities to learn about the scientific method and carry-out practical activities.

1.1 Practical skills assessed in written examinations

In the written examination papers for AS and A level, at least 15% of the marks will be from questions that assess practical skills. The questions will cover four important skill areas, all based on the practical skills that you will develop by carrying out experimental work during your course.

- Planning – your ability to solve a physics problem in a practical context.
- Implementing – your understanding of important practical techniques and processes.
- Analysing – your interpretation of experimental results set in a practical context and related to the experiments that you would have carried out.
- Evaluating – your ability to develop a plan that is fit for the intended purpose.

1.1.1 Planning

- Designing experiments
- Identifying variables to be controlled
- Evaluating the experimental method

Skills checklist

- ☐ Selecting apparatus and equipment
- ☐ Selecting appropriate techniques
- ☐ Selecting appropriate quantities of materials and substances and scale of working
- ☐ Solving physical problems in a practical context
- ☐ Applying physics concepts to practical problems

1.1.2 Implementing

- Using a range of practical apparatus
- Carrying out a range of techniques
- Using appropriate units for measurements
- Recording data and observations in an appropriate format

Skills checklist

- ☐ Understanding practical techniques and processes
- ☐ Identifying hazards and safe procedures
- ☐ Using SI units
- ☐ Recording qualitative observations accurately
- ☐ Recording a range of quantitative measurements
- ☐ Using the appropriate precision for apparatus

1.1.3 Analysis

- Processing, analysing, and interpreting results
- Analysing data using appropriate mathematical skills
- Using significant figures appropriately
- Plotting and interpreting graphs

Skills checklist

- ☐ Analysing qualitative observations
- ☐ Analysing quantitative experimental data, including
  - calculation of means
  - amount of substance and equations
- ☐ For graphs,
  - selecting and labelling axes with appropriate scales, quantities, and units
  - drawing tangents and measuring gradients

1.1.4 Evaluation

- Evaluating results to draw conclusions
- Identify anomalies
- Explain limitations in method
- Identifying uncertainties and errors
- Suggesting improvements

Skills checklist

- ☐ Reaching conclusions from qualitative observations
- ☐ Identifying uncertainties and calculating percentage errors
- ☐ Identifying procedural and measurement errors
- ☐ Refining procedures and measurements to suggest improvements

1.2 Practical skills assessed in practical endorsement

You will also be assessed on how well you carry out a wide range of practical work and how to record the results of this work. These hands-on skills are divided into 12 categories and form the practical endorsement. This is assessed for A level Physics qualification only.

The endorsement requires a range of practical skills from both years of your course. If you are taking only AS Physics, you will not be assessed through the practical endorsement but the written AS examinations will include questions that relate to the skills that naturally form part of the AS common content to the A level course.

The practicals you do as part of the endorsement will not contribute to your final grade awarded to you. However, these practicals must be covered and your teacher will go through how this is to be done in class. It is important that you are actively involved in practical work because it will help you with understanding the theory and also how to effectively answer some of the questions in the written papers.

The practical activities you will carry out in class are divided into Practical Activity Group (PAGs). PAG1 to PAG 6 will be undertaken in Year 1, PAG7 to PAG 10 in Year 2, and PAG11 to PAG 12 throughout the two-year course.

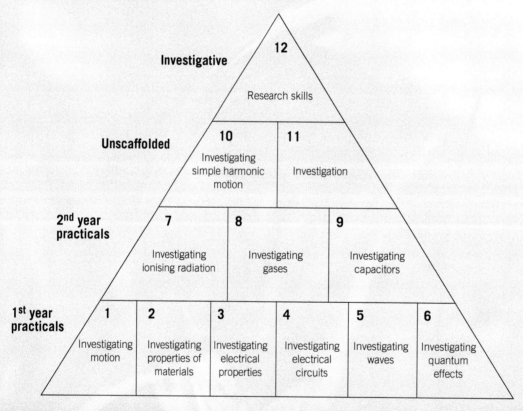

The PAGs are summarised below, together with the topic reference in the book that relates to the specific PAG.

PAG	Topic reference
1 Investigating motion	3.7, 4.4
2 Investigating properties of materials	6.1, 6.4
3 Investigating electrical properties	9.5, 9.7
4 Investigating electrical circuits	10.4, 10.6
5 Investigating waves	11.2, 11.7, 12.2–12.6
6 Investigating quantum effects	13.1
7 Investigating ionising radiation	25.4
8 Investigating gases	15.2
9 Investigating capacitors	21.4
10 Investigating simple harmonic motion	17.1
11 Investigation	Throughout
12 Research skills	Throughout

Maths skills and How Science Works across Module 1

In order to develop your knowledge and understanding in A Level Physics, it is important to have specific skills in mathematics. All the mathematical skills you will need during your physics course have been embedded into the individual topics for you to learn as you meet them. An overview is available in each of the module openers and these skills are further supported by the worked examples, summary questions, and examination-style questions.

How Science Works (HSW) is another area required for success in A Level Physics, and helps you to put science in a wider context, helping you to develop your critical and creative thinking skills in order to solve problems in a variety of contexts. Once again, this has been embedded into the individual topics covered in the books, particularly in application boxes and examination-style questions. The application and extension boxes cover some of the HSW elements.

You can find further support for maths skills and HSW on Kerboodle.

MODULE 2
Foundations of physics

Topics in this module

Introduction

Physics is not only a collection of concepts about everything from subatomic particles to the whole Universe. It is a set of different ways of thinking that have led to countless successful descriptions and explanations of the way the Universe works. Physicists have learned new ways of thinking as they look for deeper and deeper explanations of physical phenomena, searching for the most fundamental answer that can be applied across the widest range of disciplines.

Foundations of physics introduces the important ideas and conventions that permeate the fabric of physics. You will develop your skills in critical thinking, reasoning and logic, and mathematics. With these you will be able to build models to describe a wide variety of systems and to make predictions about different circumstances.

Through an exploration of **units** you will learn about the well-defined and universally understood methods used by physicists to measure physical phenomena, and methods that help physicists across the globe effectively communicate their ideas within the scientific community.

By developing your understanding of **vectors** you will build a powerful mathematical toolkit that you will use throughout your studies.

You will hone your ability to make approximations and **estimations** in order to gain a sense of magnitudes and to know what sort of answers to expect. A study of **errors, uncertainty, precision, and accuracy** develops your understanding of the limitations of experimentation. You will learn how to present your data appropriately and express numerically a level of confidence in your findings.

Some of these key ideas and skills relate to and will be developed by experimental work that you will meet across the different modules – you will find these topics covered in detail in the appendices.

Knowledge and understanding checklist

From your Key Stage 4 study you should be able to do the following. Work through each point, using your Key Stage 4 notes and the support available on Kerboodle.

☐ Use vector diagrams to illustrate forces, a net force, and equilibrium situations.

☐ Explain the vector–scalar distinction as it applies to displacement, distance, velocity, and speed.

☐ Make calculations using ratios and proportional reasoning to convert units.

Maths skills checklist

All physicists use maths. In this unit you will need to use many different maths skills, including the following examples. You can find support for these skills on Kerboodle and through MyMaths.

☐ **Recognise and make use of appropriate units in calculations.** You will need to be able to do this when identifying the correct units for physical properties and converting between units with different prefixes, for example, km and m.

☐ **Use calculators to find and use power functions.** You will need to be able to do this to calculate resultant vectors.

☐ **Use sin, cos, and tan in physical problems.** You will need to be able to do this when resolving vectors into components.

☐ **Visualise and represent two-dimensional and three-dimensional forms, including 2D representations of 3D objects.** You will need to be able to do this to solve problems involving the addition of vectors through the use of scale diagrams.

☐ **Use Pythagoras' theorem and the angle sum of a triangle.** You will need to be able to do this to solve various problems involving vectors.

☐ **Estimate results.** You will need to be able to do this when estimating the effect of changing a value during an experiment.

☐ **Find arithmetic means.** You will need to be able to do this when presenting data in tables from various pieces of experimental work.

☐ **Identify uncertainties in measurements and use simple techniques to determine uncertainty when data is combined.** You will need to be able to do this when analysing experimental data.

FOUNDATIONS OF PHYSICS
2.1 Quantities and units
Specification reference: 2.1.1, 2.1.2

▲ **Figure 1** *The correct use of units would have prevented the destruction of the Mars Climate Orbiter*

Measurements

Measurements are very important in physics. Not only must they be recorded accurately, they must also be communicated clearly. In 1998 NASA launched the Mars Climate Orbiter, a mission costing almost £195 million. When the probe arrived at Mars a few months later, it disintegrated in the planet's upper atmosphere instead of going into orbit. The disaster had a simple cause: one of NASA's teams worked in feet and pounds, whilst the other team worked in metres and kilograms. Each team assumed that the other was using the same units.

In A Level Physics, failure to use units correctly may not cost millions of pounds but it will cost you valuable marks in the examination.

Quantities

A physical **quantity** is a property of an object or of a phenomenon that can be measured. Some quantities are just numbers. For example, proton number, efficiency, and magnification are numbers. They have a numerical magnitude or size, but no units. Many other quantities consist of numbers *and* units. For example, length is a quantity that has units. It has many different units, including metres, inches, and miles. To avoid problems like the one NASA experienced with the Mars Climate Orbiter, scientists use a standard system of units called the *Système International d'Unités* (International System of Units), abbreviated to **SI**.

SI base units

SI is built around seven **base units**, six of which are shown in Table 1. The seventh unit, the unit for luminous intensity (the candela, cd), is not assessed in the A Level Physics course.

▼ **Table 1** *SI base units*

Quantity	Base unit	Unit symbol
length	metre	m
mass	kilogram	kg
time	second	s
electric current	ampère	A
temperature	kelvin	K
amount of substance	mole	mol

Symbols

A unit symbol is written in lower case, for example, m rather than M for metres, unless the unit is named after a person. In that situation, its name still begins with a lower-case letter but its symbol has a capital letter. The unit of electric current is named after André-Marie Ampère, so its name is the ampère (often just amp) and its symbol is A.

Prefixes

SI uses prefixes to show multiples and fractions of units (Table 2). For example, km stands for kilometre. The **prefix** is the 'kilo', and the unit is the 'metre'.

Notice that, apart from k for kilo, the prefixes for multiples all have initial capitals. Similarly, the prefixes for fractions are all lower case (μ is the lower-case Greek letter mu).

 ## Worked example: Using prefixes

a   Convert 1.25 kA into A.

$1.25\,\text{kA} = 1.25 \times 10^3\,\text{A}$ (or 1250 A)

b   Convert 234 μm into m.

$234\,\mu\text{m} = 234 \times 10^{-6}\,\text{m} = 2.34 \times 10^{-4}\,\text{m}$

c   Convert 0.567 s into ms.

There are 10^3 ms in 1 s. To change from seconds to milliseconds, you have to *multiply* by a factor of 10^3.

Therefore, $0.567\,\text{s} = 0.567 \times 10^3 = 567\,\text{ms}$

Summary questions

1   A student records the following figures in his notes: 60 cm and 40 ms.
   a   Name the two quantities being measured.   (*2 marks*)
   b   Change these measurements into their base units.   (*2 marks*)

2   **a**   A collision between two molecules lasts for about 100 picoseconds. Write this time in seconds.   (*1 mark*)
   b   A chemical bond is approximately 0.15 nanometres long. Write this length in metres.   (*1 mark*)
   c   The Sun's core has a temperature of approximately 16 megakelvin. Write this temperature in kelvin.   (*1 mark*)

3   Convert the following measurements to their base units. Write your answers in standard form.
   a 200 pm;   **b** 0.40 Mm;   **c** 35 μs;   **d** 0.25 mA;   **e** 756 ns.   (*5 marks*)

4   There are 86 400 s in a day. Alternatively you could say there are 86.4 ks in a day.
   a   The distance by train from London to Edinburgh is 5.34×10^5 m. What is this distance in km?
   b   The diameter of the Earth is 1.274×10^7 m. What is this diameter in Mm?
   c   The thickness of a human hair is about 7.5×10^{-5} m. What is this thickness in μm?
   d   The electric current in a nerve cell is about 1.4×10^{-7} A. What is this current in nA?   (*4 marks*)

▼ Table 2 *Prefixes for SI units*

Prefix name	Prefix symbol	Factor
peta	P	10^{15}
tera	T	10^{12}
giga	G	10^{9}
mega	M	10^{6}
kilo	k	10^{3}
deci	d	10^{-1}
centi	c	10^{-2}
milli	m	10^{-3}
micro	μ	10^{-6}
nano	n	10^{-9}
pico	p	10^{-12}
femto	f	10^{-15}

Study tip

Standard form is used to display very small or very large numbers in a scientific way. For scientific notation it is ideally expressed in the form $n \times 10^m$, where $1 < n < 10$, and m is an integer.

 ### Standard form

You can show small and large numbers in standard form.

For example, instead of writing 230 km or 230×10^3 m, we could express this distance as 2.3×10^5 m.

Write 45 ns (45×10^{-9} s) in standard form.

Study tip

Take care when you are writing prefixes and units. For example, ms means milliseconds, but Ms means megaseconds.

2.2 Derived units

Specification reference: 2.1.2

Beyond base units

The seven base units are used to measure the base quantities that they represent. However, there are many more quantities to measure than just mass, length, electric current, time, and the other three base quantities. For example, what are the units for speed and force? Quantities like these are called **derived quantities**. They use **derived units**, which can be worked out from the base units and the equations relating derived quantities to the base quantities. With derived units any quantity can be communicated.

Names and symbols

Derived units without special names

You already know some derived units. For example, the unit for speed is $m\,s^{-1}$. It comes from the equation that links average speed with two base quantities – distance and time.

$$\text{average speed} = \frac{\text{distance travelled}}{\text{time taken}}$$

Since m is the unit for distance, s is the unit for time, and we are *dividing* m by s, the derived unit for speed is m/s, written $m\,s^{-1}$ at A Level ($s^{-1} = \frac{1}{s}$). We write derived units like this because it is better for more complex units, such as the unit for specific heat capacity, $J\,kg^{-1}\,K^{-1}$, which is much clearer than $J/(kg\,K)$.

Table 1 shows some derived units without any special names.

Derived units with special names

Some derived quantities are used so often that they have special names. SI has 22 derived units with special names and symbols, but you will not need to know them all for your physics course. Table 2 shows a small selection of these units.

▼ **Table 1** *Some derived units*

Derived quantity	Derived unit
area	m^2
volume	m^3
acceleration	$m\,s^{-2}$
density	$kg\,m^{-3}$

▲ **Figure 1** *Speed is measured in $m\,s^{-1}$, a derived unit in SI*

▼ **Table 2** *Some named derived units*

Derived quantity	Unit name	Unit symbol	Unit expressed in other SI units
force	newton	N	$kg\,m\,s^{-2}$
pressure	pascal	Pa	$N\,m^{-2}$
energy or work done	joule	J	$N\,m$
power	watt	W	$J\,s^{-1}$
electric potential difference	volt	V	$J\,C^{-1}$
electric resistance	ohm	Ω	$V\,A^{-1}$
electric charge	coulomb	C	$A\,s$
frequency	hertz	Hz	s^{-1}

SI units can be combined to form a huge range of other derived units. You may be familiar with some of these already. For example, the moment of a force is measured in newton metres, N m.

Temperature

The SI base unit for temperature is the kelvin, K. In everyday life you are likely to use a different unit for temperature, a derived unit called the degree Celsius, °C. To convert from °C to K you add 273, so 20°C is 293 K and 100°C is 373 K.

A difference of 1°C is the same as a difference of 1 K, so temperature *differences* do not need conversion. For example, if you warm some water from 20°C to 100°C its temperature increases by 80°C, which is also 80 K.

1   Converting from K to °C is equally simple. Convert 298 K to °C.

2   The degree Fahrenheit, °F, is a non-SI unit for temperature. To convert from °F to °C you subtract 32, multiply by 5 then divide by 9. For example, $68°F = (68 - 32) \times \frac{5}{9} = 20°C$. Deduce the temperature that has the same value, whether given in °F or in °C.

Summary questions

1   The unit of mass is the kg. Acceleration has the derived unit $m\,s^{-2}$. The force acting on an object can be determined using the equation force = mass × acceleration. Determine the derived unit for force in base units.   (*2 marks*)

2   Use the equations given to determine the derived unit of each quantity in base units.

   a $\text{force constant} = \dfrac{\text{force}}{\text{extension}}$

     Extension is the change in length. Determine the derived unit for force constant.   (*2 marks*)

   b   work done = force × distance moved in direction of force

     Determine the derived unit for work done.   (*2 marks*)

   c $\text{pressure} = \dfrac{\text{force}}{\text{cross-sectional area}}$

     Determine the derived unit for pressure.   (*2 marks*)

3   State the difference between 1 N m, 1 nm, 1 mN and 1 MN.   (*3 marks*)

4   In electrical work, it is useful to define a quantity known as *number density* of free electrons. Number density of free electrons is the number of electrons per unit volume. What is the unit for number density in base units?   (*2 marks*)

2.3 Scalar and vector quantities

Specification reference: 2.3.1

▲ **Figure 1** *Flyboarders can hover up to 15 m above the water*

▼ **Table 1** *Some scalar quantities and units*

Scalar quantity	SI unit
length	m
mass	kg
time	s
speed	$m\,s^{-1}$
temperature	K, °C
volume	m^3
energy	J
potential difference	V
power	W

Going up

Flyboarding is a sport in which the rider stands on a board with a long hose attached that hangs into a lake. Water from the lake is forced through the hose and into jets under the board. The water rushes out of the jet nozzles, pushing the rider into the air. Skilled flyboarders can perform all sorts of aerial acrobatics, thanks to practice in judging scalar and vector quantities.

Scalar quantities

A **scalar quantity** has magnitude (size) but no direction. For example, the *distance* between a flyboarder and the surface of the water is a scalar quantity, and so is his *mass* and the *time* he can stay in the air. Table 1 shows some examples of scalar quantities with their SI units.

Adding and subtracting scalar quantities

Scalar quantities can be added together or subtracted from one another in the usual way. For example, if your mass is 55 kg and you pick up a 5 kg bag, your new total mass is (55 + 5) = 60 kg. If you sharpen a 16 cm pencil and remove 1 cm as you do so, the new length of the pencil is (16 − 1) = 15 cm.

Scalar quantities must have the same units when you add or subtract them. If you time something in an experiment you cannot add together 1 *minute* and 30 *seconds* as (1 + 30). Instead, you would convert the time from minutes into seconds and then add the times: (60 + 30) = 90 s. Alternatively, you could work in minutes to get a time of (1 + 0.5) = 1.5 minutes.

Multiplying and dividing scalar quantities

Scalar quantities can also be multiplied together or divided by one another. However, in this case the units can be the same or different, unlike adding and subtracting. It is important that you work out the final units correctly.

 Worked example: Lighter than air

A balloon is inflated with $6.1 \times 10^{-3}\,m^3$ of helium. Its mass increases by 0.98 g. Calculate the density of helium.

Step 1: The equation for density is

$$density = \frac{mass}{volume}$$

Step 2: Consider the units of the equation.

You are dividing together two scalar quantities. The SI base unit for mass is the kg. Volume has the unit m^3. The mass must be converted into kg before substitution; mass = $9.8 \times 10^{-4}\,kg$. →

Step 3: Substitute the values into the equation and calculate the density.

$$\text{density} = \frac{9.8 \times 10^{-4}}{6.1 \times 10^{-3}} = 0.16 \, \text{kg m}^{-3}$$

Vector quantities

A **vector quantity** has magnitude *and* direction. For example, the weight of a flyboarder is a vector quantity, and so is the force from the rushing water from the jet nozzles. Table 2 shows some examples of vector quantities and their SI units.

Distance and displacement

Distance and displacement are both measured in m, but distance is a scalar quantity and displacement is a vector quantity. This is illustrated in Figure 2.

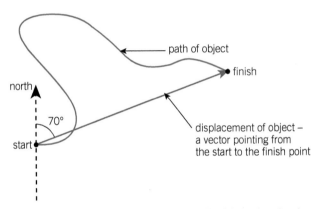

▲ **Figure 2** *Distance travelled is the length of the red path, whereas the magnitude of the displacement is the length of the blue arrow and the direction of the displacement is 70° off due north*

▼ **Table 2** *Some vector quantities and units*

Vector quantity	SI unit
displacement	m
velocity	m s^{-1}
acceleration	m s^{-2}
force	N (kg m s^{-2})
momentum	kg m s^{-1}

Synoptic link

You find out more about vector quantities when studying motion, forces, and momentum in Chapters 3, 4, and 7 of this book.

Synoptic link

In Chapter 3, you will come across two important vector quantities – velocity and acceleration.

Summary questions

1 Explain what is wrong with the following calculation:
   mass$_1$ = 150 g, mass$_2$ = 0.500 kg; total mass = 150 + 0.500 = 150.5 g   *(2 marks)*

2 Compare and contrast distance and displacement.   *(2 marks)*

3 You can calculate power by dividing energy by time. Explain whether power is a scalar or a vector quantity.   *(2 marks)*

4 Figure 2 shows the path of a beetle that takes 20 s to travel from the start to the finish.
   The diagram is drawn to 1:1 scale. Determine:
   a the distance travelled, using a length of string;   *(1 mark)*
   b the magnitude of the displacement;   *(1 mark)*
   c the average speed of the beetle.   *(2 marks)*

5 Explain why the magnitude of the displacement of an object can never be greater than the distance travelled by the object.   *(1 mark)*

2.4 Adding vectors

Specification reference: 2.3.1

Learning outcomes

Demonstrate knowledge, understanding, and application of:

→ addition of two vectors with scale drawings and with calculations.

▲ **Figure 1** *What effect will the flowing water have on the dog's progress across the river?*

▲ **Figure 2** *Representing a vector quantity, in this example a force of 5.0 N*

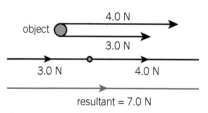

▲ **Figure 3** *Two parallel forces acting on an object are shown at the top, with the corresponding vector diagrams below*

Going against the flow

Many dogs love to jump into rivers to fetch sticks thrown for them. When a dog swims back to a point on the river bank, it has to swim against the current. The velocity of the flowing water and the velocity of the dog's paddling are vector quantities, so it is possible to work out the overall or **resultant** velocity of the dog by adding the two vectors together.

Vectors in one dimension

As you have already seen with displacement in Topic 2.3, a vector quantity is represented by a line with a single arrowhead:

● the length of the line represents the magnitude of the vector, drawn to scale

● the direction in which the arrowhead points represents the direction of the vector.

For example, Figure 2 shows a line representing a single vector. It is drawn to a scale of $1.0 \, \text{cm} \equiv 1.0 \, \text{N}$, so a line 5.0 cm long represents a force of 5.0 N.

Parallel vectors

Where two vectors are **parallel** (they act in the same line and direction), you just add them together to find the **resultant vector**. The direction of the resultant is the same as the individual vectors but its magnitude is greater. For example, if two forces of 3.0 N and 4.0 N act in the same direction on an object, the resultant force is 7.0 N.

Antiparallel vectors

Where two vectors are **antiparallel** (they act in the same line but in opposite directions), you call one direction positive and the opposite direction negative (it does not matter which), and then add the vectors together to find the resultant. The magnitude and direction of the resultant will depend on the magnitude of the two vectors.

▦ Worked example: Vectors in opposite directions

Two forces act in opposite directions on an object, as shown in Figure 4. Calculate the magnitude and direction of the resultant force.

Step 1: Assign positive and negative values to the vectors.

Assume that the positive direction is towards the right, so the two forces are −3.0 N and +4.0 N.

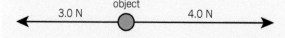

▲ **Figure 4** *Two forces acting in opposite directions*

→

Step 2: Calculate the resultant force.

resultant $= -3.0 + 4.0 = +1.0\,\text{N}$ towards the right

Two perpendicular vectors

Perpendicular vectors act at right angles to each other. Figure 5a represents two perpendicular forces of magnitudes $4.0\,\text{N}$ and $3.0\,\text{N}$ acting on an object.

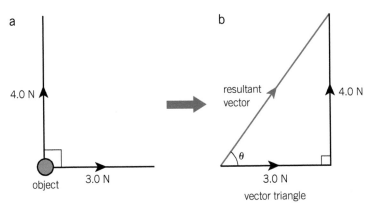

▲ **Figure 5** *Two perpendicular forces: (a) the two forces acting on the object; (b) the vector triangle used to determine the resultant vector*

The resultant vector can be found either by calculation or by a scale drawing of a **vector triangle**. Follow the rules below when adding any two vectors.

1   Draw a line to represent the first vector.

2   Draw a line to represent the second vector, starting from the *end* of the first vector.

3   To find the resultant vector, join the start to the finish. You have created a vector triangle (Figure 5b).

The method can be used to determine the resultant vector for any two vectors – displacements, velocities, accelerations, and so on. The angle between the vectors need not be 90°; any triangle works.

In this case, since the angle is 90°, you can also determine the magnitude of the resultant force F using **Pythagoras' theorem**.

$$F^2 = 4.0^2 + 3.0^2$$

$$F = \sqrt{4.0^2 + 3.0^2} = \sqrt{25}$$

$$F = 5.0\,\text{N}$$

To find the direction of the resultant force, you can calculate the angle θ made with the $3.0\,\text{N}$ force.

$$\tan\theta = \frac{\text{opp}}{\text{adj}} = \frac{4.0}{3.0} = 1.333$$

$$\theta = 53°$$

Summary questions

1   The steps on an escalator move upwards at $0.5\,\text{m s}^{-1}$. Calculate the resultant vertical velocity of a person:

   a   standing still on the escalator;   *(1 mark)*

   b   walking upwards at $2.0\,\text{m s}^{-1}$;   *(1 mark)*

   c   walking downwards at $1.0\,\text{m s}^{-1}$.   *(1 mark)*

2   The diagrams in Figure 6 represent forces acting on an object. For each one, draw a vector triangle and therefore determine the magnitude and direction of the resultant force.   *(10 marks)*

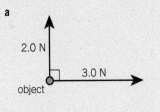

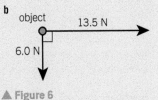

▲ **Figure 6**

3   A river flows due north at $0.90\,\text{m s}^{-1}$. A dog swims at $0.30\,\text{m s}^{-1}$. Calculate the magnitude and direction of the resultant velocity when the dog swims:

   a   due north;   *(2 marks)*

   b   due south;   *(2 marks)*

   c   due east.   *(3 marks)*

Learning outcomes

Demonstrate knowledge, understanding, and application of:

→ resolution of a vector into two perpendicular component vectors.

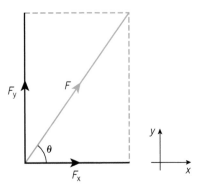

▲ **Figure 1** *Pilots must compensate for the effect of crosswinds during take-off and landing*

▲ **Figure 2** *Resolving a force F into components F_x and F_y*

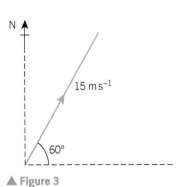

▲ **Figure 3**

Crosswinds

The wind can be helpful to aircraft. A tailwind, blowing in the same direction as the aircraft is travelling, reduces the journey time and saves fuel. On the other hand, a headwind can increase the journey time and waste fuel. Crosswinds can blow an aircraft off course unless the pilot takes them into account. An understanding of vectors is helpful in situations like these.

Resolving a vector into two components

You already know how to add together two perpendicular vectors to find a resultant vector. You can reverse this procedure to split a vector into two perpendicular components. This is called **resolving the vector**. It can be done using a scale drawing, but more often vectors are resolved by calculation.

To resolve a force F into the x and y directions, the two **components** of the force are

● $F_x = F \cos\theta$

● $F_y = F \sin\theta$

where θ is the angle made with the x direction. These equations can be used with any vector in the place of x.

 Worked example: A crosswind

At an airport, a horizontal wind is blowing at $15\,\mathrm{m\,s^{-1}}$ at an angle of $60°$ north of east (Figure 3). Calculate the components of the wind velocity in the north and east directions.

Step 1: Select the equations for resolving vectors.

● $v_x = v \cos\theta$

● $v_y = v \sin\theta$

Step 2: Substitute the values into the equations and calculate the components.

velocity component due east = $v_x = 15 \times \cos 60° = 7.5\,\mathrm{m\,s^{-1}}$

velocity component due north = $v_y = 15 \times \sin 60° = 13\,\mathrm{m\,s^{-1}}$

You can quickly check your answer using Pythagoras' theorem.

$$v^2 = v_x^2 + v_y^2 = 7.5^2 + 13^2 = 56.25 + 169$$

$$v = 15\,\mathrm{m\,s^{-1}}$$

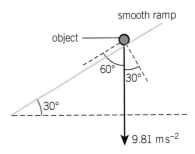
▲ Figure 4

 ## Worked example: Going down

A freely falling object has a vertical acceleration of $9.81\,\text{m s}^{-2}$. The object is placed on a smooth ramp that makes an angle of 30° to the horizontal (Figure 4). Calculate the component of the acceleration a down the ramp.

Step 1: Select the equation.

acceleration component down the ramp = $a\cos\theta$ where θ is the angle a makes to the slope.

Step 2: Substitute the values into the equations and calculate the component.

component = $9.81 \times \cos 60° = 4.91\,\text{m s}^{-2}$

You could have used $9.81 \times \sin 30°$ instead. The answer will be the same because $\sin 30°$ is the same as $\cos 60°$.

Study tip

Always check that your calculator is in the correct mode – in this case degrees – when you resolve vectors.

Summary questions

1 A force of 10 N acts on an object at an angle θ to the horizontal. Calculate the horizontal component of the force when $\theta = 0$, $\theta = 45°$, and $\theta = 90°$. Comment on your answers. *(4 marks)*

2 A parascender is attached by a rope to a boat travelling at a constant velocity (Figure 5a). The rope is angled at 35° to the surface of the sea, and the tension in the rope is 1650 N. Calculate the horizontal and vertical components of the tension in the rope. *(2 marks)*

3 A sailing boat is travelling north. It is moving because of a force due to the wind, which is 350 N blowing towards 40° east of north (Figure 5b). Calculate the components of the force from the wind:
   a towards the north (the direction in which the boat is moving); *(1 mark)*
   b towards the east (perpendicular to the direction in which the boat is moving). *(1 mark)*

a
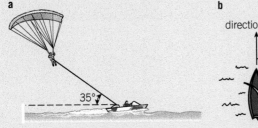

b
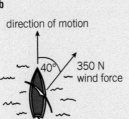
direction of motion
350 N wind force

c
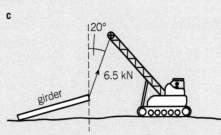
20°
6.5 kN
girder

▲ Figure 5

4 One end of a steel girder is lifted off the ground by a crane. The cable is at 20° from the vertical and the tension in the cable is 6.5 kN (Figure 5c). Calculate the vertical and horizontal components of this force. *(2 marks)*

2.6 More on vectors

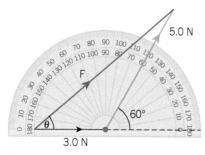

▲ **Figure 1** *Tugboats towing an oil platform*

Tugboats

A tugboat is a small but powerful boat that pushes or pulls larger vessels such as barges and tankers. Tugboats manoeuvre these large ships through crowded waterways and harbours. Larger, ocean-going tugboats can tow damaged ships to safety. Sometimes even the most powerful tugboats need to work in pairs or groups. Tugboat captains must understand the vectors involved so that the towed vessel travels in the right direction.

Adding non-perpendicular vectors

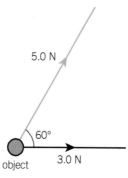

▲ **Figure 2** *Two non-perpendicular forces acting on an object*

There are several techniques you can use to add together two non-perpendicular vectors. They all rely on constructing a clear vector triangle. We will apply each of the techniques in turn to the following problem in order to demonstrate how to use them.

Two forces, of 5.0 N and 3.0 N, act on a single point at 60° to each other (Figure 2). What is the magnitude and direction of the resultant force?

Technique 1 – Scale diagram

Choose an appropriate scale for the drawing of your vector triangle. Use the rules outlined in Topic 2.4 to construct your vector triangle (Figure 3).

▲ **Figure 3** *A vector triangle drawn to scale*

Carefully measure the length of the resultant vector: it is 7.0 cm. With 1.0 cm representing 1.0 N in the diagram, the resultant force must equal 7.0 N. The angle made by the resultant and the 4.0 N force is 38°.

Technique 2 – Calculations using cosine and sine rules

Figure 4 shows a sketch of the vector triangle. The angles and magnitudes of the vectors are all shown. The resultant force is F.

You can use the cosine rule ($a^2 = b^2 + c^2 - 2bc \cos\theta$) to determine the magnitude of the resultant force.

$$F^2 = 3.0^2 + 5.0^2 - 2 \times 3.0 \times 5.0 \times \cos 120°$$

$$F = \sqrt{49} = 7.0\,\text{N}$$

The angle θ can be found using the sine rule $\dfrac{a}{\sin A} = \dfrac{b}{\sin B}$.

$$\frac{5.0}{\sin\theta} = \frac{7.0}{\sin 120}$$

$$\sin\theta = \frac{5.0 \times \sin 120}{7.0} = 0.6186$$

$$\theta = 38°$$

The magnitude of the resultant force is 7.0 N at an angle of 38° relative to the 3.0 N force.

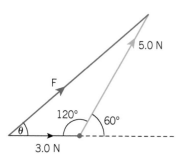

▲ **Figure 4** *A vector triangle with angles and forces shown*

Technique 3 – Calculations using vector resolution

This technique relies on choosing convenient perpendicular axes. One of the vectors is resolved along each axis so that the magnitude of the resultant vector can be determined using Pythagoras' theorem (Figure 5).

total force in x direction $= 3.0 + 5.0\cos 60° = 5.5\,\text{N}$

total force in y direction $= 5.0\sin 60° = 4.33\,\text{N}$

resultant force $F = \sqrt{5.5^2 + 4.33^2} = 7.0\,\text{N}$

$$\theta = \tan^{-1}\frac{4.33}{5.5} = 38°$$

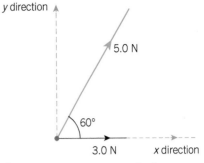

▲ **Figure 5** *Two non-perpendicular vectors shown as part of a right-angled triangle*

Subtracting vectors

Two vectors are represented by **X** and **Y**. To subtract **Y** from **X**, you simply reverse the direction of **Y** and then add this new vector to **X** (Figure 6).

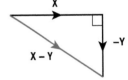

▲ **Figure 6** *Subtracting vectors*

Summary questions

1 Three forces act on an object (Figure 7). Calculate the magnitude and direction of the resultant force. *(4 marks)*

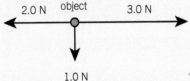

◀ **Figure 7**

2 Two tugboats are pulling a ship, each with a force of 8.0 kN, and with an angle of 40° between the cables (Figure 8). Calculate the magnitude and direction of the resultant force.

a Overhead view $F_2 = 8.0\,\text{kN}$

▲ **Figure 8**

(4 marks)

3 Three tugboats are towing an object at sea. The forces and angles between the cables are shown in Figure 9. Calculate the magnitude and direction of the resultant force on the object.

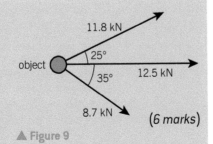

▲ **Figure 9**

(6 marks)

4 Figure 10 shows two vectors, **A** and **B**. Determine the magnitude and the direction of the resultant vector **A** – **B**. *(4 marks)*

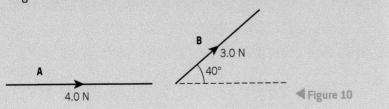

◀ **Figure 10**

Chapters in this module

Introduction

Force and motion are tightly knitted together. They form a central part of every physicist's understanding of the Universe around us.

In this module you will learn how to mathematically model the motion of objects and will develop your understanding of the effects forces have on objects. You will also learn about the important connection between force and energy.

Motion explores the key ideas used to describe and analyse motion in both one and two dimensions, including the motion of Olympic swimmers, sprinting cheetahs, and parachutists jumping from the very edge of space.

Forces in action develops ideas about the effect of forces on objects. In this chapter you will learn how the motion of an object changes when it experiences a resultant force, and how several balanced forces are essential in contexts including rock climbing and bridge building.

Work, energy, and power explores the important link between work done and energy. You will learn how to apply the important principle of conservation of energy to situations from wind turbines to roller coasters.

Materials introduces several ideas that are essential in engineering. In this chapter you will learn how to classify different materials according to their properties, and the mathematics of the differences between a bungee cord and the latest aluminium alloy.

Laws of motion and momentum will enable you to combine the ideas developed in the previous chapters. You will learn how Newton's laws are used to predict the motion of all colliding or interacting objects, from astronauts in the International Space Station to the humble supermarket shopping trolley.

Knowledge and understanding checklist

From your Key Stage 4 study you should be able to do the following. Work through each point, using your Key Stage 4 notes and the support available on Kerboodle.

☐ Relate changes and differences in motion to appropriate distance–time and velocity–time graphs.

☐ Apply formulae relating distance, time, and speed for uniform motion, and for motion with uniform acceleration.

☐ Recall examples of ways in which objects interact and describe how such examples involve interactions between pairs of objects that produce a force on each object.

☐ Apply Newton's first law to explain the motion of objects and apply Newton's second law in calculations relating forces, masses, and accelerations.

☐ Describe and calculate changes in energy and explain the definition of power as the rate at which energy is transferred.

☐ Calculate energy efficiency for any energy transfer, and describe ways to increase efficiency.

Maths skills checklist

All physicists use maths. In this unit you will need to use many different maths skills, including the following examples. You can find support for these skills on Kerboodle and through MyMaths.

☐ **Change the subject of an equation, including nonlinear equations.** You will need to be able to do this to solve mathematical problems when dealing with energies.

☐ **Use an appropriate number of significant figures.** You will need to be able to do this in solving a variety of problems in the motion topic, including projectiles.

☐ **Plot two variables from experimental or other data and use $y = mx + c$.** You will need to be able to do this when studying Hooke's law.

☐ **Calculate the gradient from a graph (including tangents).** You will need to be able to do this in experiments to determine g by free fall.

☐ **Understand the possible physical significance of the area between a curve and the x-axis and be able to calculate it or estimate it by graphical methods.** You will need to be able to do this when studying motion graphs.

3 MOTION

3.1 Distance and speed

Specification reference: 3.1.1

Average speed

Laws limit the speed at which vehicles can travel on roads. Speed cameras help the police enforce these limits. Average speed check areas use cameras along the journey (Figure 1). Automatic numberplate recognition technology identifies individual vehicles, so the time they take to travel between cameras can be measured. If the distance between the cameras is also known, the average speed can be calculated.

Calculating average speed

The **average speed** v of an object can be calculated from the distance x travelled and the time t taken using the equation

$$\text{average speed} = \frac{\text{distance travelled}}{\text{time taken}}$$

You can write this in algebraic form as

$$v = \frac{\Delta x}{\Delta t}$$

The Greek capital letter Δ (delta) means 'change in'. From the SI base units for distance and time, the unit of speed is $m\,s^{-1}$.

 Worked example: Fine or not?

A car travels 2.5 km in 1 minute 22 seconds. The average speed limit for the road is 50 mph ($22\,m\,s^{-1}$). Has the driver exceeded this average speed limit?

Step 1: Identify the equation and list the known values.

$$v = \frac{\Delta x}{\Delta t}$$

$$\Delta x = 2500\,m, \Delta t = 1 \text{ minute } 22 \text{ seconds} = 60 + 22 = 82\,s$$

Step 2: Substitute the values into the equation and calculate the answer.

$$v = \frac{2500}{82} = 30.49\,m\,s^{-1}$$
$$v = 30\,m\,s^{-1} \text{ (2 s.f.)}$$

Yes, the driver has exceeded the average speed limit.

▲ **Figure 1** *To protect the people repairing the road, your average speed on this section of the road must not be more than 50 miles per hour*

Distance–time graphs

Graphs of distance against time are used to represent the motion of objects.

- Distance is plotted on the *y*-axis (vertical axis).
- Time is plotted on the *x*-axis (horizontal axis).

In a distance–time graph, a stationary object is represented by a horizontal straight line. An object moving at a **constant speed** is represented by a straight, sloping line. The **gradient** of that line is equal to the distance travelled divided by the time taken, $\Delta x/\Delta t$, in other words, to the speed of the object.

- speed = gradient of a distance–time graph

In Figure 2

change in distance $\Delta x = 1400 - 400 = 1000\,\text{m}$

change in time $\Delta t = 70 - 20 = 50\,\text{s}$

speed $v = \dfrac{\Delta x}{\Delta t} = \dfrac{1\,000}{50} = 20\,\text{m s}^{-1}$

Instantaneous speed

A criticism of checks on average speed is that a vehicle can travel faster than the average speed allowed for part of the journey, then travel slowly to increase the total journey time sufficiently to avoid a fine.

Instantaneous speed is the speed of the car over a very short interval of time. The instantaneous speed at a particular time is found by drawing the tangent to the distance–time graph at that time, then determining the gradient of this tangent (Figure 3). The greater the gradient, the greater the instantaneous speed.

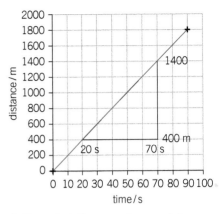

▲ **Figure 2** *A distance–time graph for an object moving at constant speed – the gradient of the graph represents the speed*

Synoptic link

You will find more information on graphs in Appendix A2, Recording results.

Study tip

When you calculate a gradient, make sure the triangle that you draw on the graph is large enough to provide an accurate answer. It is advisable to use more than half the length of the line to determine your gradient.

Summary questions

1 Calculate the average speed in m s⁻¹ for the following bicycle journeys:
   a a distance of 180 m covered in a time of 9.0 s; (*2 marks*)
   b a distance of 2.0 km covered in 6.5 minutes. (*2 marks*)

2 A snail travels 19.2 m in 1 day (24 hours).
   Calculate its average speed in m s⁻¹. (*2 marks*)

3 A car travels for 19 s at an average speed of 31 m s⁻¹.
   How far does it travel? (*2 marks*)

4 An aircraft travels at an average speed of 240 m s⁻¹ for 12 000 km.
   Calculate the time taken in seconds and in hours. (*3 marks*)

5 A lorry travels on a motorway for 2.0 minutes at a constant speed of 25 m s⁻¹. It then struggles on a hill and travels 800 m in 50 s. Calculate:
   a the total distance travelled; (*2 marks*)
   b the average speed of the lorry. (*2 marks*)

6 Use Figure 3 to determine the instantaneous speed of the object at time *t* = 80 s. (*3 marks*)

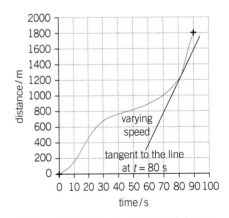

▲ **Figure 3** *A distance–time graph for an object moving at varying speed – you can determine the instantaneous speed from the gradient of the tangent to the graph*

3.2 Displacement and velocity

Specification reference: 3.1.1

▲ **Figure 1** *Each swimmer has an average speed, but if they do two laps of the pool they all have an average velocity of zero*

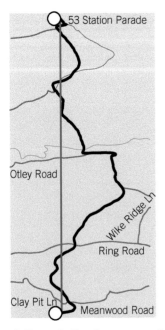

▲ **Figure 2** *The distance travelled by road on a journey from Harrogate to Leeds is different from the displacement*

Scalar and vector quantities

A 100 m swimming race in an Olympic-sized pool takes two laps. During the London Olympic Games in 2012, the American swimmer Missy Franklin beat a previous world record for the backstroke with her time of 58.33 s, giving her an average speed of $1.71\,\text{m s}^{-1}$. However, she finished where she started, which means that, despite all the hard work she did in racing 100 m, her total displacement and average velocity were zero.

The vector nature of velocity

Displacement s is a vector quantity, unlike distance, which is scalar. Displacement has both magnitude and direction. Speed is a scalar quantity calculated from distance, but velocity is a vector quantity calculated from displacement. The **average velocity** v of an object can be calculated from the change in displacement and the time taken.

$$\text{average velocity} = \frac{\text{change in displacement}}{\text{time taken}}$$

You can write this in algebraic form as

$$v = \frac{\Delta s}{\Delta t}$$

where Δs is the change in displacement and Δt is the time taken. The SI unit for velocity is m s^{-1}.

The worked example below shows that average speed and average velocity are very different quantities.

 Worked example: Speed and velocity

a Leeds is about 21 km south of Harrogate, but the distance by road is about 24 km (Figure 2). It takes 37 minutes to travel from Harrogate to Leeds by road. Calculate the average speed and the average velocity.

Step 1: Identify the equations needed.

$$\text{average speed } v = \frac{\Delta x}{\Delta t}, \text{ average velocity } v = \frac{\Delta s}{\Delta t}$$

Step 2: Substitute the values in SI units into the equations and calculate the answers.

$$\text{time taken } \Delta t = 60 \times 37 = 2220\,\text{s}$$

$$\text{average speed } v = \frac{\Delta x}{\Delta t} = \frac{24\,000}{2220} = 10.8\,\text{m s}^{-1} = 11\,\text{m s}^{-1}\ (2\text{ s.f.})$$

$$\text{average velocity } v = \frac{\Delta s}{\Delta t} = \frac{21\,000}{2220} = 9.5\,\text{m s}^{-1}\ (2\text{ s.f.})$$

The magnitude of the average velocity is $9.5\,\text{m s}^{-1}$ and its direction is due south from Harrogate.

b What would happen to the magnitude of the average velocity if the journey was from Harrogate to Leeds and then back to Harrogate?

The overall change in displacement would be zero and therefore the average velocity would be zero.

Displacement–time graphs

Graphs of displacement against time are used to represent the motion of objects.

- Displacement is plotted on the *y*-axis (vertical axis).
- Time is plotted on the *x*-axis (horizontal axis).

Figure 3 shows the displacement–time graph for a car travelling along a straight road. The car is travelling at a constant velocity between $t = 0$ and $t = 20\,s$, as can be seen from the first straight-line section of the graph. The horizontal section of the graph between $t = 20\,s$ and $t = 30\,s$ shows that the displacement of the car remains constant. Therefore, the car must be stationary. After $t = 30\,s$, the graph is still a straight line but has a negative slope. The displacement of the car is getting smaller with time. The car must therefore be returning at a constant velocity.

You can determine the velocity of an object from the gradient of its displacement–time (*s–t*) graph. If the graph is not a straight line, draw a tangent to the graph, then calculate the gradient of this tangent for the instantaneous velocity, as illustrated in Figure 4.

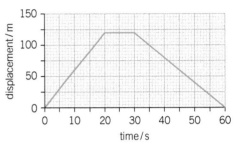

▲ **Figure 3** *A displacement–time graph for a car journey*

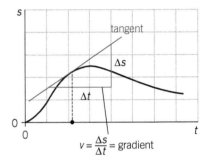

▲ **Figure 4** *Velocity can be determined from the gradient of the displacement–time graph*

 Worked example: Forwards and backwards

Use Figure 3 to determine the velocity of the car at $t = 10\,s$ and $t = 40\,s$.

Step 1: Identify the equation needed and how to obtain the values from the graph.

$$\text{velocity } v = \frac{\Delta s}{\Delta t}$$

The right-hand terms are equivalent to the change on the *y*-axis (*s*) divided by the change on the *x*-axis (*t*).

Step 2: Substitute the values into the equation and calculate the answer.

The velocity at $t = 10\,s$ can be determined from the gradient of the straight-line graph between $t = 0$ and $t = 20\,s$.

$$\text{velocity } v = \frac{\Delta s}{\Delta t} = \frac{120-0}{20-0} = 6.0\,\text{m s}^{-1}$$

Synoptic link

You will find more information about gradients and tangents in Appendix A2, Recording results.

The velocity at $t = 40\,s$ can be determined from the gradient of the straight-line graph between $t = 30\,s$ and $t = 60\,s$.

$$\text{velocity } v = \frac{\Delta s}{\Delta t} = \frac{0-120}{60-30} = -4.0\,m\,s^{-1}$$

The negative sign for the velocity shows that the car is travelling in the opposite direction to its motion between 0 and 20 s.

Summary questions

1 Describe the journey of the object with the displacement–time $(s\text{–}t)$ graph shown in Figure 5. *(3 marks)*

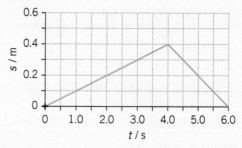

▲ **Figure 5**

2 Determine the velocity of the object in Figure 5 at time $t = 2.0\,s$ and $t = 5.0\,s$. *(5 marks)*

3 Determine the average speed of the object in Figure 5 between time $t = 0.0\,s$ and $t = 6.0\,s$. *(5 marks)*

4 A particle travels in a circular path of radius 80 cm. It starts from point **A** and takes 8.0 s to travel once round the circle (Figure 6).

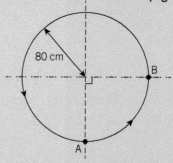

▲ **Figure 6**

Calculate the:
a average speed of the particle; *(3 marks)*
b average velocity of the particle from **A** to **B**. *(3 marks)*

3.3 Acceleration

Specification reference: 3.1.1

Bursts of acceleration

In 2013, scientists fitted wild cheetahs with collars containing a global positioning system (GPS) module and electronic motion sensors, so the animals' velocity and acceleration could be studied. The cheetahs ran at speeds of up to $26\,\mathrm{m\,s^{-1}}$ and were able to turn very quickly, with an acceleration of up to $13\,\mathrm{m\,s^{-2}}$, more than that of the fastest production car that year, the Bugatti Veyron.

Determining acceleration

The **acceleration** of an object is defined as the rate of change of velocity. In mathematical form, the acceleration a is

$$a = \frac{\Delta v}{\Delta t}$$

where Δv is the change in velocity and Δt is the time taken for the change. The unit of acceleration is $\mathrm{m\,s^{-2}}$. Since acceleration is determined from velocity, it too is a vector quantity – it has magnitude and direction. A negative acceleration is often called deceleration.

Acceleration can be determined by calculation, or from a velocity–time (v–t) graph.

▲ **Figure 1** *Cheetahs need to be able to make rapid changes in speed and direction in order to hunt*

Calculating acceleration

You can calculate acceleration if you know the change in velocity of an object and the time taken for this change.

 Worked example: 0 to 62, then slam the brakes on

a   A Bugatti Veyron can accelerate from 0 to $100\,\mathrm{km/h}$ ($27.8\,\mathrm{m\,s^{-1}}$) in $2.46\,\mathrm{s}$. Calculate its average acceleration.

Step 1: Identify the equation needed.

$$a = \frac{\Delta v}{\Delta t}$$

Step 2: Substitute the values into the equation and calculate the answer.

$$a = \frac{\Delta v}{\Delta t} = \frac{27.8 - 0}{2.46} = 11.3\,\mathrm{m\,s^{-2}}$$

b   The car takes $2.34\,\mathrm{s}$ to stop from $100\,\mathrm{km/h}$ under braking. Calculate its acceleration, assuming that this is constant during braking.

$$a = \frac{\Delta v}{\Delta t} = \frac{0 - 27.8}{2.34} = -11.9\,\mathrm{m\,s^{-2}}$$

The negative sign means that the velocity of the car is decreasing over time – it is decelerating.

▲ **Figure 2** *The Bugatti Veyron has similar acceleration to that of a cheetah*

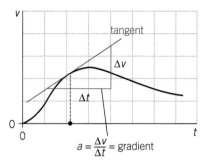

▲ Figure 3 *Acceleration can be determined from the gradient of the velocity–time graph*

Velocity–time graphs

Since $a = \frac{\Delta v}{\Delta t}$, it follows that the acceleration of an object can be determined from the gradient of a velocity–time graph. You have seen how to determine gradients for straight-line graphs and for non-linear graphs in the previous topics in this chapter; the only difference in this case is that the y-axis of the graph represents velocity v rather than displacement s.

● acceleration = gradient of velocity–time graph

Figure 4 shows how the motion of an object can be deduced from the velocity–time graph.

  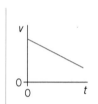

| A straight line of constant, positive gradient: constant acceleration. | A straight line of zero gradient: constant velocity or zero acceleration. | A straight line of constant negative gradient: constant deceleration. | A curve with changing gradient: acceleration is changing. |

▲ Figure 4 *Interpreting velocity–time graphs*

Summary questions

1 A racing cyclist starts from rest and reaches a velocity of $8.0\,\mathrm{m\,s^{-1}}$ after 12 s. Calculate the acceleration of the cyclist. *(2 marks)*

2 A train slows down from $40\,\mathrm{m\,s^{-1}}$ to $10\,\mathrm{m\,s^{-1}}$ over 60 s. Calculate the magnitude of the deceleration of the train. *(2 marks)*

3 Describe the journey of the object represented in the velocity–time graph shown in Figure 5. *(2 marks)*

4 a Use Figure 5 to determine the maximum acceleration of the object. Explain your answer. *(2 marks)*
  b Sketch an acceleration–time graph from $t = 0$ to $t = 3.0\,\mathrm{s}$. *(2 marks)*

5 A ball is held above the surface of water and released. It falls through the air and then through the water. Figure 6 shows the velocity–time graph for this ball.
  a Calculate the acceleration of the ball as it falls through the air. *(2 marks)*
  b Determine the magnitude of the deceleration at time $t = 1.0\,\mathrm{s}$. *(3 marks)*

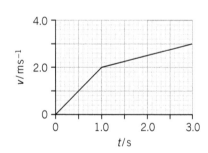

▲ Figure 5

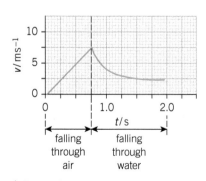

▲ Figure 6

3.4 More on velocity–time graphs

Specification reference: 3.1.1

The spy in the cab

Tiredness can affect the ability to drive safely as much as alcohol can. The law sets time limits to ensure that drivers of heavy vehicles take proper rest breaks and do not drive whilst tired. Tachographs record the speed and distance travelled by a vehicle. Modern tachographs are digital, but older ones use a stylus to record the information on a circular chart. This chart rotates once in 24 hours, providing a permanent record of the journey so that the authorities can check that the driver has taken regular breaks.

Area under the graph

In the previous topic you saw that acceleration can be determined from the gradient of a velocity–time graph. In addition, we can read the displacement of the object from the area under the graph. Figure 2 shows why this is so.

You will recall that the average velocity v is given by the equation

$$v = \frac{\Delta s}{\Delta t}$$

where Δs is the displacement in the time interval Δt. For the instantaneous velocity, assume that Δt is very small indeed – the velocity of the object is not going to change much. The change in the displacement $\Delta s \approx v\Delta t$. If you look at the graph in Figure 2, this is the area of the very thin rectangular strip marked under the graph. Therefore, the change in displacement is equal to the area of this strip. If you add similar strips for a longer interval of time, then clearly the area under the velocity–time graph is the total displacement of the object.

Calculating displacement for constant accelerations

Displacement is easy to calculate when the acceleration is constant, because the areas can be broken down into rectangles and right-angled triangles. This is illustrated in the worked example below for the short journey of a cyclist.

Learning outcomes

Demonstrate knowledge, understanding, and application of:

→ velocity–time graphs to determine displacement.

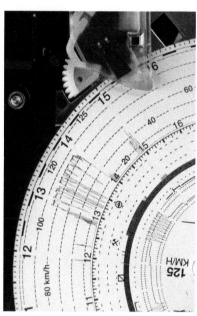

▲ **Figure 1** *An analogue tachograph record of road speed against time*

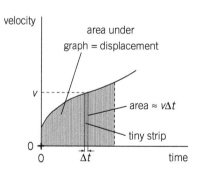

▲ **Figure 2** *The area under the velocity–time graph is equal to displacement*

 Worked example: Displacement of a cyclist

The velocity–time graph for a cyclist travelling along a straight road is shown in Figure 3.

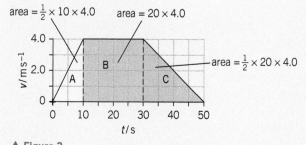

▲ Figure 3

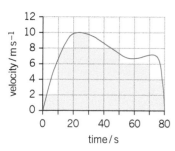

▲ **Figure 4** *Calculating the area under a non-linear velocity–time graph*

Calculate the total distance travelled by the cyclist (the cyclist's displacement) in the period of 50 s.

Step 1: Identify the method needed.

distance travelled = area between the graph and the time axis

Step 2: Calculate the answer.

distance travelled = area of triangle **A** + area of rectangle **B** + area of triangle **C**

$$= \left(\tfrac{1}{2} \times 10 \times 4.0\right) + \left(20 \times 4.0\right) + \left(\tfrac{1}{2} \times 20 \times 4.0\right)$$
$$= 20 + 80 + 40$$
$$= 140\,\text{m}$$

Step 3: There is an alternative method to determine the area under the graph. Calculate the area of the trapezium using the formula

$$\text{area} = \tfrac{1}{2} \times (\text{sum of the parallel sides}) \times \text{vertical height}$$
$$\text{distance} = \tfrac{1}{2} \times (50 + 20) \times 4.0 = 140\,\text{m}$$

Calculating displacement for changing accelerations

For non-linear velocity–time graphs, you can determine the area under the graph by counting squares. Taking Figure 4 as an example, you would start by counting the squares that are complete or nearly complete (yellow). Then count the remaining squares that lie mostly beneath the graph. Omit squares that are mostly above the graph.

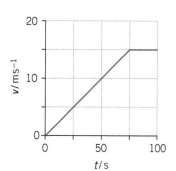

▲ **Figure 5**

Summary questions

1   The velocity–time graph for a car travelling on a straight road is shown in Figure 5. Without carrying out calculations, describe how the acceleration of the car varies between $t = 0$ s and $t = 100$ s.   (*2 marks*)

2   Use the velocity–time graph shown in Figure 5 to calculate:
   a   the total distance travelled by the car in 100 s;   (*3 marks*)
   b   the average velocity of the car.   (*2 marks*)

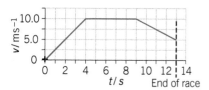

▲ **Figure 6**

3   The velocity–time graph for a sprinter running along a straight track is shown in Figure 6.
   a   Calculate the total distance travelled by the runner in 13 s.   (*4 marks*)
   b   Calculate the average velocity of the runner.   (*2 marks*)

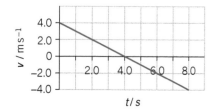

▲ **Figure 7**

4   A ball is given an initial velocity up a ramp. The subsequent motion of the ball is illustrated in the velocity–time graph in Figure 7.
   a   Describe the motion of the ball.   (*3 marks*)
   b   Calculate the total distance travelled by the ball up and down the ramp.   (*2 marks*)
   c   What are the average speed and average velocity of the ball between $t = 0$ and $t = 8.0$ s? Explain your answers.   (*4 marks*)

3.5 Equations of motion

Specification reference: 3.1.2

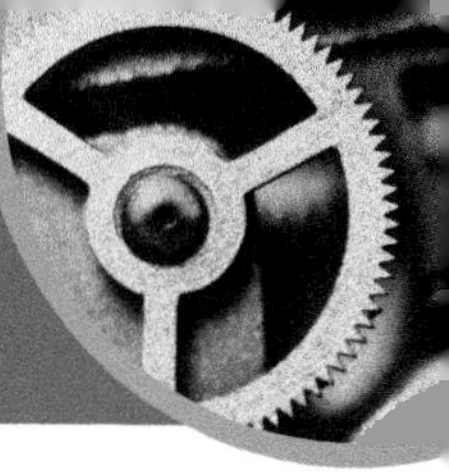

Predicting motion

We can use a knowledge of physics to predict the motion of accelerating or decelerating objects: for example, the impact speed of a small meteor about to hit the Earth, the initial speed of a car from skid marks left on the road, and the final speed of a space probe landing on a distant planet. It is amazing that all this is possible with just four equations.

Learning outcomes

Demonstrate knowledge, understanding, and application of:

→ the equations of motion for constant acceleration in a straight line.

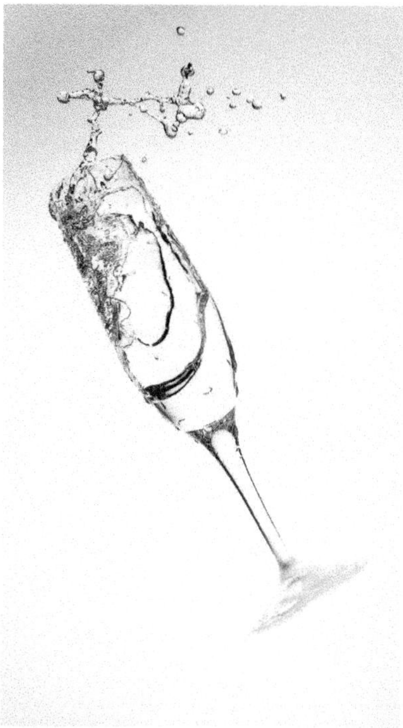

▲ **Figure 1** *Can you predict the time of fall for this glass of water?*

Equations of motion: the *suvat* equations

You need four equations to calculate quantities involving motion in a *straight line* at a constant acceleration. These equations of motion are often informally referred to as the '*suvat* equations' after the symbols for the quantities involved.

Deriving the equations of motion

Figure 2 shows the velocity–time graph for an accelerating object. The initial velocity of the object is u. After a time t the final velocity of the object is v. The object has a constant acceleration a, as you can see from the straight-line graph.

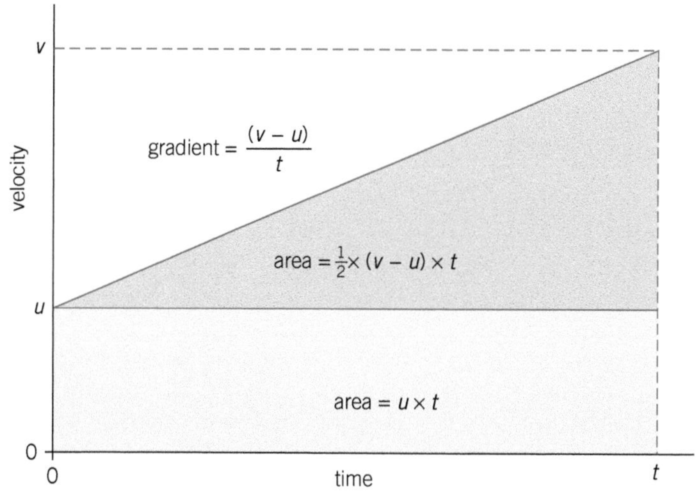

▲ **Figure 2** *Velocity–time graph showing how the* suvat *equations are derived*

▼ **Table 1** *The* suvat *quantities*

Symbol	Quantity
s	displacement (or distance travelled)
u	initial velocity
v	final velocity
a	acceleration
t	time taken for the change in velocity

Equation without *s*

From the graph in Figure 2

$$a = \frac{\Delta v}{\Delta t} = \frac{v - u}{t}$$

This can be rearranged to give

$$\boxed{v = u + at} \qquad \textbf{(1)}$$

Equation without *v*

You will recall that the area under the velocity–time graph is equal to the displacement s.

- the rectangular area $= ut$
- the triangular area $= \frac{1}{2} \times (v - u) \times t$

From Equation **1** above, $(v - u) = at$. If you substitute this into the expression for the area of the triangle, you get $\frac{1}{2} \times at \times t$. With ut for the area of the rectangle, this gives the total area s.

$$s = ut + \frac{1}{2}at^2 \qquad (2)$$

Equation without a

If you treat the area under the graph as the area of a trapezium (with u and v as the parallel sides of the trapezium, and t as the perpendicular separation between them), this becomes

$$s = \frac{1}{2}(u + v)t \qquad (3)$$

In other words, the displacement s is the average velocity, $\frac{u+v}{2}$, multiplied by the time t.

Equation without t

You need Equations **1** and **3** to derive the last useful equation of motion.

According to Equation **1** the time t is given by the equation

$$t = \frac{(v - u)}{a}$$

This equation for t can be substituted into Equation **3** to give

$$s = \frac{1}{2}(u + v) \times \frac{(v - u)}{a}$$

Rearranging this gives

$$(u + v)(v - u) = 2as$$
$$v^2 - u^2 = 2as$$
$$\boxed{v^2 = u^2 + 2as} \qquad (4)$$

> **Study tip**
>
> The four $suvat$ equations are
>
> $v = u + at$
>
> $s = ut + \frac{1}{2}at^2$
>
> $s = \frac{1}{2}(u + v)t$
>
> $v^2 = u^2 + 2as$

> **+ Equation without u**
>
> There is one more equation,
> $s = vt - \frac{1}{2}at^2$.
>
> See if you can derive it from the velocity–time graph.

> **Study tip**
>
> You can always make a good start by jotting down the $suvat$ values to see what you know.

▲ **Figure 3** *Accelerating from rest*

> 🖩 **Worked example: Car accelerating from a standing start**
>
> A car on a straight road accelerates from rest to a velocity of $12\,\mathrm{m\,s^{-1}}$ in a time of $9.0\,\mathrm{s}$. Calculate the acceleration of the car and the distance travelled in this time.
>
> **Step 1:** Write down the quantities given in the order $suvat$ and identify the equations needed.
>
> $$s = ?, \ u = 0, \ v = 12\,\mathrm{m\,s^{-1}}, \ a = ?, \ t = 9.0\,\mathrm{s}$$
>
> Use Equation **1** to calculate the acceleration a and Equation **3** to calculate s.
>
> **Step 2:** Substitute the values into the equations and calculate the answers.
>
> →

Acceleration can be calculated from Equation **1**.

$$v = u + at$$

$$12 = 0 + a \times 9.0$$

$$a = \frac{12 - 0}{9.0} = 1.33 \, \text{m} \, \text{s}^{-2}$$

The distance travelled is the displacement s along the straight road. Equation **3** can be used to calculate s.

$$s = \frac{1}{2}(v + u)t$$

$$s = \frac{1}{2} \times (12 + 0) \times 9.0$$

$$s = 54 \, \text{m}$$

 Worked example: Particle accelerating

A particle travels a distance of 16 m as it accelerates from $4.0 \, \text{m} \, \text{s}^{-1}$ to $12 \, \text{m} \, \text{s}^{-1}$. Calculate its acceleration.

Step 1: Again, start with the *suvat* values and identify the equation needed.

$$s = 16 \, \text{m}, \, u = 4.0 \, \text{m} \, \text{s}^{-1}, \, v = 12 \, \text{m} \, \text{s}^{-1}, \, a = ?, \, t = ?$$

The equation for v, u, and s is Equation **4.** We can use this to calculate the acceleration a.

Step 2: Substitute the values into the equation and calculate the answer.

$$v^2 = u^2 + 2as$$

$$12^2 = 4.0^2 + 2 \times a \times 16$$

$$a = \frac{12^2 - 4.0^2}{2 \times 16} = 4.0 \, \text{m} \, \text{s}^{-2}$$

> **Study tip**
>
> Remember that acceleration is the change in velocity with time, so although this first sentence refers to acceleration the values given are two different velocities. If you are ever confused, simply look at the units.

> **Synoptic link**
>
> You can find more information about units in Appendix A1, Physical quantities and units.

 Worked example: Falling to Earth

An apple falls from rest in a tree towards soft ground from a height of 1.50 m. Objects falling to Earth have an acceleration (free fall) of $9.81 \, \text{m} \, \text{s}^{-2}$. Calculate the time taken for the apple to reach the ground. Assume air resistance has negligible effect on the motion.

Step 1: List the *suvat* values and identify the equation needed.

$$s = 1.50 \, \text{m}, \, u = 0 \, \text{m} \, \text{s}^{-1}, \, a = 9.81 \, \text{m} \, \text{s}^{-2}, \, t = ?$$

Use Equation **2** to calculate the time t.

> **Study tip**
>
> Note that $0 \times t = 0$ and not t.

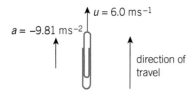

▲ **Figure 4**

Summary questions

1 A vehicle accelerated from $13.4\,m\,s^{-1}$ to $22.3\,m\,s^{-1}$ in $8.70\,s$. Calculate the distance travelled. *(3 marks)*

2 A runner changes her velocity from $3.2\,m\,s^{-1}$ to $4.2\,m\,s^{-1}$. In this time she travels $200\,m$. Calculate her acceleration. *(3 marks)*

3 In 2009 Usain Bolt sprinted $100\,m$ in $9.58\,s$, setting a record. Assuming that he travelled at a constant acceleration from start to the finish, calculate his acceleration. *(3 marks)*

4 A dragster travels 0.25 miles in a time of $4.6\,s$ along a straight track from a standing start. 1 mile $\approx 1600\,m$. Calculate:
  a its acceleration; *(3 marks)*
  b its final velocity. *(3 marks)*

5 An apple is dropped from a tall building. Calculate the distance it travels between time $t = 3.0\,s$ and $t = 5.0\,s$. Assume the acceleration of the falling apple is $9.81\,m\,s^{-2}$. *(4 marks)*

6 A car is travelling at $28\,m\,s^{-1}$. The driver applies the brakes. The car skids for a distance of $30\,m$ before stopping. Calculate the magnitude of the deceleration of the car. *(3 marks)*

Step 2: Substitute the values into the equation and calculate the answer.

$$s = ut + \frac{1}{2}at^2$$

$$1.50 = (0 \times t) + \frac{1}{2} \times 9.81 \times t^2$$

$$t^2 = \frac{2 \times 1.50}{9.81} = 0.306 \ (3 \text{ s.f.})$$

$$t = 0.553\,s$$

In this calculation, the equation is effectively $s = \frac{1}{2}at^2$, because $u = 0$.

 Worked example: What goes up

A paper clip is flicked vertically up in the air at $6.0\,m\,s^{-1}$ (Figure 4). Calculate its maximum height. Assume air resistance has negligible effect on the motion.

Step 1: List the known *suvat* values and identify the equation needed.

The paper clip will decelerate as it moves vertically, therefore, a must be *negative*. At maximum height it will stop momentarily, therefore, $v = 0$.

$$s = ?, u = 6.0\,m\,s^{-1}, v = 0\,m\,s^{-1}, a = -9.81\,m\,s^{-2}$$

The equation for v, u, and s is Equation **4**. We can use this to calculate the height s.

Step 2: Substitute the values into the equation and calculate the answer.

$$v^2 = u^2 + 2as$$

$$0 = u^2 + 2as$$

$$s = -\frac{u^2}{2a} = -\frac{6.0^2}{2 \times -9.81}$$

$$s = 1.83\,m = 1.8\,m \ (2 \text{ s.f.})$$

Stopping distances

Modern road vehicles transport people in comfort at speeds that would have astonished our great-grandparents. This speed is not always a good thing. An alert driver who has left enough room can stop in good time in good driving conditions on seeing a hazard. However, in an emergency or in poor conditions, the vehicle may still be moving when it meets an obstacle, even if the driver is braking hard.

Components of stopping distances

The **stopping distance** is the total distance travelled from when the driver first sees a reason to stop, to when the vehicle stops. It has two components:

- **thinking distance**, the distance travelled between the moment when you first see a reason to stop, to the moment when you use the brake
- **braking distance**, the distance travelled from the time the brake is applied until the vehicle stops.

Many factors influence these distances, including the speed of the vehicle, the condition of the brakes, tyres, and road, the weather conditions, and the alertness of the driver.

Thinking distance

It takes time for a driver to react to a need to stop. For a vehicle moving at constant speed

$$\text{thinking distance} = \text{speed} \times \text{reaction time}$$

 Worked example: Reaction time

In the UK Highway Code, the thinking distance at 30 mph ($13.4\,\text{m s}^{-1}$) is shown as 9.0 m. Calculate the corresponding reaction time.

Step 1: Identify the equation needed.

$$\text{reaction time} = \frac{\text{thinking distance}}{\text{speed}}$$

Step 2: Substitute the values into the equation and calculate the answer.

$$\text{reaction time} = \frac{9.0}{13.4} = 0.67\,\text{s (2 s.f.)}$$

The greater the speed or the reaction time, the further a vehicle will travel before its driver applies the brakes. Assuming a constant reaction time of 0.67 s, the thinking distance will be about 21 m at the UK national speed limit of 70 mph ($31.1\,\text{m s}^{-1}$). This is equivalent to the total length of five average cars lined up.

▲ **Figure 1** *Some motorways have chevron markings to help drivers judge the safe distance from the vehicle in front*

Study tip

Do not confuse distance and time. If you write that 'it takes longer to stop' you must make it clear whether you mean the distance or the time.

▲ **Figure 2** *Vehicles following each other too closely may not be able to stop in a short enough distance in an emergency*

Worked example: Braking distance

Step 1: Once again, start with the *suvat* quantities and identify the equation you need.

$s = 14.0\,\text{m}$

$u = 13.4\,\text{m s}^{-1}$

$v = 0$

$a = ?$

Use the equation $v^2 = u^2 + 2as$.

Step 2: Substitute the values into the equation and calculate the answer.

$$a = \frac{v^2 - u^2}{2s}$$

$v = 0$

Therefore

$$a = -\frac{u^2}{2s}$$

$$= -\frac{13.4^2}{2 \times 14.0}$$

$$= -6.4\,\text{m s}^{-2}\ (2\ \text{s.f.})$$

The magnitude of the deceleration is about $6.4\,\text{m s}^{-2}$.

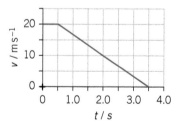

▲ **Figure 3**

Braking distance

In the UK Highway Code, the braking distance at 30 mph ($13.4\,\text{m s}^{-1}$) is shown as 14.0 m. If you assume constant deceleration from $13.4\,\text{m s}^{-1}$ to $0\,\text{m s}^{-1}$, you can use one of the equations of motion to determine the magnitude of the deceleration.

▼ **Table 1** *Thinking, braking, and overall stopping distances according to the Highway Code*

Speed / mph	20	30	40	50	60	70
Speed / m s⁻¹	8.9	13.4	17.8	22.2	26.7	31.1
Thinking distance / m	6	9	12	15	18	21
Braking distance / m	6	14	24	38	55	75
Stopping distance / m	12	23	36	53	73	96

Summary questions

1 The reaction time of a tired driver is 1.5 s. The speed of the car is $22\,\text{m s}^{-1}$. The braking distance of the car is 38 m. Calculate the stopping distance of the car. *(3 marks)*

2 According to a student, thinking distance is directly proportional to the speed of the car. Show that this is the case. *(2 marks)*

3 Use Table 1 to answer this question. A car is travelling at 70 mph ($31.1\,\text{m s}^{-1}$) on the motorway when it has to stop for an emergency. Calculate:
   a the deceleration of the car when travelling at this speed; *(4 marks)*
   b the time taken for the car to stop when the brakes are applied. *(3 marks)*

4 The velocity–time graph in Figure 3 shows the motion of a car from the instant the driver sees a hazard on the road.

   Calculate the thinking, braking, and stopping distances. Explain your answer. *(3 marks)*

5 According to a student, braking distance is directly proportional to the $(\text{speed})^2$. Show that this is the case. *(3 marks)*

3.7 Free fall and *g*

Specification reference: 3.1.2

From the edge of space

Felix Baumgartner made a record-breaking leap from the edge of space on 14th October 2012. A giant helium-filled balloon lifted his capsule 39.0 km above the surface of the Earth. Then he stepped off. Baumgartner accelerated as he fell, reaching a maximum speed of $380 \, m \, s^{-1}$ – greater than the speed of sound – after just 50 seconds. He fell 36.4 km in 4 minutes 20 seconds before deploying his parachute and landing safely just under 5 minutes later.

Acceleration due to gravity

Objects with mass exert a gravitational force on each other. The Earth is so massive that its gravitational pull is enough to keep us on its surface. An object released on the Earth will accelerate vertically downwards towards the centre of the Earth. When an object is accelerating under gravity, with no other force acting on it, it is said to be in **free fall**. The **acceleration of free fall** is denoted by the label g (not g, which means grams). Since g is an acceleration, it has the unit $m \, s^{-2}$.

Value for *g* close to Earth's surface

The value for g varies depending upon factors including altitude, latitude, and the geology of an area. For example, g is $9.825 \, m \, s^{-2}$ in Helsinki, $9.816 \, m \, s^{-2}$ in London, but only $9.776 \, m \, s^{-2}$ in Singapore. A value of $9.81 \, m \, s^{-2}$ is generally used.

Determining *g*

The basic idea behind determining g in the laboratory is to drop a heavy ball over a known distance and time its descent. The problem is that it all happens very quickly, about 0.45 s for a 1.0 m fall. Methods for measuring g are described here.

Electromagnet and trapdoor

An electromagnet holds a small steel ball above a trapdoor (Figure 2). When the current is switched off, a timer is triggered, the electromagnet demagnetises, and the ball falls. When it hits the trapdoor, the electrical contact is broken and the timer stops. The value for g is calculated from the height of the fall and the time taken.

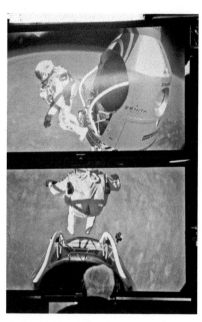

▲ **Figure 1** *Staff at the flight control centre monitoring Felix Baumgartner leaving his capsule*

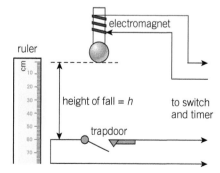

▲ **Figure 2** *Determining g using an electromagnet and timer*

Learning outcomes

Demonstrate knowledge, understanding, and application of:

→ the equations of motion for falling objects in a uniform gravitational field

→ the acceleration due to free fall *g*

→ an experiment to determine *g*.

Worked example: An experimental value for *g*

A ball drops 85.6 cm from an electromagnet to a trapdoor in 0.421 s. Use this information to determine a value for *g*.

Step 1: List the *suvat* values and identify the equation needed.

$s = 0.856$ m, $u = 0$ m s^{-1}, $a = g = ?$, $t = 0.421$ s (the distance must be converted into the SI unit m.)

Using $s = ut + \frac{1}{2} at^2$, we have

$$s = \frac{1}{2} at^2 \text{ because } u = 0$$

Step 2: Substitute the values into the equation and calculate the answer.

$$a = \frac{2s}{t^2} = \frac{2 \times 0.856}{0.421^2}$$
$$a = 9.66 \text{ m s}^{-2} \text{ (3 s.f.)}$$

The inaccuracy in this experiment is caused by the presence of air resistance and the slight delay in the release of the steel ball because of the finite time taken for the magnet to demagnetise. The accuracy may be improved by using a heavier ball and a much longer drop.

Light gates

The electromagnet and trapdoor introduce tiny delays into the timing. Instead we can use 'light gates', two light beams, one above the other, with detectors connected to a timer. When the ball falls through the first beam, it interrupts the light and the timer starts. When the ball falls through the second beam a known distance further down, the timer stops.

Taking pictures

A small metal ball is dropped from rest next to a metre rule, and its fall is recorded on video or with a camera in rapid-fire repeating mode. Alternatively, a stroboscope illuminates the scene with rapid flashes. The camera shutter is held open, producing a photograph with multiple images of the falling ball. The position of the ball at regular intervals is then determined by examining the recording.

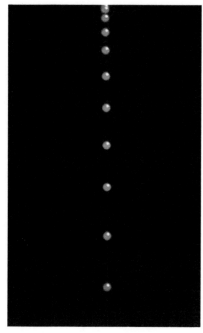

▲ **Figure 3** *The photos are taken at regular intervals and the distance between each image of the ball increases as it falls vertically towards the ground, showing that it is accelerating*

Synoptic link

You will find more information about lines of best fit in Appendix A3, Measurements and uncertainties.

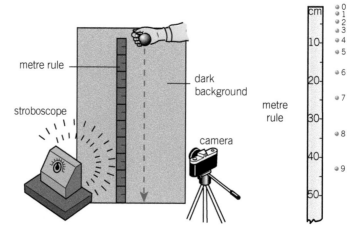

▲ **Figure 4** *Determining g using a camera and stroboscope*

Determining g by plotting a graph

The table shows data obtained from images of a ball in free fall.

The acceleration g of free fall can be determined using the equation $s = ut + \frac{1}{2}at^2$.

Since the object is dropped from rest and $a = g$

$$s = \frac{1}{2}gt^2$$

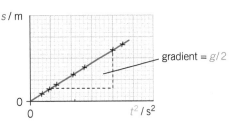

▲ **Figure 5** *Graph of the experimental results plotted as s against t²*

▼ **Table 1** *Results from the experiment in Figure 4*

Time of fall t / s	0.132	0.165	0.198	0.231	0.264	0.297	0.330	0.363	0.396	0.429
Distance fallen s / m	0.085	0.134	0.192	0.262	0.368	0.433	0.534	0.646	0.769	0.903

From the general equation for a straight-line graph, $y = mx + c$, you should get a straight line if you plot a graph of s against t^2 ($y = s$, $m = \frac{1}{2}g$, $x = t^2$, $c = 0$). The gradient, m, will be equal to $\frac{g}{2}$.

For each value of t, calculate t^2. Plot a graph of s (y-axis) against t^2 (x-axis). Draw a straight line of best fit, ignoring any anomalous points.

1 Determine the gradient of the line using a large triangle.
2 Calculate the experimental value for g.
3 What is the percentage difference between the experimental and accepted values for g?

Summary questions

1 Two different heavy objects are dropped from the same height. State the acceleration of free fall of each object. State any assumptions made. *(2 marks)*

2 A marble is dropped from a height H. It lands on the ground below after a time of 2.3 s. Calculate H. Assume the acceleration of free fall is 9.81 m s^{-2}. *(3 marks)*

3 Plan a simple experiment to estimate the acceleration g of free fall using a stopwatch and a tape measure. Explain how the experiment can be made precise. *(4 marks)*

4 A coin is dropped from the top of a bridge towards the water 9.5 m below. The coin is in free fall for a time of 1.5 s.
a Estimate the acceleration of free fall. *(3 marks)*
b Explain why the answer is not 9.81 m s^{-2}. *(1 mark)*

5 Figure 6 shows water dripping from a tap. The time between successive water drops is 0.040 s.
a Use the position of water drop **A** to determine the acceleration of free fall. *(3 marks)*
b Repeat (a) for another drop and therefore determine an average experimental value for g. *(4 marks)*

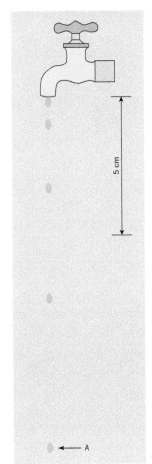

◀ Figure 6

3.8 Projectile motion

Specification reference: 3.1.3

Hitting the target

In the past, cannons were sited on clifftops to defend against attack from the sea. How far **projectiles** like cannonballs can travel depends on several factors. Ignoring the effect of air resistance, once a cannon has fired, the range depends on the height of the cannon above the sea and the initial velocity of the ball.

Independent motion

Figure 2 shows multiple images of two balls. The time interval between successive images is the same. One ball was dropped *vertically*, whilst the other was thrown *horizontally*. You will notice that both balls fall at the same rate – they are both at the same height at the same time. It does not matter whether the ball is moving horizontally. The vertical and horizontal motions of the ball are independent of each other.

Assuming no air resistance

● the vertical velocity changes due to acceleration of free fall

● the vertical displacement and **time of flight** can be calculated using equations of motion

● horizontal velocity remains constant.

Why does the horizontal velocity of the projectile remain constant? Remember that acceleration and velocity are vectors. The acceleration of free fall is vertically downwards. The component of this acceleration in the horizontal direction is zero.

$$\text{horizontal acceleration} = g\cos 90° = 0$$

The horizontal velocity is therefore unaffected by the fall.

▲ **Figure 1** *How far could a cannonball travel before landing in the sea?*

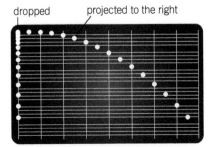

dropped   projected to the right

▲ **Figure 2** *Multiple exposures of two objects released together*

 Worked example: Cannonball

A cannonball is fired horizontally from a clifftop 44.1 m above the sea. The initial horizontal velocity of the cannonball is $304\,\text{m}\,\text{s}^{-1}$. Calculate:

a the time of flight;

b the horizontal distance it travels.

Remember that the vertical motion and the horizontal motion are independent of each other.

a There is acceleration in the vertical direction.

Step 1: Identify the equation needed and list the known values.

We can use the equation $s = ut + \frac{1}{2}at^2$ to calculate the time t of flight.

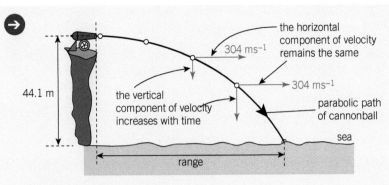

▲ **Figure 3** *Trajectory of a cannonball*

The initial vertical velocity $u = 0$ (initial vertical velocity $u = 304 \times \cos 90° = 0$).

$$s = 44.1 \text{ m}, u = 0, a = g = 9.81 \text{ m s}^{-2}, t = ?$$

Step 2: Substitute the values into the equation and calculate the answer.

$$s = \frac{1}{2} at^2$$

$$t^2 = \frac{2s}{a} = \frac{2 \times 44.1}{9.81} = 8.991 \text{ s}^2$$

$$t = 3.00 \text{ s}$$

b There is no acceleration in the horizontal direction.

Step 1: Identify the equation needed.

No acceleration, therefore
horizontal distance = horizontal velocity × time

Step 2: Substitute the values into the equation and calculate the answer.

$$\text{horizontal distance} = 304 \times 3.00 = 912 \text{ m}$$

The horizontal range of the cannonball is 912 m, almost 1 km.

Vector calculations

The path described by the cannonball in the worked example is curved because the vertical component of its velocity increases with time whilst the horizontal component is unaffected.

The magnitude of the actual velocity v of the cannonball, or any other projectile, can be calculated from the vertical and horizontal components v_x and v_y of this velocity. You just use Pythagoras' theorem (Figure 4).

$$\text{Actual velocity } v = \sqrt{v_x^2 + v_y^2}$$

The angle θ made by the velocity to the horizontal is given by

$$\theta = \tan^{-1} \frac{v_y}{v_x}$$

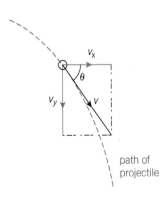

▲ **Figure 4** *The velocity v of a projectile has vertical and horizontal components*

Worked example: Arrows

An arrow is fired horizontally at $49.2\,\mathrm{m\,s^{-1}}$ from a $29.4\,\mathrm{m}$ high castle. Calculate its velocity when it hits the ground.

Vertical motion:

Step 1: Identify the equation needed and list the known values.

$$v^2 = u^2 + 2as$$

The initial vertical velocity $u = 0$, $a = g = 9.81\,\mathrm{m\,s^{-2}}$, $s = 29.4\,\mathrm{m}$

Step 2: Substitute the values into the equation and calculate the answer.

$$v^2 = 0^2 + (2 \times 9.81 \times 29.4) = 576.8\,\mathrm{m^2\,s^{-2}}$$

final vertical velocity $v_y = 24.0\,\mathrm{m\,s^{-1}}$

Horizontal motion:

Step 1: Identify the method needed and list the known values.

Use Pythagoras' theorem (Figure 5).

The horizontal component of the velocity $v_x = 49.2\,\mathrm{m\,s^{-1}}$.

Step 2: Substitute the values into the equation and calculate the answer.

$$\text{velocity of arrow} = \sqrt{24.0^2 + 49.2^2} = 54.7\,\mathrm{m\,s^{-1}}$$

$$= \tan^{-1}\frac{24.0}{49.2} = 26.0° \text{ to the horizontal.}$$

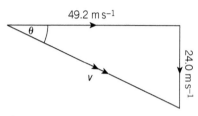

▲ **Figure 5**

Firing upwards

Figure 6 shows the path of a projectile fired at an angle θ to the horizontal. The initial velocity of the projectile is v. The motion of this projectile can still be analysed in terms of the independence of motion in the horizontal and vertical directions.

The horizontal component of the velocity is $v\cos\theta$ and the initial vertical upwards component of the velocity is $v\sin\theta$.

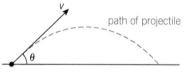

▲ **Figure 6** *The curved path described by a projectile fired upwards*

Worked example: Kicking a ball

A ball is kicked with an initial velocity of $15\,\mathrm{m\,s^{-1}}$ at an angle of $30°$ to the horizontal. Calculate the range R of this ball.

Step 1: Identify the method needed.

Work out the time of flight from the vertical motion, then the horizontal range from the time (Figure 7).

Step 2: List the values and the equation for the first part.

Vertical motion:

The time of the ball in flight can be calculated from its vertical motion. Note that the velocity of the ball when it hits the

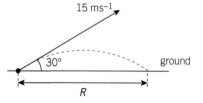

▲ **Figure 7** *How far will this ball travel?*

ground has the same magnitude as, but the opposite sign to, its upward velocity when it left the ground.

$$u = 15 \sin 30° = 7.5 \, \text{m s}^{-1}, v = -7.5 \, \text{m s}^{-1}, a = -9.81 \, \text{m s}^{-2}, t = ?$$

$$v = u + at$$

Step 2: Substitute the values into the equation and calculate the answer.

$$v = u + at$$

$$t = \frac{v \quad u}{a} = \frac{7.5 \quad 7.5}{9.81} = 1.53 \, \text{s}$$

Step 3: Use the answer for vertical motion to calculate the answer for the range.

Horizontal motion:

The horizontal component of the velocity is $15 \cos 30° = 13.0 \, \text{m s}^{-1}$. This remains constant throughout the flight. Therefore

$$\text{range } R = (15 \cos 30°) \times 1.53 = 20 \, \text{m (2 s.f.)}$$

Summary questions

1 A ball is kicked into the air. Figure 8 shows the velocity components of the ball at a particular instant. Calculate the velocity *v* of the ball.
(2 marks)

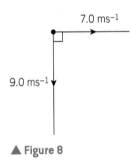

2 A cannonball is fired horizontally from a cliff 29 m above the sea. The initial horizontal velocity of the cannonball is $320 \, \text{m s}^{-1}$. Calculate:
   a the time of flight; *(3 marks)*
   b the horizontal distance it travels; *(3 marks)*
   c the speed at which it hits the sea. *(3 marks)*

▲ Figure 8

3 A cannonball is fired at $22.0 \, \text{m s}^{-1}$ at 35° to the horizontal. Calculate:
   a the maximum vertical height of the ball; *(4 marks)*
   b the horizontal distance travelled by the ball. *(5 marks)*

4 Sketch the vertical velocity–time graph for the cannonball in **3**. *(3 marks)*

Practice questions

1 **a** Copy and complete by stating the value or name of each of the remaining three prefixes.

▼ Table 1

prefix	value
micro (μ)	10^{-6}
mega (M)	
	10^{-9}
tera (T)	

(3 marks)

b Write down all the scalar quantities in the list below.

**density   weight   velocity
volume   acceleration**   *(1 mark)*

c The distance between the Sun and the Earth is 1.5×10^{11} m. Calculate the time in minutes for light to travel from the Sun to the Earth. The speed of light is 3.0×10^{8} m s^{-1}.   *(2 marks)*

May 2010 G481 Mechanics

2 **a** State the difference between a *scalar* quantity and a *vector* quantity.   *(2 marks)*

b Define *velocity* and derive its base units.
(2 marks)

c Figure 1 show the displacement *s* against time *t* for an object.

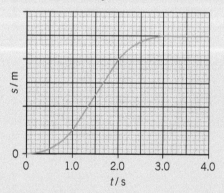

▲ Figure 1

Use the graph to describe and explain how the velocity for this object changes with time from $t = 0$ to $t = 4.0$ s.
(5 marks)

3 A student uses the apparatus shown in Figure 2 to determine the acceleration of free fall *g*.

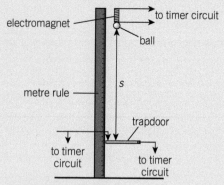

▲ Figure 2

The metal ball drops when the electromagnet is switched off. This also starts the timer. The timer stops when the ball opens the trapdoor below. The time of fall of the ball is *t*. The distance *s* between the trapdoor and the bottom of the metal ball is changed. Figure 3 shows the graph plotted by the student.

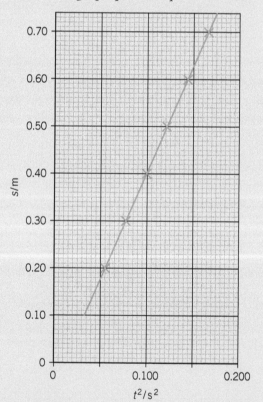

▲ Figure 3

a Use an equation of motion to explain why the graph of *s* against t^2 is a straight line.   *(2 marks)*

b Use the graph to determine the acceleration of free fall of the ball.
(3 marks)

c Calculate the percentage difference between your value in **(b)** and the accepted value for *g*. *(1 mark)*

d Explain how you can deduce that there is a systematic error in this experiment. Suggest what this error is likely to be. *(2 marks)*

4 a Define *acceleration*. *(1 mark)*

b A super-tanker cruising at an initial velocity of 6.0 m s^{-1} takes 40 minutes (2400 s) to come to a stop. The super-tanker has a constant deceleration.

    (i) Calculate the magnitude of the deceleration. *(3 marks)*

    (ii) Calculate the distance travelled in the 40 minutes it takes the tanker to stop. *(2 marks)*

    (iii) On a copy of Figure 4, sketch a graph to show the variation of distance *x* travelled by the super-tanker with time *t* as it decelerates to a stop.

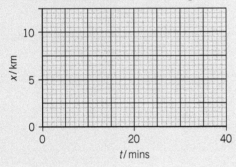

▲ Figure 4

(2 marks)
Jan 2012 G481

5 Figure 5 shows the variation of velocity *v* with time *t* for a small rocket.

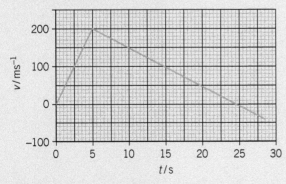

▲ Figure 5

The rocket is initially at rest and is fired vertically upwards from the ground. All the rocket fuel is burnt after a time of 5.0 s when the rocket has a vertical velocity of 200 m s^{-1}. Assume that air resistance has a negligible effect on the motion of the rocket.

(i) Without doing any calculations, describe the motion of the rocket

    1 from *t* = 0 to *t* = 5.0 s

    2 from *t* = 5.0 s to *t* = 25 s. *(3 marks)*

(ii) Calculate the maximum height reached by the rocket. *(3 marks)*

(iii) Explain why the rocket has a speed greater than 200 m s^{-1} as it hits the ground. *(1 mark)*

Jan 2013 G481

6 Figure 6 shows the path of a ball falling from the top of a table. The initial velocity of the ball is in the horizontal direction and has magnitude 2.0 m s^{-1}.

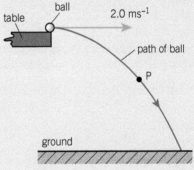

▲ Figure 6

a State the direction of the acceleration of the ball when at point **P**. *(1 mark)*

b Describe and explain the variation of the vertical component of the velocity of the ball as it travels towards the ground. *(3 marks)*

c The vertical component of the velocity at **P** is 2.9 m s^{-1}. Calculate

    (i) the velocity of the ball at **P**, *(3 marks)*

    (ii) the angle made by the velocity of the ball with the horizontal when at **P**. *(2 marks)*

FORCES IN ACTION
4.1 Force, mass, and weight
Specification reference: 3.2.1

Learning outcomes
Demonstrate knowledge, understanding, and application of:
→ the net force on an accelerating object
→ the newton
→ weight.

▲ **Figure 1** *One of the UK's three standard kilograms, a 39.17 mm high cylinder of platinum–iridium alloy stored in a bell jar at the National Physical Laboratory*

Study tip
Remember that $F \propto a$ when m is constant.

$a \propto \dfrac{1}{m}$ *when F is constant.*

Synoptic link
The equation $F = ma$ is often referred to as Newton's second law. As you will see in Topic 7.3, Newton's second law of motion, this law is defined in terms of rate of change of momentum. $F = ma$ is just a special case for constant mass.

The last artefact
The SI **base units** for length, time, current, and light intensity are directly or indirectly defined in terms of the speed of light. The base unit of mass is still defined by an artefact (an actual object). The international prototype kilogram or IPK was made in 1889. It is stored in a vault near Paris. By definition, the IPK has a mass of exactly 1 kg. Copies are used as standards around the world.

Force, mass, and acceleration
The **mass** of an object is one of its physical properties, and depends on the amount of matter it contains. A *net (resultant)* **force** acting on the object will make the object accelerate in the direction of the net force. The net force F, mass m of the object, and acceleration a of the object are related by the equation

$$F = ma$$

Force is measured in newtons (N), mass in kilograms (kg), and acceleration in metres per second squared (m s⁻²).

A force of 1 newton will give a 1 kg mass an acceleration of $1\,\mathrm{m\,s^{-2}}$ in the direction of the force.

 Worked example: High performance

An electric car has a mass of 2.1×10^3 kg. It accelerates from rest to $27\,\mathrm{m\,s^{-1}}$ in 5.4 s. Calculate the net force acting on the car.

Step 1: Identify the equation needed and list the known values.

$$m = 2.1 \times 10^3\,\mathrm{kg}, \, u = 0, \, v = 27\,\mathrm{m\,s^{-1}}, \, t = 5.4\,\mathrm{s}$$

$$F = ma$$

Step 2: Substitute the values into the equation and calculate the answer.

$$F = 2.1 \times 10^3 \times \left(\frac{27 - 0}{5.4}\right) = 1.05 \times 10^4\,\mathrm{N}$$

The net force acting on the car = $1.1 \times 10^4\,\mathrm{N}$ (2 s.f.)

Mass and weight
In physics it is important to distinguish between mass and weight. You cannot afford to confuse these two quantities (Table 1).

▼ Table 1 Mass and weight

Quantity	Unit	Comment
mass	kg	constant for a specific object or particle
weight	N $(kg\,m\,s^{-2})$	magnitude is variable – it depends on location

The **weight** of an object on the surface of the Earth is the gravitational force acting on the object. An object in free fall has an acceleration g of $9.81\,m\,s^{-2}$. The only force acting on the object is its weight W. Since $F = ma$, it follows that

$$W = mg$$

You can calculate the weight of an object on the Moon or other planets. Remember that the value of g will be different, but the mass m will be the same.

You can determine the weight of an object using a newtonmeter, which is calibrated to show the gravitational force acting on an object in newtons. A 1.0 kg object hanging from the newtonmeter will show a weight reading of about 9.8 N. The same object on the Moon would give a smaller reading of 1.6 N – the acceleration of free fall on the Moon is only $1.6\,m\,s^{-2}$. The mass remains constant, but weight is a variable.

▲ **Figure 2** This astronaut weighs less than on the Earth and can walk in a heavy suit with little effort because the value of g on the surface of the Moon is much less than $9.81\,m\,s^{-2}$

Study tip

The equations $F = ma$ and $W = mg$ are not provided in the examinations so you will have to remember them.

 Understanding mass

The definition of mass goes back to the 17th century when Isaac Newton related mass, acceleration, and force. Mass was regarded as constant for a given object or particle (and we will treat it as constant in this course).

In 1905 Albert Einstein came up with the model of relativistic mass in his special theory of relativity. He found that the mass m of a particle depends on its speed v, according to the equation

$$m = \frac{m_0}{\sqrt{1 - \left(\frac{v}{c}\right)^2}}$$

where m_o is the rest mass of the object and c is the speed of light in a vacuum $(c = 3.00 \times 10^8\,m\,s^{-1})$.

Your mass will not alter much at the speeds at which we move around. You would have to travel close to the speed of light for your mass to change significantly.

1 Sketch a graph of relativistic mass m against the speed v of a particle.
2 The rest mass of the electron is 9.11×10^{-31} kg. Calculate its mass at:
   a $0.10\,c$ (10% speed of light);   b $0.999\,c$.
3 Explain whether or not an electron can travel at the speed of light c.

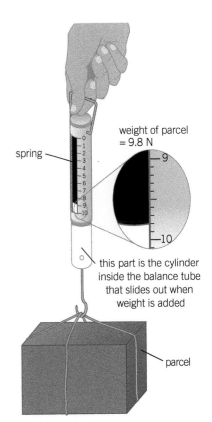

weight of parcel = 9.8 N

spring

this part is the cylinder inside the balance tube that slides out when weight is added

parcel

▲ **Figure 3** You can use a newtonmeter to determine weight

Summary questions

1 A resultant force of 500 N acts on a stationary car of mass 1200 kg. Calculate:
   a the car's acceleration; *(2 marks)*
   b its velocity after 6.0 s. *(2 marks)*

2 Calculate the mass in grams of a mobile phone of weight 1.1 N. *(2 marks)*

3 A golf ball has a mass of 46 g. It is hit with a force of 5.8 kN. Calculate the initial acceleration of the ball. *(2 marks)*

4 The weight of a car is 1.8×10^4 N. It accelerates from rest to a velocity of 28 m s^{-1} in a time of 9.6 s. Calculate the acceleration of the car. *(5 marks)*

5 The forces acting on a proton (mass 1.7×10^{-27} kg) are shown in Figure 4. Calculate the magnitude and direction of the acceleration of the proton. *(6 marks)*

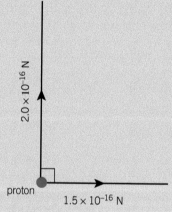

▲ **Figure 4**

6 A 8.0 g pellet travelling at 420 m s^{-1} hits a wooden crate. The pellet penetrates 98 mm into the crate. Calculate the average magnitude of the force acting on the pellet. *(3 marks)*

4.2 Centre of mass

Specification reference: 3.2.3

A balancing act

You need a sense of balance to stay upright. If you lean too far to one side you may fall over. Idol Rock in North Yorkshire (Figure 1) is an example of a natural balancing act. Over many years, the weaker layers of millstone grit have been eroded, leaving 200 tonnes of rock perched perilously on a small pyramid. As long as its **centre of mass** does not move to one side, Idol Rock will stay upright.

Where weight acts

Imagine pushing a spanner floating freely in space with your finger. The spanner will rotate, or move in a straight line, or both. The spanner can be made to move in a straight line if the force is applied along a line of action that coincides with its centre of mass. The centre of mass of an object is a point through which any externally applied force produces straight-line motion but no rotation.

Figure 2 shows an irregular object made up of identical atoms, each having a tiny weight w. The resultant gravitational force on the object, its total weight W, will act through a point often called the object's **centre of gravity**, which coincides with its centre of mass. The centre of mass of an object is an imaginary point where the entire weight of an object appears to act. The centre of mass of a uniform metre rule will be at its 50.0 cm mark. If you stand upright, your centre of mass is just behind your navel.

It is much easier to analyse and solve physics problems if we think about the weight of an object acting through its centre of mass rather than as a collection of many tiny forces on each part of the object. In Figure 3, for example, the complicated movement of the stunt rider can be represented by the smooth parabolic path described by his centre of mass.

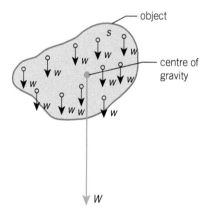

▲ **Figure 3** It is easier to analyse the motion of the stunt rider by looking at the motion of his centre of mass instead of his whole body

▲ **Figure 1** Idol Rock at Brimham Rocks near Harrogate in North Yorkshire

▲ **Figure 2** The centre of gravity, through which the object's weight acts, coincides with its centre of mass

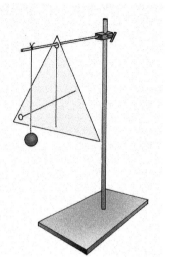

▲ **Figure 4** *Determining centre of gravity*

Finding the centre of gravity

A freely suspended object will come to rest with its centre of gravity vertically below the point of suspension. This is the idea behind a **plumb-line**, used in construction. A heavy object, the plumb-bob, is suspended from a piece of string. When the plumb-bob comes to rest, the string is vertical.

You can use a plumb-line to find the centre of gravity of an object. This can be difficult in practice with complex three dimensional objects, but it is easy to do with objects made from card.

Make small holes along the edges of the object made from card. Insert a pin through one of the holes and hold the pin firmly in a clamp. Allow the object to swing freely. It will come to rest with its centre of gravity vertically below the pin. Hang a plumb-line from the pin and draw a line along the vertical string of the plumb-line. Repeat this process for other holes. The centre of gravity will be the point of intersection of the lines.

You can check that the position of the centre of gravity is correct by removing the card and seeing whether it will balance on a pin, or your finger, at this point.

Summary questions

1 The centre of gravity of a metre rule is found to be at its 48.3 cm mark. Suggest why it is not at the 50.0 cm mark. *(1 mark)*

2 Use diagrams to show the centre of mass of:
   a a flat circular plate; *(1 mark)*
   b a rectangular table; *(1 mark)*
   c a triangular card. *(1 mark)*

3 Describe how you could determine the centre of gravity of an irregularly shaped piece of card using the edge of the ruler instead of the plumb-line method above. *(4 marks)*

4 Explain why the centre of gravity for a table-tennis ball is in the empty space inside the ball, rather than in the plastic of the ball itself. *(2 marks)*

5 Suggest how you can locate the centre of mass of an object in space where there is no detectable gravitational field. *(4 marks)*

4.3 Free-body diagrams

Specification reference: 3.2.1

Hanging around

How can we analyse the forces acting on an object? The easiest way to do this is to draw a **free-body diagram**, which isolates all the forces acting on a particular object. The photograph of the cliff climber in Figure 1 reveals many forces in operation: there are forces within the rope, forces acting on the climber, and forces acting on the cliff face too. Fortunately, we can isolate three key forces when analysing the stability of this climber – her *weight*, the *tension* in the rope, and the **normal contact force** between her shoe and the cliff face.

Some important forces

Table 1 summarises some of the forces that you will meet in your study of mechanics.

▼ **Table 1** *A summary of key forces*

Force	Comment	Force diagram
weight	the gravitational force acting on an object through its centre of mass	toy car — weight
friction	the force that arises when two surfaces rub against each other	friction — box — motion of box
drag	the resistive force on an object travelling through a fluid (e.g., air and water); the same as friction	motion — shuttle cock — drag
tension	the force within a stretched cable or rope	tension — stretched rope — tension
upthrust	an upward buoyancy force acting on an object when it is in a fluid	upthrust — toy boat
normal contact force	a force arising when one object rests against another object	normal contact force — box — ramp

Learning outcomes

Demonstrate knowledge, understanding, and application of:

→ free-body diagrams.

▲ **Figure 1** *What are the major forces acting on this climber?*

Study tip

'Normal' means 'at right angles to'.

Representing forces

In a free-body diagram

- each force vector is represented by an arrow labelled with the force it represents
- each arrow is drawn to the same scale (the longer the arrow, the greater the force).

Figure 2 shows the free-body diagram of the climber from the start of this topic.

▲ **Figure 2** *The free-body diagram for a climber*

On a slope

Figure 3 shows an object on a smooth inclined slope.

Assume that there is no friction – the only force acting on the object is its weight. This weight can be resolved into two components, parallel and perpendicular to the slope.

Force parallel to the slope = $W\sin\theta$ or $F_x = mg\sin\theta$

Force perpendicular to the slope = $W\cos\theta$ or $F_y = mg\cos\theta$

The component of the weight down the slope is responsible for the acceleration of the object down the slope. There is no acceleration of the object perpendicular to the slope. Therefore, this component of the weight must be equal to the **normal contact force** N acting on the object, that is

$$F_y = N = mg\cos\theta$$

The worked example shows how you can analyse the motion of an object down the slope.

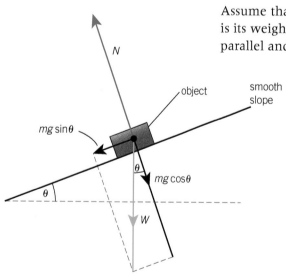

▲ **Figure 3** *Object on a slope*

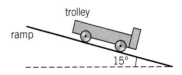
trolley
ramp
15°

▲ Figure 4 *Trolley on a slope*

 Worked example: Down the slope

An 850 g trolley is held at the top of a 1.2 m long ramp (Figure 4). The ramp makes an angle of 15° to the horizontal. The trolley is released from rest. Calculate the acceleration a of the trolley as it travels down the ramp and the time t it takes to reach the bottom of the ramp.

Step 1: Identify the equations needed.

$$\text{force on trolley down the ramp} = mg\sin\theta$$

$$F = ma$$

Step 2: Substitute the values into the equation and calculate the answer.

acceleration of trolley $a = \dfrac{F}{m} = \dfrac{mg\sin\theta}{m} = g\sin\theta$ (note that the acceleration is independent of the mass)

$$a = 9.81 \times \sin 15° = 2.54\,\text{m s}^{-2}$$

You can now use the equation of motion $s = ut + \frac{1}{2}at^2$ to calculate the time t.

$$1.2 = \tfrac{1}{2} \times 2.54 \times t^2 \ (u = 0)$$

$$t = \sqrt{\frac{2 \times 1.2}{2.54}} = 0.97\text{s (2 s.f.)}$$

Summary questions

1 Draw a labelled free-body diagram for:
   a a ball falling vertically through the air; (*1 mark*)
   b a toy boat resting on the surface of water. (*1 mark*)

2 Figure 5 shows the free-body diagram for a bag on the floor of a lift. The mass of the bag is 8.0 kg. The normal contact force is N. The lift travels vertically upwards with an acceleration of 1.5 m s⁻². Calculate the resultant force on the bag and therefore the magnitude of the force N. Explain your answer. (*4 marks*)

3 A 20 g wooden block is placed on a smooth ramp that makes an angle of 30° to the horizontal. The block is released from rest. Calculate its acceleration down the ramp. (*3 marks*)

4 Calculate the acceleration of the block in question 3 assuming a constant friction of 0.10 N acts against the motion of the block. (*4 marks*)

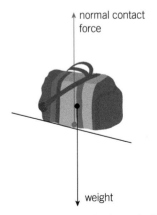
normal contact force
weight

▲ Figure 5 *A bag on the floor of a lift*

4.4 Drag and terminal velocity

Specification reference: 3.2.2

▲ **Figure 1** *A peregrine falcon*

▲ **Figure 3** *Streamlining reduces drag force*

Stooping to prey

The peregrine falcon is one of the fastest birds in the world. In level flight it can reach $30\,\mathrm{m\,s^{-1}}$, but when it goes into a controlled dive, or 'stoop', to catch its prey, it folds its wings back to minimise air resistance and hurtles vertically downwards at up to $108\,\mathrm{m\,s^{-1}}$. At this speed the bird has reached its terminal velocity, the velocity at which the drag force on it balances its weight.

Moving through a fluid

An object moving through a fluid, for example, air or water, experiences a **drag force** from the fluid. Drag is a frictional force that opposes the motion of the object. Its magnitude depends on several factors, including the speed of the object, the shape of the object, the roughness or texture of the object, and the density of the fluid through which it travels. The two most important factors that affect the magnitude of the drag force are *speed* of the object and its *cross-sectional area*.

Large cross-sectional areas result in greater drag force. For most objects, including those falling through air, the drag force is directly proportional to speed^2. This means that, for example, when the speed of an object is doubled, the drag force increases by a factor of four (Figure 2).

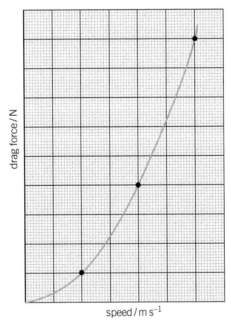

▲ **Figure 2** *The drag force–speed graph for many objects: drag force ∝ speed^2*

The drag force experienced by objects moving in air is often called **air resistance**. Modern vehicles have smooth, streamlined shapes to reduce the air resistance exerted against them whilst travelling (Figure 3). This increases the top speed of the vehicle and also reduces the amount of fuel used on a journey.

Terminal velocity

During a vertical fall through air or another fluid, the weight of the object remains constant but the drag force increases as the speed increases.

At the instant an object starts to fall, there is no drag force on the object. The total force is equal to the weight. The acceleration of the object is g, the acceleration of free fall.

As the object falls, its speed increases and this in turn increases the magnitude of the opposing drag force. The resultant (net) force on the object decreases and the instantaneous acceleration of the object becomes less than g.

Eventually the object reaches **terminal velocity**, when the drag force on the object is equal and opposite to its weight. At terminal velocity, the object has zero acceleration and its speed is a constant.

Figure 4 shows the velocity–time graph for an object falling through air and the corresponding free-body diagrams at three different times. The weight of the object is mg, the variable drag force is D, and the instantaneous acceleration is a. The gradient of the graph gives the instantaneous acceleration of the object.

Synoptic link

You will find more information about gradients of graphs in Appendix A2, Recording results.

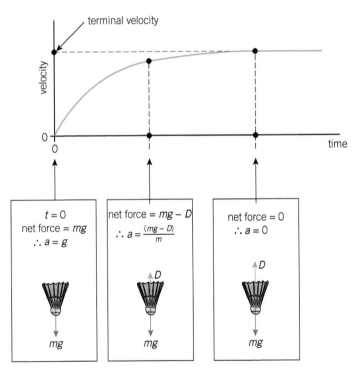

▲ **Figure 4** *Velocity–time graph for an object falling through air: the object has an acceleration of g at the start of the fall and zero acceleration when terminal velocity is reached*

Investigating motion in a fluid

You can easily investigate the motion of an object falling affected by a drag force by using a motion sensor connected to a data-logger or a laptop. The falling object is attached to a light polystyrene ball by a thin thread passed over a pulley. The object is then dropped through a cylinder of liquid such as water or glycerol, pulling the polystyrene ball vertically upwards. The motion of this ball is identical to that of the object falling through the fluid. You can generate and analyse velocity–time and acceleration–time graphs with this arrangement.

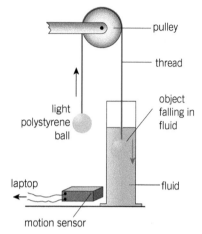

1   A student suggests that the motion sensor can be pointed directly towards the object falling in the fluid. Explain why this will not produce any useful data.
2   Describe how you could *estimate* the terminal velocity of the falling object without using the motion sensor.

▲ **Figure 5** *Investigating an object falling through a fluid*

Summary questions

1   A skydiver is falling towards the Earth at a terminal velocity of 45 m s^{-1}. Describe what she could do to change her terminal velocity.   *(1 mark)*

2   A rubber ball of mass 0.120 kg is dropped from a tall building. Calculate the magnitude of the drag force as it falls through the air at its terminal velocity. Explain your answer.   *(3 marks)*

3   At 10 m s^{-1} the drag force acting on a car is 1.0 kN. What is the drag force on the same car travelling at 30 m s^{-1}? Explain your answer.   *(3 marks)*

4   Determine the instantaneous acceleration of each object in Figure 6.   *(10 marks)*

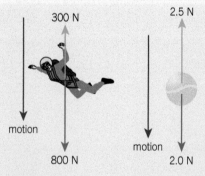

▲ **Figure 6**

5   The magnitude of the drag force D in newtons (N) acting on a 0.30 kg object falling through air is given by the expression

$$D = 0.20\,v^2$$

where v is the speed of the object in m s^{-1}. Calculate:
a   the instantaneous acceleration of the object at 1.5 m s^{-1};   *(5 marks)*
b   the terminal velocity of this object.   *(4 marks)*

4.5 Moments and equilibrium

Crossing rivers

The Tees Transporter Bridge crosses the River Tees at Middlesbrough. It is an unusual solution to building a river crossing that allows ships to pass underneath. Instead of going over the bridge, pedestrians and vehicles travel in a gondola suspended by cables from a system of wheels and rails. The engineers who designed it had to take into account the changing forces as the gondola moves across the river.

The moment of a force

The weight of the gondola shown in Figure 1 creates a turning or a twisting effect about its two supports. The engineers took this into account when thinking about the stability of the structure.

The **moment** of a force is the turning effect of a force about some axis or point. It is defined as follows (Figure 2):

moment = force × perpendicular distance of the line of
action of force from the axis or point of rotation

moment = Fx

The SI unit for the moment of a force is N m.

Perpendicular distance

It is important that you use the perpendicular distance in calculations involving moments, not just the distance from force to **pivot**. Figure 3 illustrates what is meant by the perpendicular distance. You can calculate the perpendicular distance x from the pivot using trigonometry.

$$x = 0.20 \cos \theta$$

The clockwise moment of the force must therefore be

$$\text{moment} = F \times 0.20 \cos \theta = 0.20 F \cos \theta$$

There is another method, in which you simply resolve the force F into two perpendicular directions. The perpendicular component of the force, $F \cos \theta$, has a perpendicular distance of 0.20 m from the pivot in Figure 3. Its clockwise moment about the pivot is

$$\text{moment} = F \cos \theta \times 0.20 = 0.20 F \cos \theta$$

This is exactly the same as in the first method. The other component of the force, $F \sin \theta$, has zero perpendicular distance from the pivot, so its contribution towards the moment is zero.

The principle of moments

When a body is in **equilibrium**, the net force acting on it is zero and its net moment is zero. You can use the **principle of moments** to solve problems where an object is in rotational equilibrium.

▲ **Figure 1** *The Tees Transporter Bridge, opened in 1911, is one of only twenty such bridges ever built*

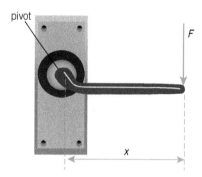

▲ **Figure 2** *Moment of a force*

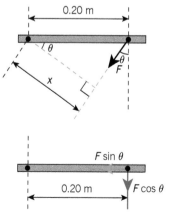

▲ **Figure 3** *Understanding perpendicular distance*

4.5 Moments and equilibrium

Study tip

When you define the moment of a force, make sure you state 'perpendicular' and not just 'distance'.

Principle of moments: For a body in rotational equilibrium, the sum of the anticlockwise moments about any point is equal to the sum of the clockwise moments about that same point.

The principle of moments was recognised almost 2000 years ago by Archimedes. Engineers still use this important principle when designing buildings and other complex structures. The worked examples below show how you can analyse problems using this principle. As you can see, it is important to draw clearly labelled free-body diagrams.

Worked example: A simple see-saw

Figure 4 shows a metre rule pivoted at its 50 cm mark. Two objects of weights 0.70 N and W are placed on the ruler as shown to balance it. Assume the weight of the ruler is negligible.

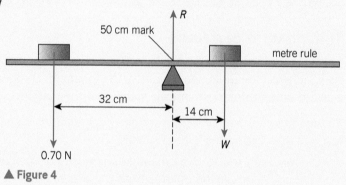

▲ **Figure 4**

Calculate the size of W and the force acting at the pivot.

Step 1: Identify the equation needed.

According to the principle of moments

sum of clockwise moments = sum of anticlockwise moments

Step 2: Substitute the values into the equation and calculate the answer.

$0.14 \times W = 0.70 \times 0.32$ (note that the line of action of the normal contact force R passes through the pivot, so its moment is zero)

Now rearrange this equation to determine W.

$$W = \frac{0.70 \times 0.32}{0.14} = 1.6 \, \text{N}$$

The size of W is 1.6 N.

Step 3: To calculate the force R at the pivot, you can now use the idea that the net force on the ruler is zero. There are two downwards vertical forces, 0.70 N and 1.6 N. This implies that there must be an upwards vertical force R at the pivot equal to sum of these two forces. Therefore

$$R = 0.70 + 1.6 = 2.3 \, \text{N}$$

Worked example: A loaded bridge

Figure 5 shows a model of a loaded section of a bridge made using a uniform wooden beam.

The wooden beam is 120 cm (1.2 m) long and has weight 15 N. One end of the beam rests on a support and the other end is fixed to a vertical string. Calculate the vertical force R at the support.

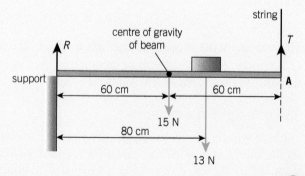

▲ **Figure 5**

Step 1: To solve this problem you need to examine where to take moments. Taking moments about the support will not be helpful. The most convenient point to take moments would be the end **A**, where the string is tied. The tension T in the string will have zero moment about **A**.

The weight of the beam acts through its centre of gravity.

Taking moments about **A**, we have

sum of clockwise moments = sum of anticlockwise moments

Step 2: Substitute the values into the equation and calculate the answer.

$R \times 1.20 = (15 \times 0.60) + (13 \times 0.40)$ (note that all perpendicular distances are from **A**)

$$R = \frac{142}{1.20} = 11.83\,\text{N} = 12\,\text{N (2 s.f.)}$$

You could now calculate the tension T from the fact that the net force on the beam is zero.

$$T = 16\,\text{N (2 s.f.)}$$

Summary questions

1 Calculate the moment from each force about the pivot in Figure 6. *(3 marks)*

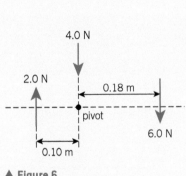

▲ **Figure 6**

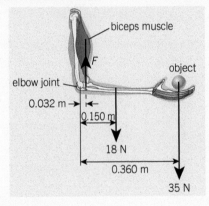

◀ **Figure 7**

2 Figure 7 shows a human forearm held horizontally and still. Calculate:
   a the clockwise moment about the elbow joint; *(3 marks)*
   b the force F in the muscle. *(3 marks)*

3 A uniform cylinder has height 10.0 cm and diameter 3.0 cm. The cylinder is placed with its circular base resting on a horizontal table. Calculate the maximum angle through which it can tip before it continues to fall by itself. *(3 marks)*

4 Calculate the magnitude of the force F in Figure 8. *(6 marks)*

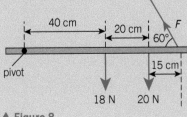

▲ **Figure 8**

4.6 Couples and torques

Specification reference: 3.2.3

Learning outcomes

Demonstrate knowledge, understanding, and application of:

→ couples and torques

→ equilibrium of objects under the action of forces and torques.

Couples

Imagine pushing the side of your calculator on the table with your finger. If the line of action of this force does not coincide with the centre of mass of the calculator, then it will both slide along the table (translate) and rotate. How can you make an object spin without any translational motion? The trick is to apply a pair of equal but opposite forces to the object. These two forces must be parallel and along different lines. Such a pair of forces is referred to as a **couple**.

Torque of a couple

Figure 2 shows a couple applied to an object. The magnitude of each force is F and the perpendicular distance between them is d. The moment of this pair of equal but opposite forces about the centre point **C** is

$$\text{moment} = \left(F \times \frac{d}{2}\right) + \left(F \times \frac{d}{2}\right) = Fd$$

The moment of a couple is known as a **torque**. The torque of a couple is defined as

$$\text{torque of a couple} = \text{one of the forces} \times \text{perpendicular separation between the forces} = Fd$$

▲ **Figure 1** An example of a couple: a pair of equal but opposite forces are applied to the pedals

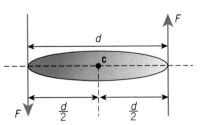

▲ **Figure 2** The moment of this couple is called a torque

🖩 Worked example: Steering wheel

Use Figure 3 to calculate the torque of the couple on this car steering wheel.

Step 1: Identify the correct equation to use.

$$\text{torque} = Fd$$

Step 2: Substitute the values in and calculate the torque.

$$\text{torque} = 20 \times 0.32$$

$$= 6.4\,\text{N m}$$

The torque applied is 6.4 N m in a clockwise direction.

Worked example: Preventing rotation

Figure 4 shows a rod of length 50 cm with a disc of radius 10 cm fixed at its centre. Two forces, each of magnitude 30 N, are applied normal to the rod at each end. Calculate the torque produced by the pair of 30 N forces and the minimum tension in the rope that would prevent the disc from rotating.

Step 1: Identify the correct equation to use.

$$\text{torque} = Fd$$

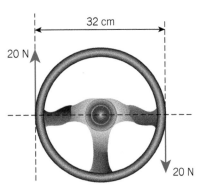

▲ **Figure 3** The couple on this steering wheel will make it turn

→

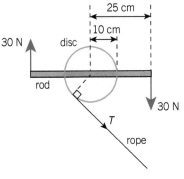

▲ Figure 4

Step 2: Substitute the values in and calculate the torque.

$$\text{torque} = 30 \times 0.25 = 7.5\,\text{N m}$$

Step 3: To prevent rotation, the moment of the tension T in the rope must be equal but opposite to this torque.

$$\text{moment} = T \times \text{distance} = T \times 0.10 = 7.5$$

$$T = \frac{7.5}{0.10} = 75\,\text{N}$$

Summary questions

1 A snooker ball is resting on a table. A single off-centre force is applied to its surface with a cue. Describe the subsequent motion of the ball. *(2 marks)*

2 The top of a kitchen tap has diameter 4.0 cm. Estimate the torque required to open such a tap using your thumb and one of the other fingers. *(3 marks)*

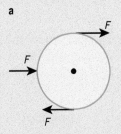
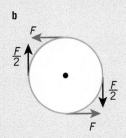

▲ Figure 5

3 Figure 5 shows two discs placed on a smooth horizontal surface.

Describe qualitatively the type of motion each disc will perform. *(4 marks)*

4 Figure 6 shows a couple acting on an object.

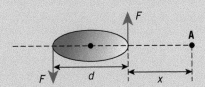

▲ Figure 6

a Determine the total moment of the couple about the point **A**. *(2 marks)*

b What can you deduce from this calculation? *(1 mark)*

Synoptic link

You will find more information on estimations in Appendix A3, Measurements and uncertainties.

4.7 Triangle of forces

Specification reference: 3.2.3

Demonstrate knowledge,
understanding, and application of:

→ the equilibrium of three
coplanar forces

→ the rule of the triangle of forces.

▲ **Figure 1** *A baby having fun with forces*

Synoptic link

You know how to add two vector
forces from Topic 2.4, Adding vectors.

Bouncing babies

Baby bouncers help babies to develop their leg muscles so that they
are ready to learn to walk. The bouncer is securely suspended
from a door frame and the baby is supported in a padded seat. The
weight acts vertically downwards and is balanced by the **tension**
in the straps.

Forces in equilibrium

A heavy ball is held at rest by two ropes (Figure 2, left). The tensions
in the ropes have magnitudes F and T. The weight of the ball is W. The
resultant of these three coplanar forces must be zero.

To add three vector forces you simply extend the procedure for
adding two vectors. Figure 2 below shows the vectors in the free-body
diagram for the ball (centre) and in a **triangle of forces** (Figure 3).

● Arrows are drawn to represent each of the three forces end-to-end.

● The triangle is closed because the net force is zero and so the
object is in equilibrium.

Different ways of thinking

The triangle of forces gives you a method for solving problems. You
can, however, interpret the equilibrium of the object in Figure 2 in
two other ways.

● The resultant of forces F and T must be equal in magnitude to the
third force W but in the opposite direction. The same is true for
any pair of forces in Figure 2.

● The resultant force vertically must be zero and the resultant
horizontal force must also be zero. Therefore, the force T can be
resolved into its vertical and horizontal components, with
$T\cos\theta = F$ and $T\sin\theta = W$.

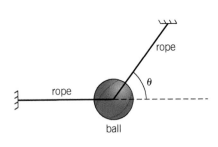

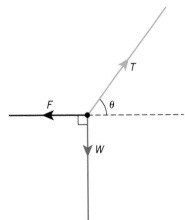

▲ **Figure 2** *All three forces acting on the ball act through the same point, as shown in
the free-body diagram in the centre*

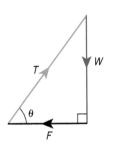

▲ **Figure 3** *Triangle of forces showing
the situation in Figure 2 – all arrows
follow one another*

62

Worked example: Resting on a slope

Figure 4 shows a block of wood weighing 1.20 N resting on a rough ramp that makes an angle of 30° to the horizontal. The normal contact force between the block and the ramp is N and the frictional force on the block is F. Draw a triangle of forces and determine the magnitude of the forces N and F.

Step 1: Draw a triangle of forces (Figure 5).

Step 2: Calculate the unknown forces.

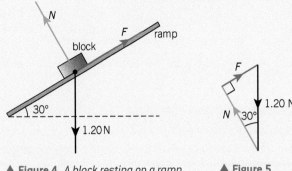

▲ **Figure 4** *A block resting on a ramp*   ▲ **Figure 5**

$$\cos 30° = \frac{N}{1.20} \text{ therefore } N = 1.20 \times \cos 30° = 1.04\,N$$

$$\sin 30° = \frac{F}{1.20} \text{ therefore } F = 1.20 \times \sin 30° = 0.60\,N$$

You can check the answers using Pythagoras' theorem.

$$0.60^2 + 1.04^2 = 1.20^2$$

Worked example: Baby bouncer

A baby and its bouncer seat weigh 120 N (Figure 6). Two straps holding the seat are suspended from a plastic bar. The tension in each strap is the same, and they meet at a point on the seat at an angle of 50°. Figure 6 shows the corresponding free-body diagram.

Step 1 (one possible method): Resolve the diagonal forces by inspection.

There is no net force in any direction as the forces balance. Therefore

$$T\cos 25° + T\cos 25° = 120$$

$$2\,T\cos 25° = 120$$

$$T = \frac{120}{2\cos 25°} = 66\,N\,(2\text{ s.f.})$$

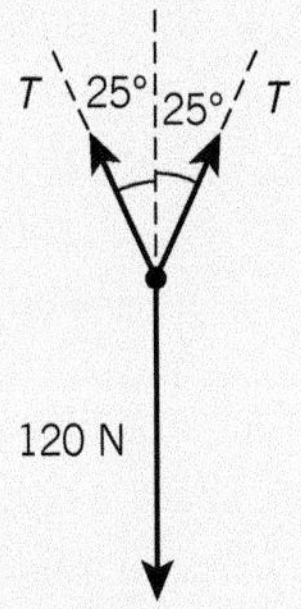

You can of course draw a triangle of forces to determine the magnitude of T, but the method above is quick and neat.

The horizontal components of T are equal and opposite and therefore balance each other.

▲ **Figure 6**

Three coplanar forces acting on an extended object

The triangle of forces method can also be applied to objects that have shape and form, from bridges to bicycles.

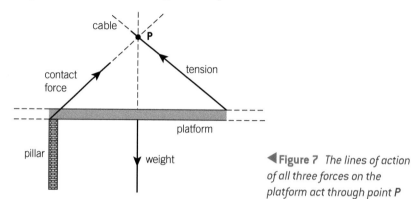

◀ **Figure 7** *The lines of action of all three forces on the platform act through point P*

Figure 7 shows the free-body diagram of a section of a bridge platform. All three coplanar forces pass through a point **P** in space, so you can draw a triangle of forces for the forces passing through **P**. This method simplifies a complex problem to the equilibrium of this imaginary point in space.

Summary questions

1 Three coplanar forces act on an object. The vectors representing these three forces form a closed triangle (triangle of forces). State the resultant force acting on the object. *(1 mark)*

2 Figure 8 shows an object in equilibrium.
  a Draw a clearly labelled triangle of forces. *(2 marks)*
  b Calculate the magnitude of the force T. *(2 marks)*
  c State and explain the magnitude of the resultant of the forces 5.0 N and 12 N. *(2 marks)*

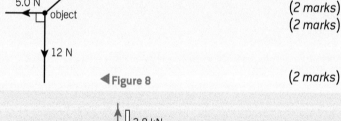

◀ **Figure 8**

3 A crane lifts a girder using a hook and two cables, as shown in Figure 9. The forces acting on the hook are shown. Calculate the magnitudes of the forces T_1 and T_2. *(4 marks)*

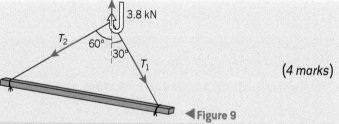

◀ **Figure 9**

4 Two newtonmeters are used to lift 500 g of slotted masses (Figure 10). The masses are at rest.
  a The angles θ_1 and θ_2 are 60° and 50°, respectively. Calculate the readings T_1 and T_2. *(4 marks)*
  b Explain why it would be impossible for both angles to be zero. *(2 marks)*

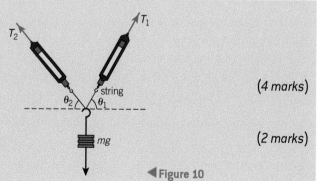

◀ **Figure 10**

4.8 Density and pressure

Specification reference: 3.2.4

Light for its size

Aerogel is one of the least dense solids known. It is so lightweight and wispy that it is sometimes called frozen smoke. Made from silica, aerogel has a density of just $1.9\,\text{kg m}^{-3}$. By comparison, the density of air is about $1.3\,\text{kg m}^{-3}$ and the density of water is $1000\,\text{kg m}^{-3}$. Its low density means that aerogel has very little mass for its volume.

Density

The **density** of a substance is defined as its mass per unit volume. You can use the following equation to calculate density ρ.

$$\rho = \frac{m}{V}$$

where m is mass and V is volume. The SI unit of density is kg m^{-3}.

 Worked example: Dense osmium

A $50.0\,\text{cm}^3$ sample of osmium has a mass of $1.13\,\text{kg}$. Calculate its density.

Step 1: Select the equation for density.

$$\rho = \frac{m}{V}$$

Step 2: Substitute in the known values in SI units and calculate the density.

$$m = 1.13\,\text{kg} \qquad V = 50 \times 10^{-6}\,\text{m}^3$$
$$(\text{note that } 1\,\text{cm}^3 = (10^{-2}\,\text{m})^3 = 10^{-6}\,\text{m}^3)$$

$$\rho = \frac{m}{V} = \frac{1.13}{50.0 \times 10^{-6}} = 2.26 \times 10^4\,\text{kg m}^{-3}$$

Osmium is 22.6 times denser than water.

Determining density

You need to know mass and volume to determine the density of a substance. The mass can be measured directly using a digital balance. For liquids, you can use a measuring cylinder to determine the volume. The volume of a regular-shaped solid can be calculated from measurements taken with a ruler, digital callipers, or a micrometer. The volume of irregular solids can be determined by displacement (Figure 2).

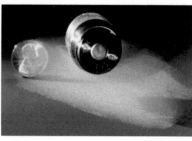

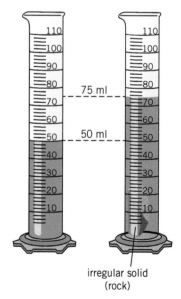

▲ **Figure 1** *A 0.70 g piece of aerogel supporting a 100 g mass – you can see a coin through it*

75 ml

50 ml

irregular solid (rock)

▲ **Figure 2** *Determining the volume of an irregular solid by displacement of a liquid: the volume is the difference between the two water levels*

Synoptic link

You will find more information on estimations in Appendix A3, Measurements and uncertainties.

Summary questions

1 Calculate the mass of air in a room of volume 140 m^3.
(2 marks)

2 A circular head of a drawing pin has a diameter of 7.5 mm. A thumb presses against the head with a force of 8.0 N. Calculate the pressure exerted on the head. *(2 marks)*

3 A measuring cylinder is filled with water up to its 70 cm^3 mark. An irregular rock of mass 0.080 kg is gently placed in the water. The new water level mark is 85 cm^3. Calculate the density of the rock. *(2 marks)*

4 Calculate the vertical force exerted by the atmosphere on the surface of a calculator of length 15 cm and width 7.5 cm.
(3 marks)

5 Calculate the average density of a neutron star of mass 3.0 × 10^{24} kg and radius 12 km.
(2 marks)

6 An 18 carat 'gold' bar of volume 3.34 × 10^{-6} m^3 comprises 58.2% gold and 41.8% copper by volume. Calculate the density of the bar. State any assumptions made.
ρ_{gold} = 1.93 × 10^4 kg m^{-3} and ρ_{copper} = 8.96 × 10^3 kg m^{-3}.
(6 marks)

Pressure

Drawing pins are designed to be pushed into noticeboards without hurting you. The pin's head has a much larger surface area than its point. This means that, when you push it into a noticeboard, the pressure exerted on your thumb is much less than the pressure exerted on the board. If you used the drawing pin the other way round it would hurt.

▲ **Figure 3** *Spreading weight over a smaller area increases the pressure exerted*

Pressure is the normal force exerted per unit cross-sectional area. You can use the following equation to calculate pressure p:

$$p = \frac{F}{A}$$

where F is normal force and A is cross-sectional area. The SI unit of pressure is N m^{-2} or pascal (Pa), where 1 Pa = 1 N m^{-2}.

We are under pressure all the time. The Earth's atmosphere exerts about 1.0 × 10^5 Pa on everything on its surface, including us. Tiny fluctuations in this pressure are responsible for the variety of weather patterns we see on the Earth.

 Worked example: Standing still

Estimate the pressure you exert on the floor when standing up.

An estimate is a rough calculation, not a guess. The numbers you use should be realistic.

Step 1: Select the equation for pressure and estimate the numbers you need for the calculation.

$$p = \frac{F}{A}$$

You can estimate the area of one shoe by measuring its width and length.

cross-sectional area of each shoe A = 0.25 m × 0.10 m
= 2.5 × 10^{-2} m^2 (2 s.f.)

The force exerted on the floor is your weight, mg. In this calculation m is estimated to be 65 kg.

Step 2: Substitute your estimates into the equation and calculate the estimated pressure.

$$p = \frac{F}{A} = \frac{65 \times 9.81}{2 \times 2.5 \times 10^{-2}} = 1.3 \times 10^4 \text{ Pa (2 s.f.)}$$

4.9 $p = h\rho g$ and Archimedes' principle

Deep water

You can safely swim underwater in the sea at depths of a few tens of metres. However, humans need a pressurised submersible to do any work at greater depth, because of the enormous pressure due to the weight of the water. Figure 1 shows a research submarine with a clear acrylic bubble housing that is 9.5 cm thick. At a depth of 610 m, the pressure acting on this submarine is about 6 000 000 Pa.

Pressure in fluids

Gases and liquids are **fluids** – substances that can flow. Gases, such as air, exert pressure on surfaces because of the constant bombardment by their molecules. Liquids also exert pressure for the same reason.

The pressure exerted by the atmosphere of the Earth varies with altitude. At sea level, atmospheric pressure is about 101 kPa. At the top of Ben Nevis (Britain's highest mountain) the pressure is only 87 kPa.

Liquids

You can calculate the pressure p exerted by a vertical column of any liquid from its weight and the cross-sectional area of the base.

$$p = h\rho g$$

where h is the height of the liquid column, ρ is the density of the liquid, and g is the acceleration of free fall (9.81 m s^{-2}).

It is important to understand how this equation is derived. Figure 2 shows a cylindrical column of liquid with height h and base of cross-sectional area A.

The pressure at the base is equal to the weight W of the column divided by A.

$$W = \text{mass of column} \times g$$

The mass of the column is the density × the volume.

$$W = (\rho V) \times g$$

The volume V of the column is Ah.

$$W = \rho \times Ah \times g$$

Finally, the pressure p is given by

$$p = \frac{\rho \times A \times h \times g}{A} = h\rho g$$

This equation shows that pressure does not depend on the cross-sectional area. It also clearly shows that $p \propto h$, so water pressure increases with depth. The term ρ shows that denser liquids will exert greater pressure.

Learning outcomes

Demonstrate knowledge, understanding, and application of:

→ $p = h\rho g$

→ upthrust and Archimedes' principle.

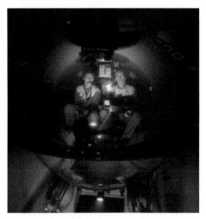

▲ **Figure 1** *This submarine has a spherical shape to resist the pressure of the surrounding water, which increases with its depth*

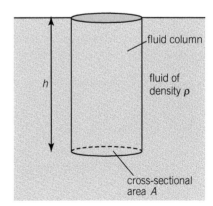

▲ **Figure 2** *Pressure in a column of liquid*

67

The pressure in a fluid at any particular depth has an unusual feature – it is the same in all directions.

 Worked example: Feeling the pressure

Calculate the total pressure acting on a submarine at a depth of 20 m.

atmospheric pressure = 101 kPa

   density of seawater = $1.03 \times 10^3\,\mathrm{kg\,m^{-3}}$

Step 1: Select the equation needed to calculate pressure in a fluid.

Take care! The pressure on the submarine will be the sum of the atmospheric pressure and the pressure due to the water. Therefore

pressure = atmospheric pressure + $h\rho g$

pressure = $1.01 \times 10^5 + (20 \times 1.03 \times 10^3 \times 9.81)$

pressure = 3.03×10^5 Pa

The pressure exerted by the seawater is twice that of the atmosphere.

Upthrust

Try pushing a small piece of wood into water and then letting it go. The wood will immediately pop out of the water, bob up and down on the surface, and remain afloat. The buoyant force on the submerged wood can be explained in terms of the pressure differences at its upper and lower surfaces.

Figure 3 shows the position of a submerged rectangular block of wood of cross-sectional area A. The density of the fluid is ρ. This block of wood will displace the fluid.

force at the top surface = $h\rho gA$

force at the bottom surface = $(h + x)\rho gA$

resultant upward force = $(h + x)\rho gA - h\rho gA = x\rho gA$

This resultant force is called **upthrust**.

upthrust = $Ax\rho g$

The volume of the block of fluid displaced is Ax and it has mass $(Ax)\rho$. So the upthrust is equal to the weight $(Ax\rho)g$ of the fluid displaced by the block of wood. This idea was known to Archimedes almost 2000 years ago and is still important to engineers, who design ships and structures under water. It applies to fully or partially submerged objects.

Archimedes' principle: The upthrust exerted on a body immersed in a fluid, whether fully or partially submerged, is equal to the weight of the fluid that the body displaces.

An object will sink if the upthrust is less than the weight of the object. For a floating object, such as a ship or a person in water, the upthrust

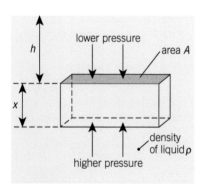

▲ **Figure 3** *Upthrust is due to pressure differences*

must equal the weight of the object. This in turn means that the weight of a floating object must be equal to the weight of the fluid it displaces.

Icebergs

It is said that nine-tenths of an iceberg lie hidden underwater. At 0°C the density of ice is about $900 \, \text{kg m}^{-3}$, whereas the density of water is about $1000 \, \text{kg m}^{-3}$. A cube of ice with sides 1.0 m will therefore have a weight of $(900 \times 9.81) = 8.83 \times 10^3 \, \text{N}$. When it floats, it will displace water of weight $8.83 \times 10^3 \, \text{N}$. This has a volume of $(8.83 \times 10^3)/(1000 \times 9.81) = 0.90 \, \text{m}^3$. So the $1.0 \, \text{m}^3$ cube of ice will sink until it has displaced $0.90 \, \text{m}^3$ of water, that is, nine-tenths of the cube is below the surface of the water.

▲ **Figure 4** *Most of an iceberg is underwater*

➕ Deep-water divers

Barotrauma is the physical damage to body tissues caused by a difference in the pressure within the body and that in the surrounding fluid. Divers swimming in deep water can sometimes suffer ear or lung damage.

A descent of 10 m in water doubles the external pressure on a diver. The record for deep-water scuba diving stands at about 330 m. Divers attempting such records are highly trained.

1 Explain, with the help of a calculation, how the external pressure on a diver at a depth of 10 m is doubled.
2 Estimate the vertical force acting on a diver's hand at a depth of 330 m. Assume the cross-sectional area of the hand is $1.2 \times 10^{-2} \, \text{m}^2$.
3 Suggest why a scuba diver who breathes air from the tank at 10 m depth and then ascends without exhaling can cause lung damage.

▲ **Figure 5** *The differences in pressure at different depths pose a danger to divers*

Summary questions

Assume the density of water $= 1.0 \times 10^3 \, \text{kg m}^{-3}$.

1 The density of mercury is $1.35 \times 10^4 \, \text{kg m}^{-3}$. Calculate the pressure exerted by a column of mercury at a depth of 0.765 m. *(2 marks)*

2 Show that the pressure exerted at a depth of 610 m in water is about 6 million pascals. *(2 marks)*

3 A table-tennis ball is held under water and then released. Describe and explain the subsequent motion of the ball. *(5 marks)*

4 A metal bar is suspended from a newtonmeter. The reading is 1.54 N in air but only 1.34 N when the bar is fully submerged in water. Calculate: a the upthrust on the bar; b the density of the bar. *(4 marks)*

5 A cube of a substance of density ρ_s floats when placed in water. Show that the fraction of the cube under water is $\dfrac{\rho_s}{\rho}$, where ρ is the density of water. *(4 marks)*

Practice questions

1   **a**   Define the *newton* and derive its base units.   *(3 marks)*

    b   Figure 1 shows a rocket on the surface of the Earth.

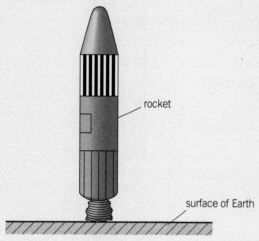

▲ Figure 1

The mass of the rocket is 3.0×10^6 kg. At lift off, the vertical upward thrust on the rocket is 34 MN.

    (i)   Calculate the initial vertical acceleration of the rocket.   *(3 marks)*

    (ii)   The upward thrust on the rocket remains the same. Explain why after some time, the vertical acceleration is much larger than the value calculated in **(i)**.   *(2 marks)*

2   Figure 2 shows the vertical forces acting on a helium-filled weather balloon just before lift off.

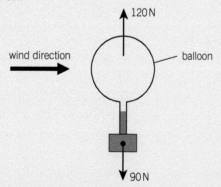

▲ Figure 2

The balloon experiences an upward vertical force (upthrust) equal to 120 N. The weight of the balloon and its contents is 90 N. The magnitude of the horizontal force provided by the wind is 18 N.

    a   Determine the magnitude of the resultant force acting on the balloon and the angle this resultant force makes with the horizontal.   *(4 marks)*

    b   As the balloon rises through the air, it experiences a drag force. State *two* factors that affect the magnitude of the drag force on this balloon.   *(2 marks)*

May 2012 G481

3   Figure 3 shows a lamp supported by two cables. The weight of the lamp is 24 N.

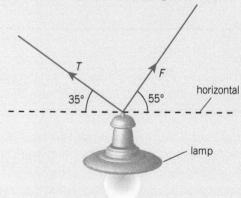

▲ Figure 3

The lamp is in equilibrium. The tensions in the cables are *T* and *F*.

    a   Without any calculations, explain the value of the resultant force due to *T* and *F*.   *(2 marks)*

    b   Calculate the magnitude of the forces *T* and *F*.   *(4 marks)*

    c   The angle made by the force *T* with the horizontal is decreased. Explain the effect this has on the tension *T*.   *(2 marks)*

4   **a**   Define *density*.   *(1 mark)*

    b   Figure 4 shows the variation of density of the Earth with **depth** from the surface.

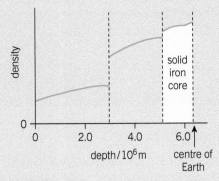

▲ Figure 4

(i) Suggest how Figure 4 shows that the Earth consists of a number of distinct layers. (*1 mark*)

(ii) Geophysicists believe that the central core of the Earth is solid iron and nickel. This central core is surrounded by a layer of molten metal. The central core starts at a **depth** of 5.1×10^6 m. The solid iron core accounts for 18% of the mass of the Earth. The mass of the Earth is 6.0×10^{24} kg and its radius is 6.4×10^6 m. Calculate the mean density of the central core of the Earth. Volume of a sphere $= \frac{4}{3}\pi r^3$
(*3 marks*)
May 2011 G481

5 a The atmosphere of the Earth exerts pressure on all objects on its surface. At a depth d in water, the total pressure is P. On a copy of the axes below, sketch a graph to show the variation of P with d. (*2 marks*)

b Figure 5 shows an object held under water.

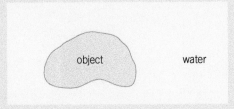

▲ Figure 5

The density of the object is $400 \,\text{kg m}^{-3}$ and it has a volume of $6.0 \,\text{cm}^3$. The density of water is $1000 \,\text{kg m}^{-3}$.

(i) Explain why the object experiences an upthrust. (*3 marks*)

(ii) Calculate initial upwards acceleration of the object when it is released. (*5 marks*)

6 a State what is meant by the *centre of gravity* of an object. (*1 mark*)

b Define *moment of a force*. (*1 mark*)

c Figure 6 shows a baby's mobile toy.

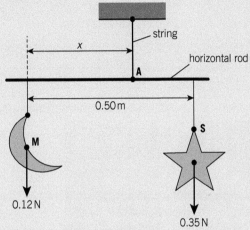

▲ Figure 6

The toy consists of a horizontal rod from which two objects shaped as a crescent moon **M** and a star **S** are suspended from lengths of string. The weight of the rod is negligible and it is pivoted about an axis passing through point **A** perpendicular to the plane of the diagram. The weights of **M** and **S** and the separation between the strings are shown in Figure 6. The distance between the string on the left and point **A** is x. The arrangement shown in Figure 6 is in equilibrium.

(i) State **two** conditions necessary for the rod to be in equilibrium. (*2 marks*)

(ii) By taking moments about **A** determine the distance x. (*3 marks*)

(iii) Determine the magnitude of the tension in the string attached to **A**. (*1 mark*)
May 2013 G481

5 WORK, ENERGY, AND POWER
5.1 Work done and energy

Specification reference: 3.3.1

Learning outcomes

Demonstrate knowledge, understanding, and application of:

→ work done by a force
→ the unit joule
→ $W = Fx \cos \theta$ for the work done by a force
→ transfer of energy equal to work done.

▲ **Figure 1** *How much work is done raising the boxes to this height?*

Moving up and along

Moving an object can be difficult if it has a large mass or it must be moved a long distance. Fork-lift trucks are used in warehouses and other places where a load may need to be stacked or moved. A system of hydraulics powered by a motor in the base of the machine provides the force to move the platform. The energy needed depends upon the force and the distance over which this force acts.

Work done and energy

In everyday conversation, the word **work** can mean any kind of physical or mental activity. Work done in physics has a very precise definition. It involves a force F and distance x moved in the direction of the force.

work done = force × distance moved in the direction of the force

$$W = Fx$$

Work done therefore has the unit Nm, or joule (J). 1 J = 1 Nm. 1 J is the work done when a force of 1 N moves its point of application 1 m in the direction of the force.

You do no work when holding a heavy book in your hand. You exert a force on the book, but there is no movement. You do work when lifting the book vertically. You apply a force on the book and it moves through a certain distance. Energy is transferred when you do work. The book has gained gravitational potential energy. If you let the book fall, then its weight will do work, which is transferred into kinetic energy. In fact

work done = energy transferred

This is not surprising, because energy is defined as the capacity to do work.

 Worked example: Work done

How much work is done by:

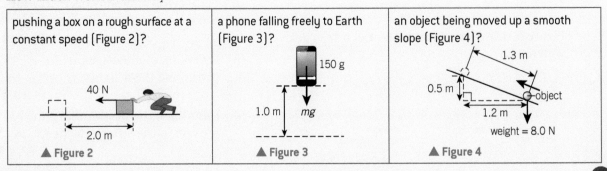

pushing a box on a rough surface at a constant speed (Figure 2)?	a phone falling freely to Earth (Figure 3)?	an object being moved up a smooth slope (Figure 4)?
▲ Figure 2	▲ Figure 3	▲ Figure 4

| $W = Fx$

 $W = 40 \times 2.0 = 80\,J$

 The work done **on** the box is transferred into thermal energy of the box and the surface below. | The force acting on the object is its weight mg.

 $W = Fx$

 $W = (0.150 \times 9.81) \times 1.0$
 $\quad = 1.5\,J\ (2\ s.f.)$
 The work done **by** the force of gravity on the object is transferred to kinetic energy. | The distance travelled in the direction of the force (the weight) is 0.5 m, and not 1.3 m nor 1.2 m.

 $W = Fx$

 $W = 8.0 \times 0.5 = 4.0\,J$
 The work done **against** the force of gravity is transferred into gravitational potential energy. |

Work done at an angle to motion

Quite often, a force is applied at an angle to the direction in which an object can move, as in Figure 5. How do you calculate the work done by the force?

The component of the force F in the direction of motion is $F\cos\theta$. Therefore

$$\text{work done } W = (F\cos\theta) \times x$$

Or simply

$$W = Fx\cos\theta$$

The work done for the object moving up the slope in Figure 4 can be calculated using this equation. You just need to know the angle θ between the force and direction of motion. Try it for yourself with this alternative equation – you will get the same answer.

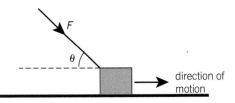

▲ **Figure 5** *Force exerted on an object at an angle to the direction of motion*

> **Study tip**
>
> When a force acts in the direction of travel, $\theta = 0°$ and $\cos 0° = 1$. This means you can use a simpler expression, $W = Fx$.

Summary questions

1 Calculate the work done when a force of 24 N moves an object a distance of 0.50 m in the direction of the force. *(2 marks)*

2 The thrust from a hovercraft fan is 430 N. Calculate the work done when the thrust moves the hovercraft by 1.0 km. *(2 marks)*

3 Calculate the work done by a person of mass 60 kg to climb to the top of a 5.8 m wall. *(2 marks)*

4 A shopper pushes a 38 kg shopping trolley at a constant speed up a car park ramp. The ramp is at 10° to the horizontal and is 3.1 m long. Calculate the work done to push the trolley to the top of the ramp. *(3 marks)*

5 A person exerts a force of 65 N at 52° to the horizontal floor to push a box 5.0 m across a floor at constant speed. Calculate the work done on the box. Explain, in terms of energy, what happens to the work done on the box. *(3 marks)*

6 A bullet travels straight through a piece of wood of thickness 30 mm. The change in the kinetic energy of the bullet is 1.4 kJ. Calculate the average force exerted by the wood on the bullet. *(4 marks)*

Learning outcomes

Demonstrate knowledge, understanding, and application of:

→ the principle of conservation of energy

→ energy in different forms, transfer and conservation.

▲ **Figure 1** *In this design for a perpetual motion machine, weights falling on one side of the wheel make weights on the other side move upwards, supposedly keeping the wheel turning forever. In reality this will not happen – why not?*

▲ **Figure 2** *The spinning blades of a wind turbine have kinetic energy, and they gained gravitational potential energy when they were lifted into place by a crane*

Perpetual motion

There have been many attempts to build a perpetual motion machine – a device which, once started, will continue to move and do work without any further input of energy. None of them work and all eventually stop. In fact, perpetual motion is impossible according to the **principle of conservation of energy**. Would-be inventors should know that a patent will never be granted because their designs violate this well-established physical law.

Energy

Energy is the capacity for doing work. It is a scalar quantity, with magnitude but not direction. The SI unit for energy is the joule (J), the same unit as for work done.

Forms of energy

The energy of systems with mass can be classified as **kinetic energy** and **gravitational potential energy**.

● Kinetic energy is the energy due to the movement of an object.

● Gravitational potential energy is the energy due to the position of an object in the Earth's gravitational field.

The term *potential* is often used in physics to mean 'hidden' or 'stored'. Table 1 summarises some other forms of energy you have come across in your physics lessons.

▼ **Table 1** *Forms of energy*

Form of energy	Description	Examples
kinetic energy	energy due to motion of an object with mass	moving car moving atoms
gravitational potential energy	energy of an object due to its position in a gravitational field	child at the top of a slide water held in clouds
chemical energy	energy contained within the chemical bonds between atoms – it can be released when the atoms are rearranged	energy stored within a chemical cell energy stored in petrol and released when it is burnt
elastic potential energy	energy stored in an object as a result of reversible change in its shape	a stretched guitar string a squashed spring
electrical potential energy	energy of electrical charges due to their position in an electric field	electrical charges on a thundercloud static charge on a charged balloon

▼ **Table 1** *Continued*

Form of energy	Description	Examples
nuclear energy	energy within the nuclei of atoms – it can be released when the particles within the nucleus are rearranged	energy from fusion processes in the Sun energy from nuclear fission reactors
radiant (or electromagnetic) energy	energy associated with all electromagnetic waves, stored within the oscillating electric and magnetic fields	energy from the hot Sun energy from an LED
sound energy	energy of mechanical waves due to the movement of atoms	energy emitted when you clap output energy from your headphones
internal (heat or thermal) energy	the sum of the random potential and kinetic energies of atoms in a system	a hot cup of tea has more thermal energy than a cold one

▲ **Figure 3** *In a fire, chemical energy stored in the substances that make up the building and its contents is converted to thermal energy and radiant energy*

The principle of conservation of energy

Energy can be converted from one form to another. For example, an archer's bow usefully converts elastic potential energy into kinetic energy in an arrow. The bow also converts some of the elastic potential energy into thermal energy and sound energy, but the *total* final energy is always equal to the *total* initial energy.

The **principle of conservation of energy** states that the total energy of a closed system remains constant: energy can never be created or destroyed, but it can be transferred from one form to another.

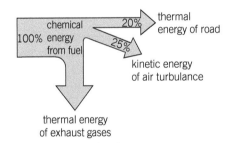

▲ **Figure 4** *Energy conversions from fuel burnt*

Summary questions

1   a   State what is meant by the term *potential* in physics. *(1 mark)*
    b   State the energy changes taking place when you rub your hands together. *(1 mark)*

2   A lamp converts 20 J of electrical energy into 5 J of light energy and one other form of energy. Suggest what this other form of energy is, and calculate its quantity. *(2 marks)*

3   Describe the useful energy conversions that happen in:
    a   a filament lamp; *(1 mark)*
    b   the headphones connected to a mobile phone. *(1 mark)*

4   A car is travelling on a level road at constant speed. Figure 4 is a visual representation of the energy conversions of the chemical energy in the fuel – the diagram is called a Sankey diagram.
    a   Explain why the diagram does not show the kinetic energy of the car. *(1 mark)*
    b   Calculate the percentage of total energy wasted as thermal energy. *(1 mark)*

5.3 Kinetic energy and gravitational potential energy

Specification reference: 3.3.2

▲ **Figure 1** *The loop in this rollercoaster plunges thrill-seekers vertically downwards, converting GPE into KE*

Don't look down

Roller coasters have been thrilling people for at least two hundred years. In a typical rollercoaster, the cars are hauled up a steep slope by a motor, and then the force of gravity takes over. As the cars hurtle along the track, gravitational potential energy (GPE) and kinetic energy (KE) are interchanged repeatedly, enough to excite or terrify the passengers without causing injury. It is possible to analyse the motion of the rollercoaster from its GPE and KE.

Kinetic energy

Kinetic energy is energy associated with an object as a result of its motion. You can calculate the KE E_k of an object in linear motion from its mass m and speed v using the equation

$$E_k = \frac{1}{2}mv^2$$

Notice that for objects travelling at the same speed, the KE is directly proportional to the mass. For a given object, the KE is directly proportional to the square of its speed.

 Worked example: A meteor hitting the Earth

The Chelyabinsk meteor was a near-Earth asteroid that crashed into Russia in February 2013. It had a mass of $1.2 \times 10^7\,\text{kg}$ and a speed on impact of $19\,\text{km}\,\text{s}^{-1}$. Calculate its KE, and compare this energy with the KE of a $1.2 \times 10^4\,\text{kg}$ truck travelling at $20\,\text{m}\,\text{s}^{-1}$.

Step 1: Identify the equation needed and list the known values.

$$E_k = \frac{1}{2}mv^2$$

Meteor: $m = 1.2 \times 10^7\,\text{kg}$, $v = 19 \times 10^3\,\text{m}\,\text{s}^{-1}$ ($1.9 \times 10^4\,\text{m}\,\text{s}^{-1}$)

Truck: $m = 1.2 \times 10^4\,\text{kg}$, $v = 20\,\text{m}\,\text{s}^{-1}$

Step 2: Substitute the values into the equation and calculate the answer.

Meteor: $E_k = \frac{1}{2}mv^2 = \frac{1}{2} \times 1.2 \times 10^7 \times (1.9 \times 10^4)^2$ (note that you must convert the speed into $\text{m}\,\text{s}^{-1}$)

$E_k = 2.17 \times 10^{15}\,\text{J} \approx 2.2 \times 10^{15}\,\text{J}$

Truck: $E_k = \frac{1}{2}mv^2 = \frac{1}{2} \times 1.2 \times 10^4 \times 20^2$

$E_k = 2.4 \times 10^6\,\text{J}$

The ratio of the kinetic energies is $\dfrac{2.17 \times 10^{15}}{2.4 \times 10^6} = 9.0 \times 10^8$ (2 s.f.)

So, the KE of the meteor on impact was equivalent to the KE of 900 million trucks!

Deriving $E_k = \frac{1}{2}mv^2$

You can derive the equation for KE by using ideas developed in this book already.

Figure 2 shows a constant force F acting on an object of mass m. The object is initially at rest. The acceleration of the object is a. After a distance s it has a speed v.

The distance s travelled by the object can be determined from the equation of motion $v^2 = u^2 + 2as$.

$$s = \frac{v^2 - u^2}{2a} = \frac{v^2}{2a} \quad (u = 0)$$

The work done by the force is entirely transferred to the KE of the object. Therefore

$$\text{work done} = E_k = Fs$$

The force F is given by $F = ma$. Therefore

$$E_k = ma \times s = ma \times \frac{v^2}{2a}$$
$$\therefore E_k = \frac{1}{2}mv^2$$

(where $\therefore$ denotes *therefore*).

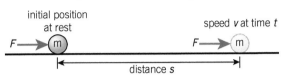

initial position at rest
speed v at time t
F m
F m
distance s

▲ **Figure 2** *Gaining kinetic energy*

1 Use the equation for E_k to derive the SI base units for kinetic energy.
2 Derive an equation for E_k in terms of speed v, weight W of an object, and acceleration of free fall g.

Gravitational potential energy

Gravitational potential energy is the capacity for doing work as a result of an object's position in a gravitational field. You can calculate the change in GPE E_p of an object in a **uniform gravitational field** from its mass m and the change in height h using the equation $E_p = mgh$

where g is acceleration of free fall.

You can use this equation for all objects close to the Earth's surface, where g may be assumed to be constant and to have a value of $9.81\,\text{m s}^{-2}$.

GPE is *gained* when an object gets higher, and is *lost* when an object gets lower.

You can explain the origin of the equation for GPE using the idea of work done by a force. Imagine lifting a mass vertically upwards through a height h at constant speed (so its KE does not change). You have to apply a force equal to its weight mg. The work done W by this force is transferred into GPE. Therefore

$E_p = W = \text{force} \times \text{distance moved in the direction of force}$

$E_p = (mg) \times h \qquad E_p = mgh$

Synoptic link

You can use work done = change in KE or $Fs = \frac{1}{2}mv^2$ to explain why the braking distance (s) of a particular car is directly proportional to (initial speed)^2 – see Topic 3.6, Car stopping distances.

Study tip

Remember that things fall without any extra energy being supplied, so GPE is dissipated.

 ## Worked example: Skydiver

A skydiver with a mass of 80 kg falls through a distance of 2.0 km at a terminal velocity of $45.0\,\text{m s}^{-1}$. Calculate the loss of GPE and explain what happens to the energy lost.

Step 1: Identify the equation needed.

$$\text{loss in GPE} = E_p = mgh$$

▲ **Figure 3** *Parachutists gain gravitational potential energy on the way up in an aircraft, and lose it as they fall to the ground*

Step 2: Substitute the values into the equation and calculate the answer.

$E_p = 80 \times 9.81 \times 2000 = 1.6 \times 10^6 \, \text{J}$ (remember h must be in metres)

There is no change in the KE of the skydiver, so the GPE is transferred to the thermal energy of the surrounding air.

Energy exchange

There are many situations where KE and GPE are exchanged. For example, as an object falls, its GPE decreases and its KE increases. You can see this happening for a waterfall or when you drop a pen. An object falling from rest will lose GPE. From the principle of conservation of energy, the object will gain an equal amount of KE. Therefore

$$mgh = \frac{1}{2}mv^2$$

where v is the final speed of the object.

The mass m on both sides of the expression cancels out. Therefore

$$gh = \frac{1}{2}v^2 \qquad v^2 = 2gh \qquad v = \sqrt{2gh}$$

The equation above is only valid if there are no resistive forces involved. The mass of the object has no bearing on its final speed. You should not be too surprised by this because you already know that the acceleration of free fall is the same for all objects. An object dropped from a height of only 7.0 m will hit the ground at a speed of $31 \, \text{m s}^{-1}$ (70 mph) – you can use the equation above to check this.

▲ **Figure 4** *The speed of the roller coaster at the bottom is given by $v = \sqrt{2gh}$, as long as there are no frictional losses*

Summary questions

1 Calculate the kinetic energy of a 1500 kg object travelling at $10 \, \text{m s}^{-1}$. *(2 marks)*

2 Calculate the gain in the gravitational potential energy of a plane of mass $= 9.4 \times 10^4 \, \text{kg}$ climbing a vertical distance of 1500 m. *(2 marks)*

3 A 120 g ball is dropped from rest from a height of 90 cm. Its rebound height is 70 cm. Calculate the energy lost to the ground. *(3 marks)*

4 A roller coaster of mass 400 kg begins at rest, then drops 55 m into a tunnel. Assume that air resistance has negligible effect on its motion.
   a Calculate the initial GPE of the roller coaster. *(2 marks)*
   b Calculate its KE as it enters the tunnel. *(1 mark)*
   c Use your answer to (b) to calculate its speed. *(3 marks)*

5 Victoria Falls (Mosi-oa-Tunya) is a waterfall on the Zambezi River at the border of Zambia and Zimbabwe. It has a height of about 110 m. Calculate the maximum speed of the water at the bottom. State any assumptions made. *(3 marks)*

6 A book of mass 0.80 kg falls from a height of 1.2 m. Its speed is $3.2 \, \text{m s}^{-1}$ when it hits the ground. Calculate the work done against the drag force. *(3 marks)*

7 A bullet of mass 30 g is fired at a block of wood at a speed of $240 \, \text{m s}^{-1}$. The bullet penetrates a distance of 8.5 cm into the wood. Calculate the average resistive force exerted on the bullet by the wood block. *(3 marks)*

5.4 Power and efficiency

Specification reference: 3.3.3

Electric mountain

The Dinorwig Power Station in north Wales is an unusual hydroelectric scheme. Built into a mountain in Snowdonia National Park, it has two reservoirs, one more than 500 m above the other. Several kilometres of water-filled tunnels pass through the mountain, carrying water between the reservoirs. A huge cavern excavated deep inside the mountain contains six reversible turbines.

When there is a high demand for electricity from the National Grid, water flows from the upper reservoir, driving the turbines and generators. During off-peak times, the turbines pump water back up from the lower reservoir. Dinorwig is the largest pumped storage facility in Europe, but although it is a brilliant solution to providing electricity at short notice, it is not 100% efficient.

Learning outcomes

Demonstrate knowledge, understanding, and application of:

→ power; the unit watt; $P = \dfrac{W}{t}$

→ $P = Fv$

→ efficiency of mechanical systems; $efficiency = \dfrac{useful\ output\ energy}{total\ input\ energy} \times 100\%$.

Energy and power

For a sprinter or a racing car, we are interested not only in the amount of energy but also in the *rate* at which energy is transferred. A powerful car is the one with the largest value for the rate of energy transfer.

Power is the rate of work done.

As an equation, this is written as

$$P = \frac{W}{t}$$

where P is the power and W is the work done in a time t. Since work done is equal to energy transfer, we can also define power as the rate of energy transfer.

Power is measured in joules per second ($J\,s^{-1}$) or in watts (W). 1 W is equal to one joule per second.

▲ **Figure 1** *The upper reservoir for the Dinorwig Power Station has a low water level in the daytime*

 Worked example: Body power

A 60 kg person runs up a flight of steps in a time of 7.2 s. The gain in vertical height in this time interval is 5.0 m. Calculate the rate of work done against the force of gravity.

Step 1: Calculating the rate of work done against gravity is the same as calculating the power P. Also, the work done against the force of gravity is the same as the gain in the gravitational potential energy E_p of the person.

$$power = \frac{work\ done}{time}$$

$$P = \frac{E_p}{t} = \frac{mgh}{t}$$

Step 2: Substitute the values into the equation and calculate the answer.

$$P = \frac{60 \times 9.81 \times 5.0}{7.2} = 410\,\text{W (2 s.f.)}$$

The rate of work done against the force of gravity is about 410 W. This is the same as 410 J per second.

The actual power developed by the person is much greater than this value. Our muscles are incapable of transferring all available energy into movement – some is wasted as thermal energy within the muscles. This is why you get hot when you exercise. Muscles are not very efficient at transferring energy (see Table 1).

Power and motion

There are situation in physics where constant force has to be exerted on an object to maintain its constant speed. A good example of this is a car travelling on a level road at a constant speed (Figure 2). The net force on the car is zero. The rate of work done by the forward force provided by the car engine is equal to the rate of work done against the frictional forces acting on the car. It is possible to calculate the power P developed by the force provided by the car.

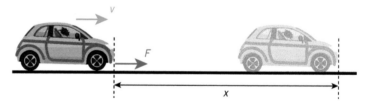

▲ **Figure 2** *The power developed by the car depends on the force F and the speed v*

A constant force F moves the car a distance x in a time t.

work done by the force $W = F \times x$

$$P = \frac{W}{t} = \frac{Fx}{t}$$

The speed v of the car is $\left(\frac{x}{t}\right)$, therefore

$$P = Fv$$

This equation is not just for cars. You can use it in a variety of situations where a constant force is necessary to maintain a constant speed. Examples include a swimmer travelling though water at constant speed (Figure 3) and a skydiver falling through the air at terminal velocity.

Efficiency

You will remember the principle of conservation of energy: the total energy of a closed system remains constant. However, this does not mean that processes and machines convert all their energy into *useful* work. In particular, thermal losses can mean that some of the input energy is not converted into useful output energy.

▲ **Figure 3** *Greater power is required to swim faster through the water*

You can calculate **efficiency** using the expression

$$efficiency = \frac{useful\ output\ energy}{total\ input\ energy} \times 100\%$$

The greater the efficiency, the greater the percentage of input energy converted. Some examples of efficiency are shown in Table 1.

▼ **Table 1** *Typical efficiencies – can you suggest why the electric heater is almost 100% efficient?*

System	filament lamp	muscles	petrol engine	solar cell	LED	diesel engine	wind-generator	electric heater
Typical efficiency /%	5	20	20	25	35	35	40	~ 100

Summary questions

1   Calculate the power of a lamp that transfers 240 J in 30 s.   *(2 marks)*

2   Calculate the energy transferred by a 2.0 kW motor in a time of 60 s.   *(2 marks)*

3   A lamp is about 5.0% efficient at producing light energy from electrical energy. Calculate the light energy produced from a 60 W lamp in a time of 1.0 hour.   *(3 marks)*

4   A car of mass 1200 kg starting from rest reaches a speed of 18 m s⁻¹ in 20 s. Calculate the average rate of work done on the car.   *(3 marks)*

5   An aircraft has four jet engines, each producing 210 kN of thrust. Calculate the total power output of the engines when the aircraft is in level flight and travelling at a constant speed of 250 m s⁻¹.   *(2 marks)*

6   A small 3.5 W electric motor raises a 15 N load through a vertical height of 1.4 m in 30 s (Figure 4). Calculate the efficiency of the motor.   *(3 marks)*

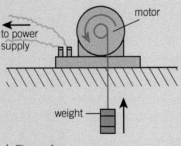

to power supply

weight

motor

▲ **Figure 4**

7   A hydroelectric power station produces 600 MW of electrical power. Water falls 50 m before passing through its turbines. The transfer of gravitational potential energy to electrical energy is 40% efficient. Calculate the rate (volume/time) at which water passes through the turbines. The density of water is 1000 kg m⁻³.   *(5 marks)*

Practice questions

1 **a** Define power.   (*1 mark*)

   b An electric motor on a mechanical crane is used to lift heavy objects, see Figure 1.

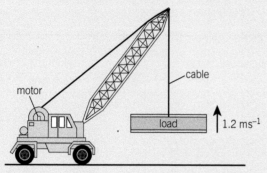

▲ Figure 1

The cable attached to the motor is used to lift a load of total mass of 1500 kg at a constant vertical velocity of 1.2 m s⁻¹.

   (i) Calculate the tension in the cable. Explain your answer.   (*3 marks*)

   (ii) Calculate the minimum output power of the motor needed to raise the 1500 kg mass.   (*3 marks*)

2 **a** Write a word equation for *kinetic energy*.   (*1 mark*)

   b A bullet of mass 3.0×10^{-2} kg is fired at a sheet of plastic of thickness 0.015 m. The bullet enters the plastic with a speed of 200 m s⁻¹ and emerges from the other side with a speed of 50 m s⁻¹.

   Calculate

   (i) The loss of kinetic energy of the bullet as it passes through the plastic   (*3 marks*)

   (ii) The average frictional force exerted by the plastic on the bullet.   (*2 marks*)

   c Plan a simple experiment to determine the kinetic energy of a student running round a track.   (*4 marks*)

3 **a** State the *principle of conservation of energy*.   (*1 mark*)

   b Define *work done* by a force and state its unit.   (*1 mark*)

   c Figure 2 shows a crater on the surface of the Earth.

▲ Figure 2

The crater was formed by a meteorite impact 50,000 years ago. The meteorite was estimated to have a mass of 3.0×10^{8} kg with an initial kinetic energy of 8.4×10^{16} J just before impact.

   (i) State one major energy transformation that took place during the impact of the meteorite with the Earth.   (*1 mark*)

   (ii) Show that the initial impact speed of the meteorite was about 2.0×10^{4} m s⁻¹.   (*2 marks*)

   (iii) The crater is about 200 m deep. Estimate the average force acting on the meteorite during the impact.   (*3 marks*)

Jan 2012 G481

4 **a** Derive the base units for work done.   (*3 marks*)

   b Figure 3 shows a person at the base of an escalator.

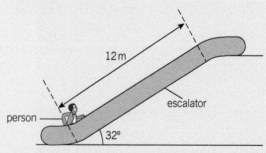

▲ Figure 3

The mass of the person is 70 kg. The escalator travels at an angle of 32° to the horizontal. The person travels a total distance of 12 m from the bottom to the top of the escalator in a time of 8.0 s.

(i) Calculate the kinetic energy of the person on the moving escalator.

(3 marks)

(ii) Calculate the gravitational potential gained by the person. *(3 marks)*

(iii) Calculate the power required to lift the person from the bottom to the top of the escalator. *(2 marks)*

5 A skydiver jumps off a helicopter and falls vertically towards the ground. The skydiver reaches a steady velocity before he opens his parachute and lands safely on the ground.

a Describe the energy changes taking place from the time the skydiver jumps off the helicopter to the instant before he opens his parachute. *(4 marks)*

b The total mass of the skydiver is 80 kg and his terminal velocity is 45 m s^{-1}. Calculate the rate of work done against drag. *(3 marks)*

6 a Explain what is meant by *energy* and relate it to *power*. *(2 marks)*

b Define the *watt*. *(1 mark)*

c A lift is used to carry people up a building. The mass of the lift is 1500 kg and it can carry a maximum of 8 people of average mass 70 kg. The vertical height travelled by the lift is 120 m and it takes 55 s.

(i) Calculate the gain in gravitational potential energy of the 8 people in the lift. *(2 marks)*

(ii) Calculate the minimum output power of the electrical motor used to operate the lift. *(3 marks)*

7 a In a downhill race, the total distance between start and finish is 5.00 km, and the total vertical drop is 520 m. The weight of a runner is 70 kg and the time taken for the descent is 15 minutes.

(i) Calculate the average kinetic energy of the runner during this race. *(3 marks)*

(ii) Calculate the total loss in the gravitational potential energy of the runner. *(2 marks)*

(iii) Explain why your answers to **(a)**(i) and **(a)**(ii) are not the same. *(1 mark)*

b (i) State the principle of conservation of energy. *(1 mark)*

(ii) Use the principle of conservation of energy to show that an object dropped through a vertical distance of 520 m can have a speed of more than 100 m s^{-1}. *(3 marks)*

8 A stunt person, initially at rest, slides down a cable attached between a tall building and the ground.

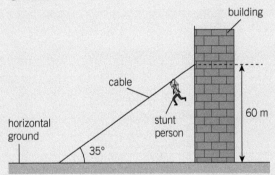

▲ Figure 4

The person has mass 72 kg. The speed of the person at the bottom of the cable is 20 m s^{-1}.

a Calculate the loss in gravitational potential energy of the person. *(2 marks)*

b Calculate the work done against resistive forces. *(3 marks)*

c Calculate the average frictional force acting on the person. *(3 marks)*

6 MATERIALS

6.1 Springs and Hooke's law

Specification reference: 3.4.1, 3.4.2

Learning outcomes

Demonstrate knowledge, understanding, and application of:

→ tensile and compressive deformation

→ Hooke's law

→ force constant k of a spring or wire and $F = kx$

→ force–extension (or compression) graphs for springs and wires

→ elastic and plastic deformation of springs.

Tensile and compressive forces

You need a pair of equal and opposite forces to alter the shape of an object. For example, you can extend the length of a rubber band by pulling its ends. Forces that produce **extension** are known as **tensile forces**, and those that shorten an object (**compression**) are **compressive forces**. Imagine sitting on a stool. Your weight downwards and the contact force upwards from the ground provide the pair of compressive forces on the stool; these will shorten the length of the stool by a tiny amount.

In the suspension bridge shown in Figure 1, the cables and the huge vertical supports hold together the road structure below. Can you identify where the tensile and compressive forces act?

Hooke's law

A helical spring undergoes **tensile deformation** when tensile forces are exerted and **compressive deformation** when compressive forces are exerted.

Figure 2 shows a simple apparatus used to investigate how the extension x of a helical spring is affected by the applied force F, and a typical graph of force against extension obtained.

▲ **Figure 1** *The Forth Road Bridge in Scotland*

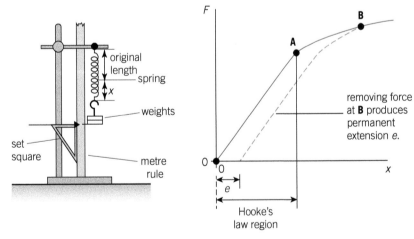

▲ **Figure 2** *Recording the force–extension graph for a spring*

▲ **Figure 3** *You can easily identify the spring that has been stretched beyond its elastic limit*

The **force–extension graph** is a straight line from the origin up to the **elastic limit** (point **A**) of the spring. In this linear region, the spring undergoes **elastic deformation**. This means that the spring will return to its original length when the force is removed. Beyond point **A**, the spring undergoes **plastic deformation**: permanent structural changes to the spring occur and it does not return to its original length when the force is removed (Figure 3).

For forces less than the elastic limit of the spring, the spring obeys **Hooke's law**: The extension of the spring is directly proportional to the force applied. This is true as long as the elastic limit of the spring is not exceeded.

Force constant k

For a spring obeying Hooke's law, the applied force F is directly proportional to the extension x. Therefore

$$F \propto x$$

or

$$F = kx$$

where k is called the **force constant** of the spring (SI unit newton per metre, $N\,m^{-1}$). This is a measure of the **stiffness** of a spring. A spring with a large force constant is difficult to extend and you would refer to it as a stiff spring. You can also use the equation $F = kx$ for a compressible spring: x then represents the compression of the spring.

You can determine the force constant k from the gradient of the linear region of the force–extension graph.

Hooke's law also applies to wires under tension and concrete columns under compression. You can even model the behaviour of atoms in a solid using Hooke's law. In fact the equation $F = kx$ can be applied to almost any object that can be elastically squashed or extended.

Synoptic link

You will find more information about gradients of straight-line graphs in Appendix A2, Recording results.

 Worked example: Force constant of a wire

A shelf of mass 14.00 kg is supported by four identical wires. The original length of each wire was 1.800 m. When attached to the shelf, the length of each wire is 1.804 m. Calculate the force constant of each wire.

Step 1: Select the correct equation to calculate the force F acting on each wire.

The weight W of the shelf can be calculate using $W = mg$; assume that this weight is shared equally amongst the four wires.

$$F = \frac{\text{weight}}{4} = \frac{14.00 \times 9.81}{4} = 34.3\,N$$

Step 2: Determine the extension x of each wire.

$$x = \text{new length} - \text{original length}$$
$$= 1.804 - 1.800 = 0.004\,m$$

Step 3: Select the correct equation to calculate k.

$$F = kx$$

$$34.3 = k \times 0.004$$

$$k = \frac{34.3}{0.004} = 9 \times 10^4\,N\,m^{-1}\,(1\text{ s.f.})$$

The force constant of each wire is $9 \times 10^4\,N\,m^{-1}$ (1 s.f.).

Investigating Hooke's law

You can investigate Hooke's law using a spring and some standard masses (see Figure 2).

Attach the spring at one end using a clamp, boss, and clamp stand secured to the bench using a G-clamp or a large mass. Set up a metre rule with a resolution of 1 mm close to the spring. Suspend slotted masses from the spring and, as you add each one, record the total mass added and the new length of the spring.

You can improve the accuracy of the length measurements using a set square, and by taking readings at eye level to reduce parallax errors. You might also measure the mass of each slotted mass using a digital balance. To obtain reliable results, aim to take at least six different readings and to repeat each one at least once.

Synoptic link

You will find more information about accuracy of readings in Appendix A3, Measurements and uncertainties.

▲ **Figure 4** *Every year in summer the swans born on a section of the river Thames are caught, weighed and measured, and ringed before being released to grow to maturity*

Force meters

Hooke's law is used in the design and calibration of simple force meters or newtonmeters, often used for weighing. Such spring-loaded scales are useful in situations where scales must be mobile or robust and easy to repair. They are used to monitor babies' growth in clinics in developing countries, for example. In Figure 4 a cygnet – a young swan – is weighed in a wildlife survey.

1 The extension of the spring in the force meter is 12 mm when a cygnet with a mass of 4.0 kg is weighed. Predict the extension of the spring when a cygnet of 6.0 kg is weighed. State any assumptions made.
2 Determine the force constant of the spring in $N\,m^{-1}$.

Summary questions

1 Figure 5 shows the force F against compression x graphs for two metal rods **A** and **B**. Compare the behaviour of the rods.

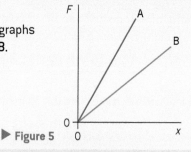

▶ **Figure 5** (2 marks)

2 The elastic limit of a wire is at 5.0 N. When a force of 2.5 N is applied the wire has an extension of 4 mm. Sketch a force–extension graph for this wire. (3 marks)

3 A spring is compressed by 5 mm by a force of 4.0 N.
   a Calculate the force constant of the spring. (2 marks)
   b Calculate the force applied when its compression is 32 mm. State any assumptions made. (3 marks)

4 Table 1 shows the results obtained for a spring in the experiment described above. Repeat readings could not be taken in this experiment because the spring was stretched beyond its elastic limit.

▼ Table 1 *Results from an investigation into the extension of a spring*

Mass attached to the spring / g	Length of spring, L / 10^{-2} m
100	4.3
200	8.6
300	13.0
400	17.1
500	21.6
600	28.1
700	37.0

a Copy the table of results. Add a column for the force F/N acting on the spring by calculating $F = mg$, where m is the mass in kg and $g = 9.81$ m s^{-2}. *(1 mark)*

b Plot a graph of force F against length L of the spring. *(3 marks)*

c Explain how the graph shows that the spring obeys Hooke's law and state the value of the force at the elastic limit. *(3 marks)*

d Determine the force constant of the spring in N m^{-1}. *(2 marks)*

5 A 200 mm long spring is suspended vertically. The length of the spring increases to 294 mm when a mass of 280 g is attached to it.

a Calculate the force constant of the spring. *(3 marks)*

b A second, identical spring is suspended alongside the first spring and both are then attached to a rod of negligible mass (Figure 6).

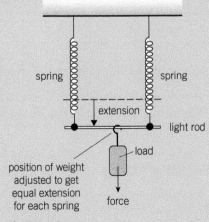

spring   spring
extension
light rod
position of weight
adjusted to get
equal extension
for each spring   load
force

▲ Figure 6

i Calculate the combined force constant of this parallel arrangement. Explain your answer. *(3 marks)*

ii Calculate the combined force constant when the same springs are joined end-to-end in a series arrangement. Explain your answer. *(3 marks)*

6.2 Elastic potential energy

Specification reference: 3.4.2

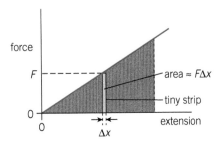

▲ **Figure 1** *The stretched elastic of a catapult stores energy and transfers it to kinetic energy when released*

▲ **Figure 2** *Force–extension graph and work done*

Synoptic link

In Topic 5.1, Work done and energy, you learnt the definition for work done by a force.

Stored energy

Imagine holding a rubber band between your fingers and applying a force by stretching it. The rubber band will extend in the direction of the force. The work done on the rubber band is transferred into stored energy in the band. You can show that a stretched rubber band has stored energy when you suddenly let it go and it flies across the room.

Clockwork toys use coiled springs to store and then release energy. The stretched strings of a guitar store energy – when a string snaps, this energy is transferred to kinetic energy and the broken string flies away at $15\,\text{m}\,\text{s}^{-1}$. Figure 1 shows another store of energy.

Work done and springs

When a material is compressed or extended without going beyond its elastic limit, the work done on the material can be fully recovered. If the material has gone through plastic deformation, then some of the work done on the material has gone into moving its atoms to new permanent positions. This energy is not recoverable.

How can you determine the energy stored in an elastic material? A good start is the force–extension graph for a spring (Figure 2).

The small amount of work done ΔW by a force F in extending the spring by a small length Δx is given by the equation

$$\Delta W \approx F \times \Delta x$$

If Δx is very small indeed, the force F acting on the spring will change very little over the range Δx. In the graph in Figure 2, $F\Delta x$ is the area of the thin rectangular strip, which is equal to ΔW. If you add similar strips for the entire extension of the spring, the area under the graph is the total work done on the spring.

area under a force–extension graph = work done

The work done on the spring is transferred to **elastic potential energy** within the spring. This energy is fully recoverable because of the elastic behaviour of the spring.

Elastic potential energy

You can derive an equation for the elastic potential energy E for a spring from the area under the force–extension graph (Figure 3).

$$E = \text{area under graph} = \text{area of shaded triangle}$$

$$E = \frac{1}{2}Fx$$

where F is the force producing an extension x.

You can also interpret the equation above as 'work done = average force × final extension'.

A spring obeys Hooke's law, $F = kx$. Substituting this equation into $E = \frac{1}{2}Fx$ gives us another useful equation for elastic potential energy.

$$E = \frac{1}{2}Fx = \frac{1}{2}(kx) \times x$$

$$E = \frac{1}{2}kx^2$$

For a given spring, E is directly proportional to extension², so doubling the extension quadruples the energy stored.

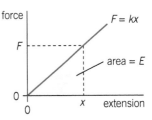
▲ **Figure 3** *Force–extension graph for a spring or wire*

 Worked example: Firing a spring

A compressible spring of force constant $k = 50\,\mathrm{N\,m^{-1}}$ and mass 4.0 g is placed around a short horizontal rod (Figure 4). The spring is compressed by 8.0 cm and then released.

a   Calculate the elastic potential energy in the spring when compressed.

b   Calculate the speed of the spring immediately after it has fully extended. State any assumptions made.

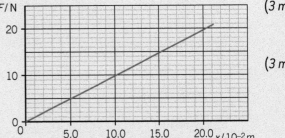
▲ **Figure 4**

a   **Step 1:** Write down all the quantities given in this question in SI units.

$$k = 50\,\mathrm{N\,m^{-1}}, m = 4.0 \times 10^{-3}\,\mathrm{kg}, x = 0.08\,\mathrm{m}$$

Step 2: Select the equation for the energy stored in the spring and calculate it.

$$E = \frac{1}{2}kx^2 = \frac{1}{2} \times 50 \times 0.08^2 = 0.16\,\mathrm{J}$$

b   **Step 3:** Assume all the elastic potential energy in the spring is transferred to its kinetic energy. Therefore

$$\text{kinetic energy} = \frac{1}{2}mv^2 = 0.16\,\mathrm{J}$$

Step 4: Rearrange this equation for v and then substitute the mass of the spring to calculate the speed.

$$v = \sqrt{\frac{2 \times \text{kinetic energy}}{m}} = \sqrt{\frac{2 \times 0.16}{4.0 \times 10^{-3}}} = 8.9\,(2\,\text{s.f.})$$

The speed of the spring is $8.9\,\mathrm{m\,s^{-1}}$ (2 s.f.).

Summary questions

1   A rubber band is extended by 20 cm by a force of 18 N. Estimate the energy stored in this stretched rubber band. *(2 marks)*

2   The energy stored in a stretched cable is 1.5 J when it is extended by 2.0 mm. Calculate the force constant of the wire. *(3 marks)*

3   Figure 5 shows a force–extension graph for a spring. Calculate the work done on the spring when its extension changes from 5.0 cm to 15.0 cm. *(3 marks)*

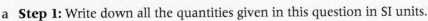
▶ **Figure 5**

4   A spring of mass 8.0 g has a force constant of 120 N m⁻¹. It is placed upright on a horizontal table and compressed by 4.0 cm. It is then released and it jumps vertically above the table. Calculate the maximum height gained by the spring above the table. State any assumptions made. *(5 marks)*

6.3 Deforming materials

Specification reference: 3.4.1, 3.4.2

▲ **Figure 1** *Bungee cord must be strong and elastic*

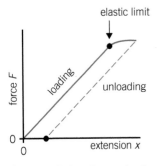

▲ **Figure 2** *Loading and unloading curves for a metal wire*

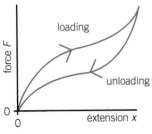

▲ **Figure 3** *Loading and unloading curve for a rubber band*

Bungee cords

Bungee jumpers leap from heights whilst securely attached to a long bungee cord. At first the jumper free falls, but then the slack is taken up, the cord stretches, and the jumper slows to a halt. As elastic potential energy in the cord is converted into kinetic energy and gravitational potential energy, the jumper accelerates upwards again. For safety, operators of bungee jumps need to take into account factors such as the weight of the jumper, the height of the jump, and the force–extension properties of the cord.

Loading and unloading

Different materials respond differently to tensile forces. Their extension increases as the force increases, then decreases again as the force is reduced. However, the **loading curve** and the corresponding unloading curve may not be the same.

Metal wire

Figure 2 shows a typical force–extension graph for a metal wire.

The loading graph in Figure 2 follows Hooke's law until the elastic limit of the wire. The unloading graph will be identical for forces less than the elastic limit. However, beyond the elastic limit it is parallel to the loading graph but not identical to it. The wire is permanently extended after the force is removed – it is longer than it was at the start. Like the springs in Topic 6.1, Springs and Hooke's law, the wire has suffered plastic deformation.

Rubber

Figure 3 shows a typical force–extension graph for a rubber band.

Rubber bands do not obey Hooke's law. The rubber band will return to its original length after the force is removed – elastic deformation – but the loading and unloading graphs are both curved and are different.

The 'loop' formed by the loading and unloading curves is called a **hysteresis loop**. You will recall that the area under a force–extension graph is equal to work done. More work is done when stretching a rubber band than is done when its extension decreases again. Thermal energy is released when the material is loaded then unloaded, represented by the area inside the hysteresis loop.

Polythene

Figure 4 shows a typical force–extension graph for a strip of polythene, the polymer used in plastic carrier bags.

A polythene strip does not obey Hooke's law. Thin strips of polythene are very easy to stretch and they suffer plastic deformation under relatively little force. As you know, shopping bags made from polythene do not return to their original size after being stretched.

Plastic deformation is not necessarily a bad thing. For example, steel sheet is pressed into car body parts, which must retain their new shape after manufacture.

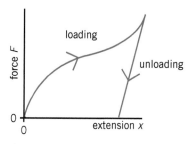

▲ **Figure 4** *Loading and unloading curve for a polythene strip*

Warming up

Rubber consists of squashed and tangled long-chain molecules. These can be untangled easily with small forces, but once straightened they require large forces to extend any further (Figure 5).

Rubber is an elastic material, but it is poor at storing energy. This makes it an ideal material for aeroplane tyres. Aeroplane tyres suffer sudden impact forces during landing. Their material makes landings smooth. The temperature of an aeroplane tyre can increase by as much as 100°C during landing.

1 Suggest how the shape of the loading curve for rubber shown in Figure 3 can be explained by the molecular structure of rubber.

2 Explain what is meant by the statement 'Rubber is an elastic material, but it is poor at storing energy'.

3 Use the force–extension graph in Figure 3 to explain why aeroplane tyres:
   a reduce the bumpiness of landings;
   b warm up during landing.

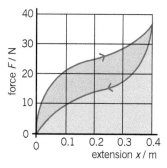

▲ **Figure 5** *The long chains of rubber molecules can easily be extended up to a certain point*

Summary questions

1 Rubber is an elastic material. Explain what this means.   (*1 mark*)

2 Explain why a rubber band does not have a force constant.   (*2 marks*)

3 Compare and contrast the behaviour of a metal wire and a strip of polythene.   (*2 marks*)

4 Figure 6 shows the force–extension graph for a length of bungee cord. Use the graph to estimate the thermal energy released when the cord is stretched by 0.4 m and returns to its original length.   (*3 marks*)

▲ **Figure 6**

6.4 Stress, strain, and the Young modulus

Specification reference: 3.4.2

▲ **Figure 1** *Spider web*

As strong as a cobweb

Spiders make their webs from a natural protein polymer called fibroin. A single strand of spider silk can be stretched up to 40% of its length before it breaks. As well as being very elastic, it is as strong as steel but has one-sixth the density.

Stretching materials

Imagine extending a wire by pulling at its ends. The extension will depend on the original length of the wire, its diameter, the tension in the wire, and of course the material of the wire. There are two helpful terms for the behaviour of materials under tensile forces: **tensile stress** and **tensile strain**.

Tensile stress

Tensile stress is defined as the force applied per unit cross-sectional area of the wire.

$$\text{tensile stress} = \frac{\text{force}}{\text{cross-sectional area}}$$

You can write this as

$$\sigma = \frac{F}{A}$$

where σ (Greek letter sigma) is the tensile stress (units pascals), F is the applied force, and A is the cross-sectional area.

Tensile strain

Tensile strain is defined as the fractional change in the original length of the wire.

$$\text{tensile strain} = \frac{\text{extension}}{\text{original length}}$$

You can write this as

$$\varepsilon = \frac{x}{L}$$

where ε (Greek letter epsilon) is the tensile strain, x is the extension, and L is the original length. Tensile strain is the ratio of two lengths, so has no units. Sometimes strain is written as a percentage, for example, 6.4% instead of 0.064.

Stress–strain graph for a metal

Figure 2 shows a typical stress against strain graph for mild (low-carbon) steel, a **ductile** material. A ductile material can easily be drawn into a wire or hammered into thin sheets.

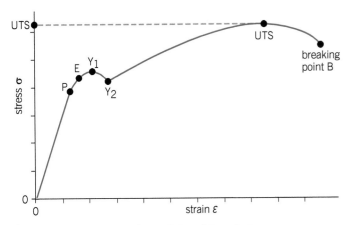

▲ Figure 2 *A stress–strain graph for mild steel wire*

In this graph, the stress is directly proportional to the strain from the origin to P, the **limit of proportionality**. The material obeys Hooke's law in this linear region. E represents the elastic limit. Materials may obey Hooke's law up to this limit but not always. Elastic deformation occurs up to the elastic limit, and plastic deformation beyond it. Y1 and Y2 are upper and lower **yield points**, where the material extends rapidly. This part of the curve is typical of mild steel but may be absent from other ductile materials.

The stress at the point labelled UTS represents the material's **ultimate tensile strength**. This is the maximum stress that a material can withstand when being stretched before it breaks. Beyond this point, the material may become longer and thinner at its weakest point, a process called necking. The material eventually snaps at its breaking point, labelled B. The stress value at the point of fracture is known as the **breaking strength** of the material.

A **strong material** is one with a high ultimate tensile strength. For example, copper is stronger than lead, but mild steel is stronger than copper.

The Young modulus E

Within the limit of proportionality, stress is directly proportional to strain. The ratio of stress to strain for a particular material is a constant and is known as its **Young modulus**, E. That is

$$\text{Young modulus} = \frac{\text{tensile stress}}{\text{tensile strain}}$$

or

$$E = \frac{\sigma}{\varepsilon}$$

The unit of the Young modulus is the same as that for stress, $N\,m^{-2}$ or Pa.

The Young modulus E of a material is the gradient of the linear region of the stress–strain graph ($y = mx + c$ is the same as $\sigma = E\varepsilon$). It depends only on the material, not its shape and size. An experiment carried out on a block of copper and a thin copper wire will give the same value for the Young modulus.

Synoptic link

For more information on gradients of straight-line graphs, see Appendix A2, Recording results.

You can compare the stiffness of materials by comparing their Young modulus values. A material with a large Young modulus is stiffer than one with a smaller Young modulus (Table 1).

▼ **Table 1** *Young modulus of some materials*

Material	E / Pa
polystyrene	~ 3×10^9
lead	1.8×10^{10}
aluminium	7.0×10^{10}
mild steel	2.1×10^{11}
graphene	1.1×10^{12}
diamond	1.2×10^{12}

 Worked example: Crane cable

An object of weight 790 N is suspended vertically from a crane on a steel cable 5.0 m long and 6.0 mm in diameter. The Young modulus of the material of the cable is 2.0×10^{11} Pa. Calculate the extension of the cable.

Step 1: Select the equation for the stress in the cable and calculate it. Remember to convert the diameter into metres when calculating the cross-sectional area.

$$\sigma = \frac{F}{A} = \frac{F}{\pi r^2} = \frac{790}{\pi \times (3.0 \times 10^{-3})^2} = 2.794 \times 10^7 \, \text{Pa}$$

Note: Although the data is given to two significant figures, the value for σ above is an intermediate value in the calculation, so it is best to retain as many significant figures as you can.

Step 2: Rearrange the equation for Young modulus to calculate the strain of the cable.

$$\varepsilon = \frac{\sigma}{E} = \frac{2.794 \times 10^7}{2.0 \times 10^{11}} = 1.397 \times 10^{-4} \, \text{Pa}$$

Step 3: Calculate the extension x. Rearrange the equation first to make x the subject.

$$\varepsilon = \frac{x}{L}$$

$$x = \varepsilon L = 1.397 \times 10^{-4} \times 5.0 = 7.0 \times 10^{-4} \, \text{m} \ (2 \text{ s.f.})$$

The extension of the cable is about 0.7 mm.

Determining the Young modulus of a wire

You can determine the Young modulus of a material in the form of wire by measuring its diameter, applying various loads to it, and measuring its length each time.

Figure 3 shows a simple method in which a wire of starting length greater than 1.00 m is clamped securely at one end, passed over a pulley, and loaded with slotted masses at the other end. You should wear eye protection in case the wire breaks.

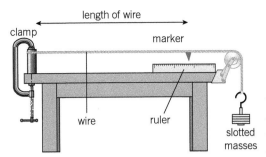

Figure 3 *Stretching a wire in the lab – the marker (tape) helps you to determine the length accurately*

The diameter d of the wire can be measured using a micrometer. The cross-sectional area A of the wire can be calculated from $A = \frac{\pi d^2}{4}$. You can obtain a more accurate diameter by averaging measurements from several places along the wire, and also by monitoring any changes during the experiment. The tensile force F acting on the wire can be calculated from the hanging mass m using $F = mg$, where g is the acceleration of free fall.

After applying each additional mass, the extension is calculated (x = extended length – original length L). You can improve accuracy by taking readings for at least six different masses, and repeating them.

The stress and strain values for each load are calculated and used to plot a stress–strain graph. The Young modulus of the material can be determined from the gradient of the linear section of the graph.

More stress–strain graphs

Not all materials behave in the same way. You can infer a great deal about the properties of a material from its stress–strain graph.

You have already seen the stress–strain graph for a ductile material (Figure 2). The stress–strain graphs for glass and cast iron, both **brittle** materials, are shown in Figure 4. A brittle material shows elastic behaviour up to its breaking point, without plastic deformation.

Polymeric materials are materials that consist of long molecular chains. (You have already met an example of a polymer, rubber, in Topic 6.3, Deforming materials). These behave differently depending on their molecular structure and temperature. Both rubber and polythene can stretch a great deal before breaking, but rubber shows elastic behaviour and polythene shows plastic behaviour (Figure 5).

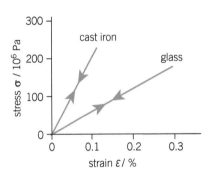

▲ Figure 4 *Stress–strain graphs for two brittle materials – ultimate tensile strength is the same as breaking strength for a brittle material*

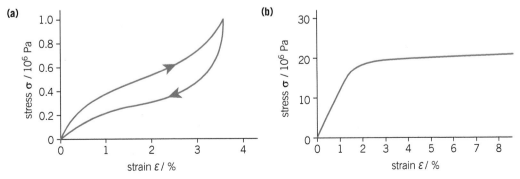

▲ Figure 5 *Stress–strain graphs for polymers: (a) rubber and (b) polythene*

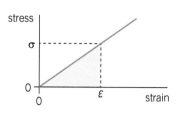

▲ **Figure 6** *Stress–strain graph for an elastic material*

 Storing energy per unit volume

For a wire or a spring, you can use the area under a force–extension graph to determine work done and therefore, as you saw in Topic 6.2, Elastic potential energy, to determine the elastic potential energy. Figure 6 shows the stress–strain graph for an elastic material.

1 Show that the area under the graph is the energy stored per unit volume by the material.
2 Show that the base units for *energy per unit volume* and *tensile stress* are the same.
3 A single strand of spider silk can store 120 MJ of energy per unit volume. Use this and the information given in the introduction to estimate the typical stress for a spider's silk. State any assumptions made.

Aeroplanes are engineered to exploit the useful properties of these materials. Their wings are made of an aluminium alloy that is both strong and stiff (high Young modulus). The rotor blades in its jet engines are made from ceramics that can withstand high temperatures and are very strong. Ceramics are brittle and show no plastic deformation. The tyres of planes are made from rubber, which is a polymeric material. Rubber has elastic properties and is an excellent shock absorber (Figure 7).

▲ **Figure 7** *A modern aeroplane*

Summary questions

1 Explain what is meant by ultimate tensile strength of a material. *(1 mark)*

2 Use Figure 4 to describe the properties of cast iron. *(1 mark)*

3 A metal wire has diameter 0.20 mm. A force of 6.3 N changes its length from 1.035 m to 1.048 m. Calculate the Young modulus of the metal. *(3 marks)*

4 The ultimate tensile strength of a metal is 220 MPa. Calculate the maximum force that can be applied on a wire of diameter 1.2 mm made from this metal. *(3 marks)*

5 a Show that the Young modulus E of a material can be calculated using the equation

$$E = \frac{FL}{Ax}$$

where F is the force applied, L is the original length of the wire, x is the extension of the wire, and A is the cross-sectional area of the wire. *(2 marks)*

   b Use the equation in (a) to determine E given the data: diameter of wire = 0.84 mm, original length of wire = 2.500 m, force applied = 300 N, extension = 1.4 cm. *(2 marks)*

   c State one significant assumption made in your calculation in (b). *(1 mark)*

Practice questions

1 a Describe how you can determine the force constant of an extendable helical spring in the laboratory. *(4 marks)*

b (i) Sketch a graph to show the variation of elastic potential energy E for a spring with its extension x. *(2 marks)*

(ii) The energy stored in a spring is 0.10 J when it has an extension of 6.0 cm. Calculate the energy stored when the extension is 9.0 cm. Explain your answer. *(3 marks)*

2 A glider of mass 0.180 kg is placed on a horizontal frictionless air track. One end of the glider is attached to a compressible spring of force constant 50 Nm^{-1}. The glider is pushed against a fixed support so that the spring compresses by 0.070 m, see Figure 1. The glider is then released.

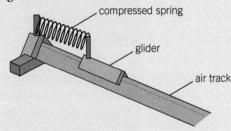
compressed spring
glider
air track

▲ **Figure 1**

(i) Calculate the horizontal acceleration of the glider **immediately** after release. *(3 marks)*

(ii) After release, the spring exerts a force on the glider for a time of 0.094 s. Calculate the average rate of work done by the spring on the glider.

(2 marks)

May 2012 G481

3 A metal wire is suspended from a tall ceiling. The wire has diameter of (0.90 ± 0.01) mm and length 2.500 m. A mass of 8.00 kg hung from its lower end produces an extension of 4.0 mm.

a Calculate the absolute uncertainty in the value of the cross-sectional area. *(3 marks)*

b Calculate the Young modulus of the metal. State any assumptions made. *(4 marks)*

c (i) On a copy of the axes below, sketch the stress against strain graphs for a ductile metal and a brittle metal. *(2 marks)*

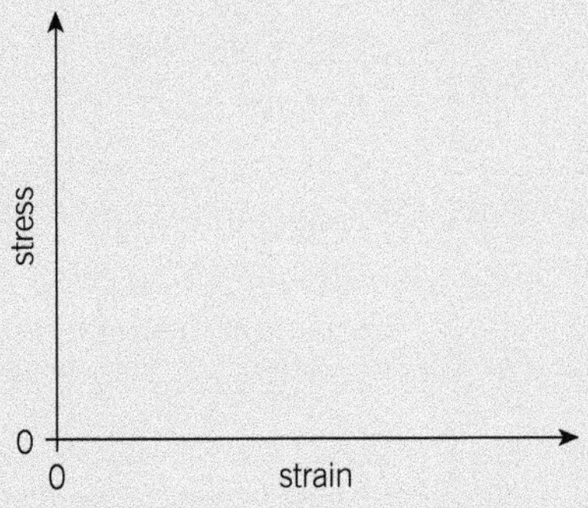

stress

0
0 strain

▲ **Figure 2**

(ii) Describe the behaviour of the ductile material as it is stretched. *(2 marks)*

4 a Define tensile stress and tensile strain. *(2 marks)*

b A metal wire of length 1.80 m and cross-sectional area 1.92×10^{-7} m^2 is extended by a force of 12.0 N. The metal has a Young modulus of 2.00 GPa.

(i) Calculate the extension of the wire. *(3 marks)*

(ii) A second wire made from the same metal has the same length but twice the diameter. State and explain whether the extension of this second wire under the same force is greater, the same or less than the first wire. *(3 marks)*

5 a Figure 3 shows the force F against extension x graph for a spring.

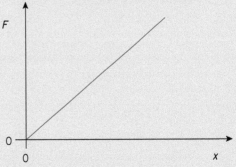
F

0
0 x

▲ **Figure 3**

State what is represented by

(i) the gradient of the graph *(1 mark)*

(ii) the area under the graph. *(1 mark)*

b A spring has a force constant of $160\,\text{N m}^{-1}$. The energy stored in the spring is used to propel an object of mass $80\,\text{g}$. The spring is compressed by $7.2\,\text{cm}$ and then released.

  (i) Calculate the energy stored in the spring. (*2 marks*)

  (ii) Calculate the initial speed of the object leaving the spring, assuming 60% of the energy stored in the spring is transferred as kinetic energy of the object. (*3 marks*)

6 a A spring shows elastic behaviour when it is subjected to forces. State what is meant by the term *elastic*. (*1 mark*)

b Figure 4 shows two springs A and B, connected in series and supporting a weight of $16\,\text{N}$. The force constant of each spring is shown on the figure.

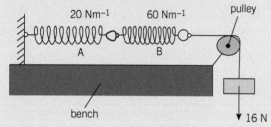

▲ Figure 4

  (i) Determine the total extension of the two springs. State any assumptions made. (*3 marks*)

  (ii) Determine the force constant of the combination of these springs. (*2 marks*)

7 a State *Hooke's law*. (*1 mark*)

b Define the force constant of a spring. (*1 mark*)

c Describe how you can determine the force constant of an extendable spring in the laboratory. In your description pay particular attention to

  • how the apparatus is used

  • what measurements are taken

  • how the data is analysed. (*4 marks*)

d Figure 5 shows the variation of force F with extension x for a spring.

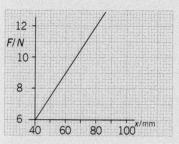

▲ Figure 5

Use Figure 5 to answer to following questions.

  (i) Explain how the graph shows that Hooke's law is obeyed. (*2 marks*)

  (ii) Determine the force constant of the spring. (*2 marks*)

  (iii) Determine the energy stored in the spring when its extension is $80\,\text{mm}$. (*3 marks*)

8 a Define *tensile stress* and *tensile strain*. (*1 mark*)

b Derive the units for stress in base units. (*3 marks*)

c A group of students are carrying out an experiment to determine the Young modulus of a metal wire. The wire has original length of $1.640\,\text{m}$ and it is suspended vertically from a support. The wire is loaded in steps of $10.0\,\text{N}$ up to $50.0\,\text{N}$ and then unloaded. Table 1 shows the experimental results from the group.

▼ Table 1

	loading the wire	unloading the wire
force F/N	extension x/mm	extension x/mm
0.0	0.00	0.00
10.0	0.50	0.49
20.0	1.01	1.00
30.0	1.49	1.50
40.0	2.00	1.99
50.0	2.52	2.52

  (i) Use Table 1 to describe the behaviour of the wire when forces up to $50.0\,\text{N}$ are applied to it. (*2 marks*)

  (ii) Name the most likely instrument used to determine the extension of the wire. (*1 mark*)

  (iii) The cross-sectional area of the wire

is $3.2 \times 10^{-7}\,m^2$. Use the value of the extension for the force of 50.0 N to calculate a value for the Young modulus of the metal. *(3 marks)*

(iv) Describe how the table of results can be used to plot a graph and hence determine a precise value for the Young modulus of the metal. *(3 marks)*

9 **a** Define the *Young modulus*. *(1 mark)*

   b Figure 6 shows a violin.

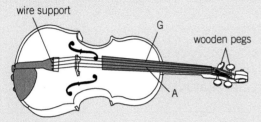

▲ Figure 6

Two of the wires used on the violin, labelled A and G, are made of steel. The two wires are both 500 mm long between the pegs and support. The 500 mm length of wire labelled G has a mass of $2.0 \times 10^{-3}\,kg$. The density of steel is $7.8 \times 10^3\,kg\,m^{-3}$.

   (i) Show that the cross-sectional area of wire G is $5.1 \times 10^{-7}\,m^2$. *(2 marks)*

   (ii) The wires are put under tension by turning the wooden pegs shown in Figure 6. Calculate the force on the wire when the extension is 0.4 mm. The Young modulus of steel is $2.0 \times 10^{11}\,Pa$. *(3 marks)*

   (iii) Wire A has a diameter that is half that of wire G. Determine the tension required for wire A to produce an extension of $16 \times 10^{-4}\,m$. *(1 mark)*

   (iv) State the law that has been assumed in the calculations in (ii) and (iii). *(1 mark)*

   June 2007 2821

10 Figure 7 shows the force F against extension x for a metal wire.

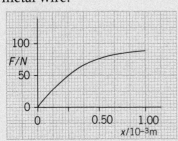

▲ Figure 7

a State the value of the force and extension at the elastic limit of the wire. *(1 mark)*

b Calculate the elastic potential energy for the wire when its extension is 0.25 mm. *(3 marks)*

c The Young modulus of the metal is $1.2 \times 10^{11}\,Pa$ and the length of the wire is 1.82 m.

Use Figure 7 to determine the cross-sectional area of the wire. *(3 marks)*

11 Figure 8 shows an arrangement used by a student to investigate the energy stored in a compressible spring.

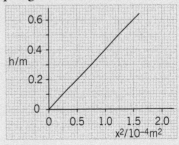

▲ Figure 8

a The spring is compressed by a distance x and then released. It climbs a vertical height h along the length of the rod. Show that h is directly proportional to x^2. *(3 marks)*

b Figure 9 shows a graph of h against x^2 for the spring.

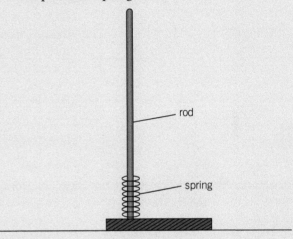

▲ Figure 9

   (i) Use Figure 9 to predict the height h when $x = 2.0\,cm$. State any assumptions made. *(4 marks)*

   (ii) The mass of the spring is 8.0 g.

   Use Figure 9 to determine the force constant of the spring. *(3 marks)*

7 LAWS OF MOTION AND MOMENTUM

7.1 Newton's first and third laws of motion

Specification reference: 3.5.1

▲ **Figure 1** *A trainee astronaut in 'zero gravity' – the padded interior reduces the risk of injury*

Weightless seconds

Figure 1 shows a NASA astronaut training in simulated 'zero gravity' on an aircraft. The aircraft climbs, gradually reducing engine thrust, and then points downwards with the thrust increasing again. This manoeuvre allows trainees to experience about 25 s of free fall. The ideas developed by Sir Isaac Newton more than three centuries ago can predict and explain this and other effects of motion and forces.

Newton's laws of motion

First law

An asteroid moving in deep space will keep moving at constant velocity. There is no force acting on it to alter its motion.

A cyclist travelling on a straight road at constant velocity experiences several forces, including contact force, weight, and air resistance. Again, no resultant force acts on the cyclist and the acceleration is zero.

You have already met **Newton's first law of motion**. A formal statement of this law is given below.

Newton's first law of motion: An object will remain at rest or continue to move with constant velocity unless acted upon by a resultant force.

If an object's velocity changes, then you know a resultant force must be acting on the object. Remember that velocity is a vector quantity, so an object's velocity changes if its speed and/or direction changes.

Third law

Tap the screen of your mobile phone. Your finger will feel a force. You also exert a force on the screen. In fact, the forces acting on your finger and on the screen have the same magnitude, but they are in opposite directions. The same happens when you stand on the ground, clap your hands, or use a hammer to hit a nail. These interactions between objects are summed up in **Newton's third law of motion**. (You will study Newton's second law of motion in Topic 7.3, Newton's second law of motion.)

Newton's third law of motion: When two objects interact, they exert equal and opposite forces on each other.

When two objects interact, the pair of forces produced will always be equal and opposite. The forces acting on the interacting objects are always of the same type.

Electrons have a negative charge; they exert an *electrostatic* repulsive force on each other (Figure 2). The two electrons experience the same magnitude of force but in opposite directions. Similarly, the unlike poles of the two magnets exert a pair of *magnetic* forces. Once again the forces are equal and opposite. When you jump off a wall and fall towards the Earth, you and the Earth are interacting. The *gravitational* force exerted by the Earth, your weight, is equal and opposite to the gravitational force that you exert on the Earth. Notice how each interaction in Figure 2 involves a pair of forces of the same type each acting on a different object. In Newton's Third Law you never see both forces acting on the same object.

Types of interaction

All interactions can be explained in terms of four fundamental forces – gravitational, electromagnetic, strong nuclear, and weak nuclear. The two nuclear forces have an extremely short range and so very little impact on the things we observe in daily life. You are familiar with gravitational force. When you push your hands together, the contact force you feel is due to the electrostatic repulsive forces between the electron clouds around the atomic nuclei in your hands. Next time you walk, remember that you are exerting a backward force on the ground and the ground is exerting a forward force on you, both forces of electrical origin.

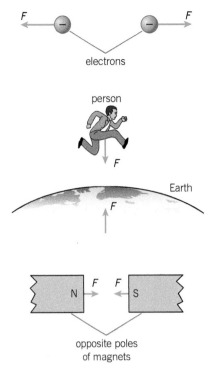

▲ **Figure 2** *Interacting objects exert forces of the same type on each other*

Summary questions

1 A car is travelling at constant velocity. State the resultant force on the car. *(1 mark)*

2 A person of weight 600 N is standing still on the ground. State and explain the force exerted by the person on the Earth. *(2 marks)*

3 Two magnets repel each other. State and explain the resultant force acting on the magnets. *(2 marks)*

4 Explain how Newton's laws can be used to explain how a person runs on a road. *(2 marks)*

5 Dalmatian pelicans are massive birds (Figure 3). One such bird weighs 150 N.
   a State the acceleration of free fall of this bird. *(1 mark)*
   b Calculate the acceleration of the Earth towards this bird. (The mass of the Earth is 5.97×10^{24} kg.) Explain your answer. *(3 marks)*

▲ **Figure 3** *This flying Dalmatian pelican is interacting with the Earth*

7.2 Linear momentum

Specification reference: 3.1.2, 3.5.1, 3.5.2

Learning objectives

Demonstrate knowledge, understanding, and application of:

→ linear momentum; $p = mv$; vector nature of momentum

→ principle of conservation of momentum

→ collisions and interactions of bodies in one dimension

→ perfectly elastic and inelastic collisions.

Mission to Mars

The Mars Science Laboratory spacecraft was launched on top of an Atlas V rocket in November 2011. It landed on Mars on August 2012 on a mission to determine if Mars is, or was ever, able to support microbial life.

Because the mass of the rocket carrying it decreased as its fuel was consumed and expelled, the equation $F = ma$ cannot be used to predict the motion of the rocket. We need a new quantity – **linear momentum** – to analyse motion of objects such as rockets.

Momentum

The linear momentum (or simply momentum) p of an object depends on its mass m and its velocity v.

$$\text{momentum} = \text{mass} \times \text{velocity}$$

or

$$p = mv$$

The SI unit of momentum is kg m s^{-1}. Momentum is a vector quantity because it is a product of a scalar (mass) and a vector (velocity).

Figure 2 shows two identical cars, each of mass 1200 kg, travelling at the same speed of $20\,\text{m s}^{-1}$ in opposite directions. The car moving to the right has a velocity of $+20\,\text{m s}^{-1}$ and the car moving left has a velocity of $-20\,\text{m s}^{-1}$. Therefore

● momentum of car moving to right
$$= mv = 1200 \times (+20) = +2.4 \times 10^4\,\text{kg m s}^{-1}$$

● momentum of car moving to left
$$= mv = 1200 \times (-20) = -2.4 \times 10^4\,\text{kg m s}^{-1}$$

It does not matter which direction is taken as positive. The important idea is that one of the values for momentum is negative.

Conservation of momentum

What happens when two or more objects collide or interact? The objects transfer both momentum and kinetic energy between themselves, but the total momentum does not change, provided that no external forces act on the interacting objects. The group of interacting objects is referred to as a **closed system**. The **principle of conservation of momentum** is expressed as follows:

For a system of interacting objects, the total momentum in a specified direction remains constant, as long as no external forces act on the system.

▲ **Figure 1** An Atlas V rocket burning the fuel it carries

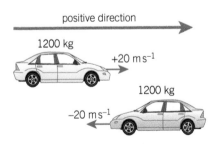

▲ **Figure 2** Both cars have the same magnitude of momentum, $2.4 \times 10^4\,\text{kg m s}^{-1}$, but the green car has a negative momentum

This means that when objects collide, the total momentum before and after the collision is the same. You can use this principle to predict the motion of interacting objects, which could be atoms bouncing off each other, cars crashing, and even colliding distant galaxies.

 ## Worked example: Air track collision

Two gliders are on a linear air track. Glider **A** is travelling at $0.20\,\text{m s}^{-1}$ and has mass $0.10\,\text{kg}$. It is hit by glider **B** of mass $0.15\,\text{kg}$ travelling at $0.40\,\text{m s}^{-1}$ in the opposite direction. They stick together. Calculate their new velocity v.

Step 1: Write down the information given for each glider before and after the collision. Alternatively, you can do a quick sketch to help you to visualise the problem, as shown in Figure 3.

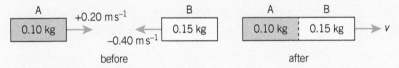

▲ **Figure 3** *Before and after sketches*

The velocity, and therefore momentum, of one glider must be negative because of its direction of travel.

Step 2: Write an equation for this collision using the principle of conservation of momentum.

total momentum before = total momentum after

$$(0.10 \times 0.20) + (0.15 \times -0.40) = (0.10 + 0.15)v$$

Step 3: Solve this equation to calculate v.

$$0.020 - 0.060 = 0.25v$$
$$-0.040 = 0.25v$$
$$v = \frac{-0.040}{0.25} = -0.16\,\text{m s}^{-1}$$

The velocity of the joined gliders is $-0.16\,\text{m s}^{-1}$.

Note: The negative sign shows that the gliders move in the direction in which glider **B** was originally travelling.

Zero momentum?

A gun recoils when a bullet is fired. The total momentum of this system remains the same and is equal to zero. The momentum of the gun and the momentum of the bullet have the same magnitude but act in opposite directions. The same physics can be used to explain a recoiling radioactive nucleus when it emits an alpha-particle, and an exploding firework (Figure 4).

▲ **Figure 4** *The total momentum of this exploding firework is zero*

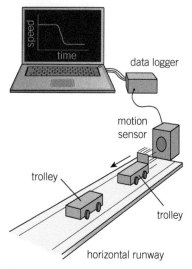

data logger

motion
sensor

trolley

trolley

horizontal runway

▲ Figure 5 *Investigating linear momentum using trolleys and a motion sensor*

▲ Figure 6 *An example of an inelastic collision*

Investigating momentum

There are several ways to investigate momentum in the laboratory. A linear air track is ideal because a cushion of air minimises the friction between the gliders and track, but trolleys and a horizontal runway also work (Figure 5). The velocity of each object is determined with a motion sensor and a laptop, light gates and a digital timer, ticker timers, or simply a stopwatch to measure the time taken to cover a known distance.

Elastic and inelastic collisions

Momentum and total energy are both conserved when two objects collide. All the kinetic energy could be retained by the objects, or the kinetic energy could be transformed into other forms such as heat and sound. In a collision of two cars, the kinetic energy could be completely transformed as the cars crumple and deform (Figure 6).

There are two types of collisions: **perfectly elastic** and **inelastic**. Table 1 summarises the key characteristics of these two types.

▼ Table 1 *Summary of the two types of collisions*

Type of collision	Momentum	Total energy	Total kinetic energy
perfectly elastic	conserved	conserved	conserved
inelastic	conserved	conserved	not conserved

Summary questions

1 State two quantities conserved in all collisions. *(2 marks)*

2 Two identical objects, each of mass 2.0 kg, collide. After the collision, the loss of momentum for one of the objects is 120 kg m s^{-1}. Calculate the change in momentum of the other object and the change in its velocity. *(3 marks)*

3 A stationary cannon of mass 1200 kg fires a 20 kg shell at a velocity of 300 m s^{-1}. Calculate the recoil speed of the cannon. Explain your answer. *(3 marks)*

4 A bumper car of mass 300 kg moving at 2.5 m s^{-1} collides head-on with another bumper car of mass 400 kg moving at 4.0 m s^{-1} in the opposite direction. The 400 kg bumper car stops after the collision.
   a Calculate the final velocity of the 300 kg bumper car. *(3 marks)*
   b Calculate the change in the kinetic energy of the bumper cars. *(3 marks)*
   c State and explain the type of collision between these bumper cars. *(2 marks)*

7.3 Newton's second law of motion

Specification reference: 3.5.1

Air bags

Air bags in modern cars are designed to reduce the chance of injury to occupants in an accident. Sensors detect the rapid changes in acceleration associated with a crash and trigger an explosive chemical reaction, which fills the air bag with nitrogen gas in just 30 ms to provide a cushion for the occupant's head and upper body. This reduces the momentum of the occupant to zero more gradually than sudden deceleration by contact with the dashboard or windscreen, and therefore reduces the force and minimises injuries.

Learning outcomes

Demonstrate knowledge, understanding, and application of:

→ Newton's second law of motion

→ net force = rate of change of momentum; $F = \frac{\Delta p}{\Delta t}$.

Newton's second law

Newton's second law of motion links the idea of net force acting on an object and rate of change of its momentum. A formal statement of the law is given below.

Newton's second law: The net (resultant) force acting on an object is directly proportional to the rate of change of its momentum, and is in the same direction.

Therefore

$$\text{net force} \propto \text{rate of change of momentum}$$

or

$$F \propto \frac{\Delta p}{\Delta t}$$

where F is the net force acting on the object and Δp is the change in momentum over a time interval Δt. The expression above can be rewritten as an equation with a constant of proportionality k.

$$F = \frac{k\Delta p}{\Delta t}$$

The value of k is made equal to 1 by defining the unit of force, the newton, as the force required to give a 1 kg mass an acceleration of $1\,\text{m s}^{-2}$.

Newton's second law of motion can therefore be written mathematically as

$$F = \frac{\Delta p}{\Delta t}$$

This astonishingly useful equation can predict the motion of any object subjected to forces, even when its mass changes with time, like a rocket.

▲ **Figure 1** *Crash-test dummies are used to test air bags*

Study tip

Remember that the letter delta Δ is a shorthand for *change in*.

 Worked example: Crash-test dummy

In a crash test at $14\,\text{m s}^{-1}$, a 4.5 kg dummy head hits the steering wheel and comes to rest in 9.1 ms. Calculate the net force acting on the head in the impact. →

Synoptic link

The equation $F = ma$ was introduced in Topic 4.1, Force, mass, and weight.

Summary questions

1 In a period of 5.0 s the change in momentum of a car is 1.2×10^4 kg m s^{-1}. Calculate the net force acting on the car. *(2 marks)*

2 A force of 150 N acts on a ball for a time of 0.025 s. Calculate the change in momentum of the ball. *(2 marks)*

3 The mass of a rocket is 2.3×10^6 kg and its velocity is 1.2×10^3 m s^{-1}. After burning fuel for 200 s, its mass has decreased to 1.0×10^6 kg and its velocity has increased to 3.4×10^3 m s^{-1}. Calculate the force acting on the rocket. *(2 marks)*

4 A 150 g ball travelling at 15 m s^{-1} hits a wall at right angles and rebounds at the same speed. The ball is in contact with the wall for 0.025 s. Calculate the force exerted by the wall on the ball. *(3 marks)*

5 A hosepipe squirts water at a rate of 2.5 kg s^{-1}. The speed of the water in the hosepipe is 4.0 m s^{-1}. Calculate the force needed to push the water out of the hosepipe. *(2 marks)*

Step 1: Write down all the quantities given in SI units, and select the equation you need.

initial velocity $u = 14$ m s^{-1}, final velocity $v = 0$ m s^{-1}, $\Delta t = 9.1 \times 10^{-3}$ s, $m = 4.5$ kg

$$F = \frac{\Delta p}{\Delta t}$$

Step 2: Determine the change in momentum Δp.

Δp = final momentum − initial momentum
$$\Delta p = (4.5 \times 0) - (4.5 \times 14) = -(4.5 \times 14)$$

Step 3: Substitute the values above into the equation to calculate F.

$$F = \frac{\Delta p}{\Delta t} = \frac{-(4.5 \times 14)}{9.1 \times 10^{-3}} = -6.9 \times 10^3 \text{ N}$$

Note: The magnitude of the force is 6.9 kN, roughly 10 times your weight. The minus means that the net force is opposite to the initial velocity, so the net force causes a deceleration.

$F = ma$ – a special case

Figure 2 shows a constant force F acting on an object of constant mass m. The initial velocity of the object is u and after a time t it has a final velocity v. According to Newton's second law

$$F = \frac{\Delta p}{\Delta t} = \frac{mv - mu}{t} = m\left(\frac{v - u}{t}\right)$$

The term in brackets is the acceleration a of the object. Therefore

$$F = ma$$

This equation is just a special case of Newton's second law when the mass m of the object remains constant during the period of acceleration. In the worked example above, you could have used $F = ma$ to determine the force on the dummy's head.

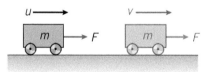

▲ **Figure 2** *The constant force F applied for a time t increases the momentum of the object*

▲ **Figure 3** *Two interacting objects*

Why is momentum conserved in collisions?

The principle of conservation of momentum is a natural consequence of Newton's laws. Figure 3 shows two interacting objects. According to Newton's third law, each experiences an equal but opposite force F.

The net force acting on the objects in this closed system is zero. According to Newton's second law $\frac{\Delta p}{\Delta t} = 0$. The change in momentum Δp of both objects must be zero; therefore, the total momentum of the objects does not change. Momentum is always conserved.

7.4 Impulse

Specification reference: 3.5.1

Squash

Squash is a fast-paced racket game played in a walled court. The 40 mm diameter rubber balls are hit at speeds of up to $76\,\mathrm{m\,s^{-1}}$. A cold squash ball does not bounce well, so it must be warmed up before play by hitting it around the court. The forces exerted on a squash ball change during impact with a racket (Figure 1), and can be analysed using **force–time graphs**.

Impulse of a force

Forces accelerating or decelerating an object usually change over time, for example, kicking a ball or crashing a car into a barrier. This type of motion can be analysed using the idea of **impulse**.

According to Newton's second law of motion

net force = rate of change of momentum

$$F = \frac{\Delta p}{\Delta t}$$

Rearranging this equation gives

$$F \times \Delta t = \Delta p$$

The product of force and time is equal to the change in momentum Δp.

Impulse of a force is defined as the product of force and the time for which this force acts on an object.

Therefore

impulse of a force = change in momentum

The unit of impulse is $\mathrm{N\,s}$ or $\mathrm{kg\,m\,s^{-1}}$.

Force–time graphs

Figure 2(a) is the force–time graph for an object experiencing a *constant* force F for a time t.

▲ **Figure 1** *A ball stretches the racket's strings and changes shape itself when it is hit*

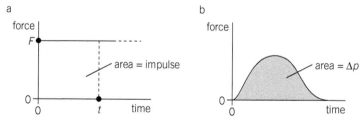

▲ **Figure 2** *Force–time graphs for (a) constant force; (b) changing force*

As you can see from Figure 2(a), the area under the graph is equal to Ft, which is the impulse of the force or the change in momentum of the object. In fact, the area under a force–time graph is always equal to the change in momentum, even when the force is changing (Figure 2(b)).

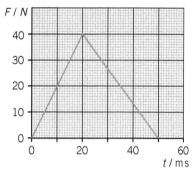

▲ Figure 3

Worked example: Hitting a squash ball

A stationary squash ball of mass 0.025 kg is hit with a racket. Use the force–time graph for the ball in Figure 3 to determine the final velocity v of the ball.

Step 1: The area under the graph is equal to the impulse of the force. Calculate the area of the triangle.

$$\text{impulse} = \text{area} = \frac{1}{2} \times 40 \times 50 \times 10^{-3} \text{ (remember: } 1 \text{ ms} = 10^{-3} \text{ s)}$$

$$\Delta p = 1.0 \text{ N s or kg m s}^{-1}$$

Step 2: The impulse of the force is equal to the change in momentum. Use the mass of the ball to calculate the final velocity.

$$\Delta p = mv - mu = 0.025v - 0 \text{ (because } u = 0)$$

$$0.025v = 1.0$$

$$v = \frac{1.0}{0.025} = 40 \text{ m s}^{-1}$$

The final velocity of the ball is 40 m s^{-1}.

Photons have momentum

You may have heard of photons, subatomic particles that travel at the speed of light and have no mass. According to quantum physics, as well as having energy, photons have momentum. The momentum p of a photon is given by the equation $p = \dfrac{h}{\lambda}$, where h is the Planck constant and λ is the wavelength of the photon.

Photons exert a tiny force when they collide with objects. Our Sun is a source of photons. These photons exert a radiation pressure of about 9.1 μPa at the Earth.

1 State and explain how the momentum of a photon depends on its wavelength.

2 Calculate the momentum of a visible light photon of wavelength 500 nm. The Planck constant $h = 6.63 \times 10^{-34}$ J s.

3 Estimate the area required to produce a force of 1 N at the Earth from the radiation pressure from the Sun.

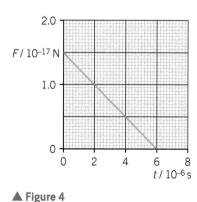

▲ Figure 4

Summary questions

1 State the two SI units for impulse and momentum. *(2 marks)*

2 A constant resultant force of 200 N acts on a car for 5.0 s. Calculate the impulse of the force. *(2 marks)*

3 A stationary 0.050 kg ball experiences an impulse of 1.1 N s. Calculate its final velocity. *(2 marks)*

4 The force F against time t graph for a proton is shown in Figure 4.
   a Calculate the change in momentum of the proton. *(2 marks)*
   b The initial velocity of the proton is 5.0×10^4 m s^{-1}. Calculate its final velocity. The mass of a proton is 1.7×10^{-27} kg. *(3 marks)*

Collisions big and small

When a snooker ball collides obliquely with an identical ball, they move off on paths at an angle of 90° to each other. This is an example of a collision in two dimensions. As with all collisions, both total energy and linear momentum are conserved.

Similar events also occur at microscopic levels. Figure 1 shows a collision between a helium nucleus and a proton. The angle after the collision is not 90° because the particles have different masses. Physicists can use particle tracks to determine the momentum of particles.

Conservation of momentum

In collisions and interactions, linear momentum is conserved in all directions. You can use your knowledge of vector triangles and of resolving vectors to analyse a variety of problems.

Adding momentum

Figure 2 shows an object **A** moving to the right with momentum p. It collides with a stationary object **B**. After the collision, **A** and **B** move off in different directions with momenta p_1 and p_2, respectively, as shown. Since linear momentum must be conserved, the vector sum p_1 and p_2 (total final momentum) must be equal to p (initial momentum). You can draw a vector triangle to add the vectors p_1 and p_2 together (Figure 2).

Learning outcomes

Demonstrate knowledge, understanding, and application of:

→ collisions and interactions of bodies in two dimensions.

Study tip

You will not be assessed on two-dimensional collisions until A Level.

▲ **Figure 1** *After a collision between a helium nucleus and a proton, the proton shoots off to the right (red track) and the helium nucleus moves off to the left (yellow track)*

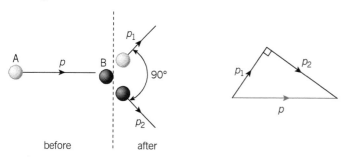

▲ **Figure 2** *Objects A and B before (left) and after the collision, and the vector triangle (right)*

Worked example: Snooker balls

A 160 g white ball travelling at $4.0\,\text{m s}^{-1}$ hits a stationary 170 g black ball (Figure 3). After the impact, the balls move apart at approximately 90° to each other, with the white ball travelling at $2.5\,\text{m s}^{-1}$. Calculate the magnitude of the final velocity of the black ball.

Step 1: Select the equation you need and calculate the momentum of each ball.

$$p = mv$$

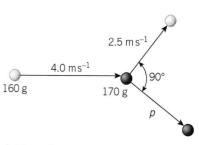

▲ **Figure 3**

initial momentum of white ball = $160 \times 10^{-3} \times 4.0 = 0.64\,\text{kg m s}^{-1}$

initial momentum of black ball = 0

final momentum of white ball after impact = $160 \times 10^{-3} \times 2.5 = 0.40\,\text{kg m s}^{-1}$

Step 2: Draw a vector triangle for the momenta after the impact (Figure 4). These must add up to the initial momentum of $0.64\,\text{kg m s}^{-1}$.

Step 3: Use Pythagoras' theorem to determine the final momentum p of the black ball.

$$p^2 = 0.64^2 - 0.40^2 = 0.25 \qquad p = 0.50\,\text{kg m s}^{-1}$$

Step 4: Use the equation $p = mv$ to calculate the velocity of the black ball.

$$0.50 = 0.170 \times v \qquad v = \frac{0.50}{0.170} = 2.9\,\text{m s}^{-1}$$

The magnitude of the final velocity of the black ball is $2.9\,\text{m s}^{-1}$.

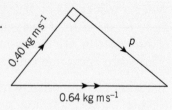

▲ **Figure 4**

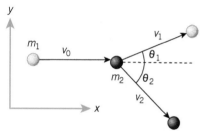

▲ **Figure 5** *The white object is originally moving in the x direction, so after the collision, the total momentum in the y direction must be zero*

Resolving momentum

Figure 5 shows a general collision involving two objects of mass m_1 and m_2. The white object travels at velocity v_0 and collides with the stationary black object. After the collision, the white object travels at angle θ_1 to its original direction with velocity v_1, and the black object travels at angle θ_2 with velocity v_2.

The momentum in any direction must be conserved. In this case, the momentum must remain the same in the x direction and y direction.

x direction: total initial momentum = total final momentum

$$m_1 v_0 = m_1 v_1 \cos\theta_1 + m_2 v_2 \cos\theta_2$$

y direction: total initial momentum = total final momentum

$$0 = m_1 v_1 \sin\theta_1 + m_2 v_2 \sin\theta_2$$

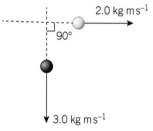

▲ **Figure 7**

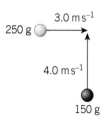

▲ **Figure 8**

Summary questions

1 Two objects of the same mass have an oblique collision. State the angle between the directions in which the objects travel after the collision. (*1 mark*)

2 Figure 6 shows the momentum of two objects **X** and **Y** before and after a collision. Explain how you can tell that the collision is incorrectly represented. (*2 marks*)

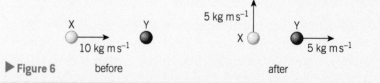

▶ **Figure 6**   before   after

3 Figure 7 shows the final momenta of two particles after a collision. Calculate the magnitude of the initial momentum of the particles. Explain your answer. (*3 marks*)

4 Two objects collide at 90° to each other and stick together (Figure 8). Calculate the magnitude of their final velocity. (*4 marks*)

Practice questions

1 a Derive the base units of momentum. *(3 marks)*

b A stationary tennis ball of mass 60 g is hit with a racquet. Figure 1 shows a graph of force F on the ball against time t of impact between the racquet and the ball.

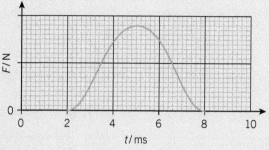

▲ **Figure 1**

The area under the graph is 3.2 N s.

(i) Calculate the final velocity of the ball. *(3 marks)*

(ii) Show that the maximum force acting on the ball is about 1 kN. *(3 marks)*

(iii) Describe and explain the motion of the racquet after it has hit the ball. *(3 marks)*

2 a Define impulse of a force. *(1 mark)*

b A driving force acts on a stationary car on a level road. Figure 2(a) shows the variation of the net force F acting on the car with time t.

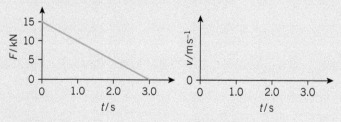

▲ **Figure 2a**   ▲ **Figure 2b**

(i) Describe how the acceleration of the car changes time $t = 0$ to $t = 3.0$ s. *(2 marks)*

(ii) On a copy of Figure 2(b), sketch a graph to show the variation of the velocity of the car with time t. You are not expected to show the values on the v-axis. *(2 marks)*

(iii) The mass of the car is 1200 kg. Calculate the final kinetic energy of the car. *(5 marks)*

3 a State Newton's second law of motion. *(1 mark)*

b Use Newton's second law of motion to show how the net force F acting on an object is related to its mass m and acceleration a. *(3 marks)*

c An 80 g ball is dropped from a height h above the ground. It has a speed of 5.4 ms⁻¹ just before it hits the ground. After hitting the ground, it has a rebound speed of 3.0 ms⁻¹.

(i) Calculate the height h of the ball. State any assumption made. *(4 marks)*

(ii) Figure 3 shows a graph of force F exerted by the ground on the ball against the time t of contact between the ball and the ground.

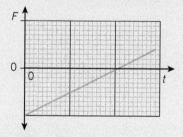

▲ **Figure 3**

On a copy of Figure 3 sketch a graph to show the variation of the force exerted by the ball on the ground. Explain your answer. *(2 marks)*

(iii) Calculate the magnitude of the change in momentum of the ball. *(2 marks)*

Jan 2013 G484

4 This question is about pressing a red hot bar of steel into a sheet in a rolling mill.

a A bar of steel of mass 500 kg is moved on a conveyor belt at $0.60\,\text{m s}^{-1}$.

Calculate the momentum of the bar, giving a suitable unit for your answer. *(2 marks)*

b From the conveyor belt, the bar is passed between two rollers, shown in Figure 4. The bar enters the rollers at $0.60\,\text{m s}^{-1}$. The rollers flatten the bar into a sheet with the result that the sheet leaves the rollers at $1.8\,\text{m s}^{-1}$.

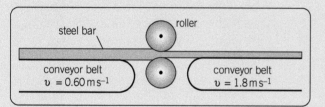

steel bar · roller · conveyor belt $v = 0.60\,\text{m s}^{-1}$ · conveyor belt $v = 1.8\,\text{m s}^{-1}$

▲ Figure 4

  (i)   Explain why there is a resultant horizontal force on the bar at the point immediately between the rollers. *(2 marks)*

  (ii)   In which direction does this force act? *(1 mark)*

  (iii)   The original length of the bar is 3.0 m. Calculate the time it takes for the bar to pass between the rollers. *(1 mark)*

  (iv)   Calculate the magnitude of the resultant force on the bar during the pressing process. *(3 marks)*

5 Figure 5 shows the masses and velocities of two objects **A** and **B** moving directly towards each other. **A** and **B** stick together on impact and move with a common velocity *v*.

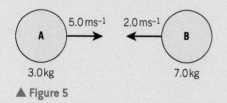

A $5.0\,\text{ms}^{-1}$ $2.0\,\text{ms}^{-1}$ B

3.0 kg   7.0 kg

▲ Figure 5

  (i)   Determine the velocity *v*, stating its magnitude and direction. *(3 marks)*

  (ii)   Determine the impulse of the force experienced by the object **A** and state its direction. *(2 marks)*

  (iii)   Explain, using Newton's third law of motion, the relationship between the impulse experienced by **A** and the impulse experienced by **B** during the impact. *(2 marks)*

Jan 2013 G484

6 a Collisions between two objects can be described as being either *elastic* or *inelastic*.

Copy and complete Table 1 by placing a tick (✔) in the relevant row(s) for each statement that is true for each type of collision.

▼ Table 1

Statement	Elastic collision	Inelastic collision
Total momentum for the objects is conserved.		
Total kinetic energy of the objects is convserved.		
Total energy is conserved.		
Magnitude of the impulse on each object is the same.		

(4 marks)

b A snooker ball is at rest on a smooth horizontal table. It is hit by a snooker cue. Figure 6 shows a simplified graph of force *F* acting on the ball against time *t*.

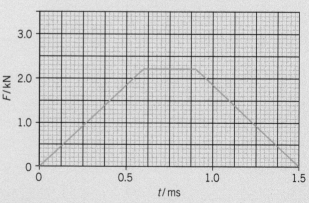

▲ Figure 6

(i) Describe how the velocity of the ball varies between $t = 0.6$ ms and $t = 0.9$ ms. *(1 mark)*

(ii) Use Figure 6 to calculate the impulse acting on the ball. *(2 marks)*

(iii) The mass of the snooker ball is 140 g. Calculate the final speed of the snooker ball as it leaves the cue. *(2 marks)*

7 a Compare and contrast elastic and inelastic collisions. *(3 marks)*

b Figure 7 shows an object. It has two sections A and B. The mass of section A is 25% of the total mass of the object. An explosion within the object ejects A and B in opposite directions.

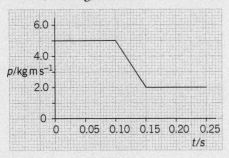

▲ **Figure 7**

(i) Use Newton's third law of motion to explain why A and B are ejected in opposite directions. *(2 marks)*

(ii) Calculate the ratio $\dfrac{\text{kinetic energy of A}}{\text{kinetic energy of B}}$. *(5 marks)*

8 a State what is meant by an *inelastic collision*. *(1 mark)*

b Two objects A and B collide. Figure 8 shows the variation momentum p of A with time t before, during, and after the collision.

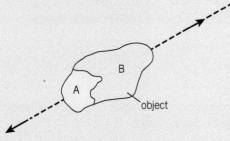

▲ **Figure 8**

Use Figure 8 to determine

(i) the magnitude of the change in momentum of B *(2 marks)*

(ii) the magnitude of the force acting on B during the collision. Explain your answer. *(3 marks)*

c Figure 9 shows the initial and final states of two trolleys involved in a collision.

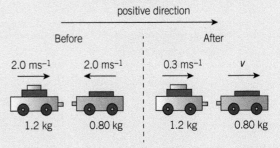

▲ **Figure 9**

The mass and the velocities are shown in Figure 9.

Calculate the velocity v of the 0.80 kg trolley after the collision. *(3 marks)*

d Figure 10 shows the initial and final states of two identical balls, X and Y, involved in a collision.

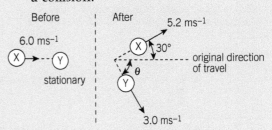

▲ **Figure 10**

Before the collision, X has a velocity of $6.0\,\text{ms}^{-1}$ and Y is stationary. After the collision, X has a velocity of $5.2\,\text{ms}^{-1}$ at an angle of 30° to its original direction of travel. Y is deflected at an angle θ and has a velocity of $3.0\,\text{ms}^{-1}$.

(i) Explain why the total momentum in the direction at right angles to the initial velocity of X is zero. *(1 mark)*

(ii) Calculate the angle θ. *(3 marks)*

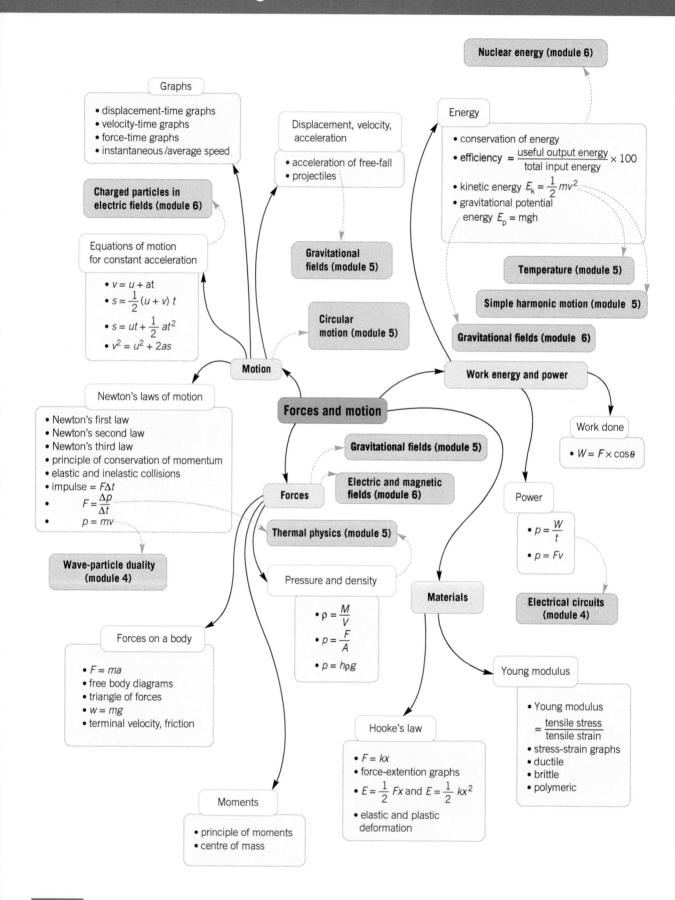

Graphs

- displacement-time graphs
- velocity-time graphs
- force-time graphs
- instantaneous/average speed

Charged particles in electric fields (module 6)

Equations of motion for constant acceleration

- $v = u + at$
- $s = \frac{1}{2}(u + v)\,t$
- $s = ut + \frac{1}{2}at^2$
- $v^2 = u^2 + 2as$

Displacement, velocity, acceleration

- acceleration of free-fall
- projectiles

Gravitational fields (module 5)

Circular motion (module 5)

Energy

- conservation of energy
- efficiency $= \dfrac{\text{useful output energy}}{\text{total input energy}} \times 100$
- kinetic energy $E_k = \frac{1}{2}mv^2$
- gravitational potential energy $E_p = mgh$

Nuclear energy (module 6)

Temperature (module 5)

Simple harmonic motion (module 5)

Gravitational fields (module 6)

Motion

Work energy and power

Newton's laws of motion

- Newton's first law
- Newton's second law
- Newton's third law
- principle of conservation of momentum
- elastic and inelastic collisions
- impulse $= F\Delta t$
- $F = \dfrac{\Delta p}{\Delta t}$
- $p = mv$

Wave-particle duality (module 4)

Forces and motion

Gravitational fields (module 5)

Electric and magnetic fields (module 6)

Forces

Thermal physics (module 5)

Pressure and density

- $\rho = \dfrac{M}{V}$
- $p = \dfrac{F}{A}$
- $p = h\rho g$

Materials

Work done

- $W = F \times \cos\theta$

Power

- $p = \dfrac{W}{t}$
- $p = Fv$

Electrical circuits (module 4)

Forces on a body

- $F = ma$
- free body diagrams
- triangle of forces
- $w = mg$
- terminal velocity, friction

Moments

- principle of moments
- centre of mass

Hooke's law

- $F = kx$
- force-extention graphs
- $E = \frac{1}{2}Fx$ and $E = \frac{1}{2}kx^2$
- elastic and plastic deformation

Young modulus

- Young modulus $= \dfrac{\text{tensile stress}}{\text{tensile strain}}$
- stress-strain graphs
- ductile
- brittle
- polymeric

Buildings

Architects have to carefully consider the physical properties of the materials when designing a building. Materials such as brick, concrete, and stone are weak in tension but strong in compression. Steel is incredibly strong in tension. Wood has interesting properties because it can withstand relatively large compressive and tensile stresses.

Figure 1 shows a conventional brick house with a tiled roof. The brick walls are in compression. A framework of wood is used to support the roof and to prevent the brick walls from buckling outwards. The strut beams in are compression and the tie beams are in tension.

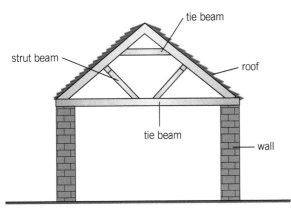

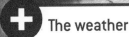

▲ **Figure 1** *Tie and strut beams*

▲ **Figure 2** *The Shard in London is 308 m tall. It is constructed around a steel framework.*

1. Describe what would happen to the structure shown in Figure 1 if the tie beams were not included.
2. Could steel cables be used in place of the tie beams?
3. Explain why the compressive stress is not constant along the length of the wall. Where would you expect the compressive stress in the wall to be a maximum?
4. Figure 2 shows The Shard in London. Discuss why very tall buildings use a steel framework rather than brick.

The weather

In 4.9 you learnt about pressure due to fluid columns and the Archimedes' principles. The measurement of atmospheric pressure is important in the study of meteorology (weather). High and low pressures govern the long and the short term weather patterns we observe. On weather maps, pressure is shown in millibars. Air pressure can be measured using instruments known as barometers.

Here are some extension tasks you can carry out to further improve your understanding of this topic. You may use the internet to carry out some of the research.

1. What is the relationship between millibars and pascals?
2. Investigate the different types of barometers and explain why many of them use mercury.
3. If you were to design you own barometer using water, how tall will it be?

4. Design an aneroid barometer using everyday items around you.
5. How does air pressure vary with altitude?

MODULE 4
Electrons, waves, and photons

Chapters in this Module

Demonstrate knowledge, understanding, and application of:

Introduction

Quantum physics is perhaps one of our greatest ever achievements. It allows us to make incredibly accurate predictions of what happens on tiny scales far smaller than an atom. In order to help you understand its key ideas, this module takes you on a journey, starting with electrons and how they behave in electrical circuits through an exploration of wave properties ending with quantum physics.

Charge and current provides an introduction to the fundamental ideas of charge and current, exploring the link between lightning strikes, the human brain, and the wonder material that is graphene.

Energy, power, and resistance develops the use of electrical symbols, along with key ideas like electromotive force, potential difference, and resistivity. You will learn about how differences in resistance help archaeologists discover ancient remains and doctors care for premature babies.

Electric circuits brings together ideas from the previous two chapters to explore the use of electrical circuits, including explanations of how potential dividers are used to make volume control dials and why a car battery can supply such a high current.

Waves 1 explores waves and their properties. You will learn about electromagnetic waves, earthquakes, and how diamonds get their sparkle.

Waves 2 includes explanations of how musical instruments produce their characteristic notes and how noise-cancelling headphones work so effectively. You will learn about the effect of interference of waves in a variety of situations.

Quantum physics introduces several truly amazing concepts, including the ideas that not only do electromagnetic waves have wave- and particle-like behaviour but this dual nature is also found to be characteristic of all particles, including electrons. Electrons can be made to diffract!

Knowledge and understanding checklist

From your Key Stage 4 study you should be able to answer the following questions. Work through each point, using your Key Stage 4 notes and the support available on Kerboodle.

- ☐ Recall that current is a rate of flow of charge and that for a charge to flow, a source of potential difference and a closed circuit are needed.

- ☐ Recall that current depends on both resistance and potential difference.

- ☐ Describe the difference between series and parallel circuits.

- ☐ Calculate the currents, potential differences, and resistances in series circuits.

- ☐ Explain the use of circuits containing components including lamps, diodes, thermistors, and LDRs.

- ☐ Describe wave motion in terms of amplitude, wavelength, frequency, and period.

- ☐ Describe how ripples on water surfaces are examples of transverse waves whilst sound waves in air are longitudinal waves, and describe the differences between transverse and longitudinal waves.

- ☐ Know that electromagnetic waves are transverse and are transmitted through space where all have the same velocity.

Maths skills checklist

All physicists need to use maths in their studies. In this unit you will need to use many different maths skills, including the following examples. You can find support for these skills on Kerboodle and through MyMaths.

- ☐ **Determine the gradient and intercept from a graph and use $y = mx + c$ to find unknown values.** You will need to be able to do this when investigating the resistivity of a wire.

- ☐ **Recognise and use expressions in decimal and standard form.** You will need to be able to do this when developing ideas around the electromagnetic spectrum.

- ☐ **Sketch relationships which are modelled by equations.** You will need to be able to do this when studying the photoelectric effect.

- ☐ **Substitute numerical values into algebraic equations using appropriate units for physical quantities.** You will need to be able to do this to determine the wavelength of light from a double slit experiment.

MyMaths.co.uk
Bringing Maths Alive

8 CHARGE AND CURRENT
8.1 Current and charge
Specification reference: 4.1.1

Learning outcomes

Demonstrate knowledge, understanding, and application of:

→ electric current as rate of flow of charge $I = \dfrac{\Delta Q}{\Delta t}$

→ the coulomb as the unit of charge

→ the elementary charge $e = 1.60 \times 10^{-19}$ C

→ net charge on a particle or an object is quantised and a multiple of e.

▲ **Figure 1** *The major path of ionised air through which billions of electrons have travelled to the ground is clearly visible here*

Synoptic link

You will recall from Topic 2.1, Quantities and units, that the ampere is one of the seven base units. It is used to define many other derived units in electricity, for example, the coulomb (A s) and the ohm (kg m^2 s^{-1} A^{-2}).

Study tip

Be sure to convert time into SI units (seconds) when using the equation for current.

A bolt from the blue

A lightning storm is one of nature's most awesome spectacles. Each bolt of lightning shows the path taken by billions and billions of electrons travelling from the cloud to the ground (or occasionally the other way around). This flow of *charged* particles produces a massive **electric current**.

Whereas current of 1.2 A is enough to charge a typical smartphone, the current in a lightning strike is often in excess of 30 000 A and heats the surrounding air to over five times the temperature of the surface of the Sun.

Defining electric current

Electric current is measured in **amperes** (or just amps for short). The ampere is one of the seven SI base units.

Electric current is defined as the rate of flow of charge, and can be calculated using the equation

$$I = \frac{\Delta Q}{\Delta t}$$

where I is the electric current in amperes, ΔQ is the charge transferred in coulombs, and Δt is the time in seconds.

In simple terms this is the amount of charge passing a given point in a circuit per unit time. One ampere (1 A) is the same as one coulomb of charge passing a given point per second (1 C s^{-1}), 12 kA is the same as 12 000 coulombs per second, and so on.

 Worked example: Current in a heater

A charge of 0.26 MC passes through a heater in 6.0 hours. Calculate the average current in the heater in that time, giving your answer to an appropriate number of significant figures.

Step 1: Identify the equation needed.

$$\text{Use } I = \frac{\Delta Q}{\Delta t}$$

Step 2: Substitute known values in SI units into the equation and calculate the answer.

$$I = \frac{0.26 \times 10^6}{6.0 \times 3600}$$

Express the answer to the correct number of significant figures.

$I = 12$ A (2 s.f.)

What is electric charge?

Electric charge is a physical property, much like mass, volume, or temperature. You can think of it as a measure of 'chargedness'. Some particles are charged, like protons and electrons, and others are not, like neutrons. Any object that is not charged is called neutral.

There are two types of charge, **positive** and **negative**. Objects with charge interact and exert forces on each other. Like charges repel each other, and opposite charges attract.

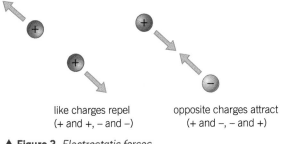

like charges repel
(+ and +, – and –)

opposite charges attract
(+ and –, – and +)

▲ **Figure 2** *Electrostatic forces*

Measuring electric charge

Electric charge is measured in **coulombs** (C). The coulomb is a derived unit, named after the French physicist Charles-Augustin de Coulomb. It is defined as the electric charge flowing past a point in one second when there is an electric current of one ampere.

From $\Delta Q = I\Delta t$, in base units, one coulomb (1 C) is equivalent to one ampere second (1 A s).

Any particle that has an **electric charge** is a charge carrier. Table 1 lists some examples of charge carriers and their charges.

You will be familiar with measuring charges as simply +1 or −2, for example, on **ions** in chemistry, but these are **relative charges**, that is, measured against the constant e. This is the **elementary charge** and is equal to 1.60×10^{-19} C. It is the same as the charge on one proton. A proton has a relative charge of $+1\,e$ and an electron $-1\,e$, not just +1 or −1.

If we know the value of the current in a metal wire, for example, we can also calculate the charge passing through it in a given time and even the number of electrons.

> **Study tip**
>
> Electric charges should not be confused with magnetic poles. There is a deep link between electricity and magnetism, but north and south poles are not the same as positive and negative charges.

▼ **Table 1** *Charges*

Charge carrier	Charge / C
proton	1.60×10^{-19}
electron	-1.60×10^{-19}
copper^{2+} ion	3.20×10^{-19}
sodium^{+} ion	1.60×10^{-19}
chloride^{-} ion	-1.60×10^{-19}

 Worked example: Electrons passing through a lamp

The current in a lamp is 6.2 A. Calculate the number of electrons passing through one point in the lamp in 2.0 minutes.

Step 1: Identify the correct equation to use.

$$I = \frac{\Delta Q}{\Delta t}$$

Rearrange to make the charge the subject. $\Delta Q = I\Delta t$

Step 2: Substitute all the values in SI units and calculate the answer.

$$\Delta Q = 6.2 \times 120 = 744\,C$$

Step 3: The number of electrons responsible for the charge of 744 C can be determined by dividing this charge by the charge e on each electron.

→

$$\text{number of electrons} = \frac{744}{1.60 \times 10^{-19}}$$
$$= 4.65 \times 10^{21} \approx 4.7 \times 10^{21}$$

The number of electrons is about 4 700 000 000 000 000 000 000.

Net charge

The charge on most objects results from either a gain or a loss of electrons by the object. If electrons have been added to the object it will be negatively charged, if electrons have been removed it will have a positive charge. The size of the charge on a particular object can be expressed as a multiple of e. The net charge on an object is given by

$$Q = \pm ne$$

where Q is the net charge on the object in coulombs, n is the number of electrons (either added or removed), and e is the elementary charge.

We describe the charge on an object as being **quantised**. This is because charge can only have certain values. These values must be integer multiples of e. For example, an object with a charge of 1.92×10^{-18} C has a charge of $+12e$.

Millikan's experiment – the discovery of the quantisation of charge

In 1909 Robert Millikan, aided by Harvey Fletcher, carried out one of the most important physics experiments of the 20th century, now simply called 'the oil-drop experiment'. He managed to levitate charged oil droplets between two oppositely charged metal plates by precisely balancing the weight of the negatively charged droplet acting downwards with an upwards attractive force from the positively charged plate.

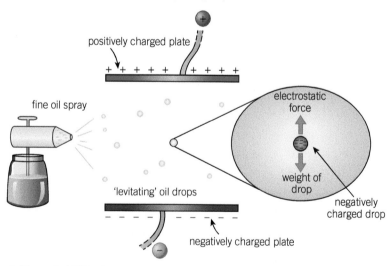

▲ **Figure 3** *Millikan's experiment balanced the forces on an oil drop to determine the charge on the drop precisely*

By taking very precise measurements he was able to determine the charge on each droplet, and did this for many droplets. He found that the charge on the drops was quantised: it did not take just any value, but only values that were multiples of elementary charge.

Using his data Millikan calculated e to be -1.59×10^{-19} C. He was within 1% of the currently accepted value.

1 Using our current understanding of charge, explain why the charge on each oil drop was quantised.

2 Millikan was not able to directly determine the mass of each oil drop. Instead he took careful measurements of the diameter of each drop and calculated its weight using the density of oil. Suggest the steps he would have taken and the formulae he would have used.

3 It is now alleged that Millikan may have ignored some of his data in order to provide a smaller range of readings. Discuss the importance of including all raw data in scientific papers.

Summary questions

1 Calculate the charge in coulombs for the following relative charges:
   a $+2.0\,e$ (1 mark)
   b $-5.0\,e$ (1 mark)
   c $-12\,e$ (1 mark)
   d $+41\,e$.   (1 mark)

2 An object acquires a charge when electrons are either removed from it or deposited on it. Determine the number of electrons deposited on an object to give a net charge of:
   a $-10\,e$ (1 mark)   b -15 C.   (3 marks)

3 A mobile phone charger draws 500 mA and takes 4.0 hours to charge a phone. Calculate the charge transferred in that time.   (3 marks)

4 Determine the value of the electric current if 5.0×10^{14} electrons pass through a wire in 1.0 s.   (2 marks)

5 A negatively charged plate gradually loses electrons to the surrounding air. If the plate loses a charge of 9000 C in 2.0 hours, calculate the average number of electrons per second that leave the plate.   (3 marks)

6 Calculate the number of electrons passing through the lamp in the second worked example if it were to be left switched on for two weeks.   (2 marks)

7 A rechargeable battery pack is labelled 5000 mA h. Calculate how much charge the pack can deliver when fully charged.   (3 marks)

8.2 Moving charges

Specification reference: 4.1.1

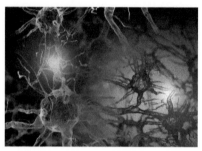

▲ **Figure 1** *Neurons within the brain conduct an electric current, but it is not just a simple flow of electrons*

A flow of charge, not just electrons

Electric currents are everywhere: not just confined to wires, but in the Earth's core, within the layers of the atmosphere, and even inside the cells in your body. We have already described an electric current as a flow of charge, a movement of charge carriers. Remember that this does not necessarily mean a flow of electrons: an electric current is a flow of any type of charge carrier – an electron is just one possibility. In metals the charge carriers are electrons, but in liquids the charge carriers tend to be **ions**.

The electric current in the nerve cells in your brain involves sodium and potassium ions passing through different parts of the membrane surrounding each cell.

Modelling electric current in metals

To help us better understand electric current, we use models to describe the movement of charge carriers through different materials. In metals an electric current is usually a flow of electrons. Because of the way atoms in metals are bonded, most electrons in metal atoms remain fixed to their atom. However, a small number of electrons from each atom are free to move (Figure 2).

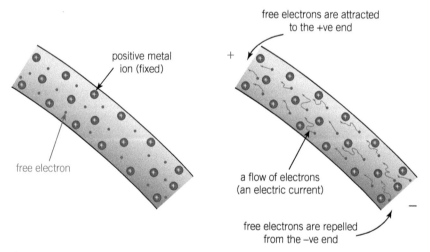

▲ **Figure 2** *A model of electric current in metals: free electrons move through the wire, randomly colliding with the positive ions as they drift past them*

The structure of a metal can be thought of as a regular crystal structure or lattice of positive ions, surrounded by a number of **free electrons** (sometimes called conduction or delocalised electrons). The positive ions are not free to move, but they do vibrate around fixed points, and they vibrate more vigorously as the temperature of the metal increases.

One way to make electrons move is to make one end of a wire positive and the other negative. The electrons in the metal of the wire will be attracted towards the positive end, and so move through the wire as an electric current. The greater the rate of charge flow, the greater the electric current in the wire. A larger current may be due to:

- a greater number of electrons moving past a given point each second (for example, a wire with a greater cross-sectional area)
- the same number of electrons moving faster through the metal.

Conventional current and electron flow

Electric current has been studied for centuries. **Conventional current** was defined long before the discovery of the electron as a current from a positive terminal towards a negative one. The direction of all electric currents is still treated as from positive to negative, regardless of the direction of movement of the charge carriers. In metals, for example, electrons travel from the negative terminal towards the positive terminal, but this flow of electrons is in the opposite direction to the conventional current (Figure 4).

▲ **Figure 3** *Modern processors contains billions of connections, each one allowing a tiny flow of charge to pass through it. In 1985 the conductors inside chips were 1500 nm across, whereas in 2014 this size has fallen to just 14 nm (only a few tens of atoms across)*

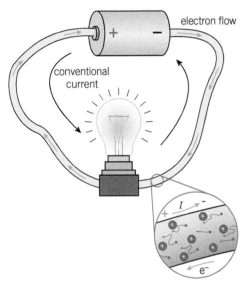

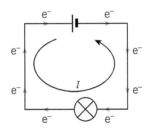

▲ **Figure 4** *Conventional current is always written from positive to negative, but in metals the electrons move the other way*

Electric current in electrolytes

It is not only metals which can conduct electricity. Several liquids can also conduct. Liquids that can carry an electric current are called **electrolytes**. In these cases the electric current is not a flow of electrons but a flow of ions. All electrolytes are either molten ionic compounds or, more commonly, **ionic solutions**. Pure water is an excellent insulator, but water from taps and rivers is an ionic solution – it contains dissolved ions, so can carry an electric current.

A common example of an ionic solution is salt (sodium chloride, NaCl) dissolved in water. The salt separates into positively charged sodium ions (**cations**), Na^+, and negatively charged chlorine ions (**anions**), Cl^- (Figure 5).

If a positive electrode (**anode**) and a negative electrode (**cathode**) are placed in the solution, ions are attracted to the electrodes. The Na^+ ions move towards the cathode and the Cl^- ions move towards the anode. This movement of ions is a flow of charge, an electric current.

Synoptic link

You will learn more about circuit diagrams in Topic 9.1, Circuit symbols.

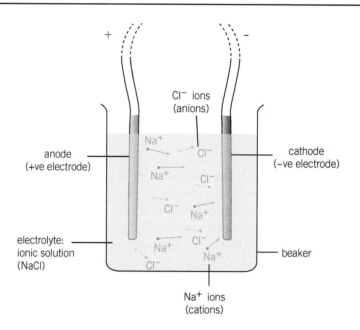

▶ **Figure 5** *In an electrolyte an electric current is a flow of ions*

▲ **Figure 6** *Ammeters come in many different forms – modern ammeters are often one function of a digital multimeter*

When the Na^+ ions reach the cathode, they accept an electron, and when the Cl^- ions reach the anode they donate an electron, so electrons can flow through the metal part of the circuit.

Measuring electric current

An **ammeter** is used to measure the electric current at any point in a circuit. It is always placed directly in series in the circuit at the point where you want to measure the current.

As ammeters are placed in series they should have the lowest possible **resistance** in order to reduce the effect they have on the current – an ammeter with high resistance would decrease the current it should be measuring. The ideal (perfect) ammeter has zero resistance, and so has no effect on the current it measures.

Summary questions

1 Outline the difference between conventional current and electron flow in a metal wire.   *(2 marks)*

2 Describe the similarities and differences between electric current in a metal wire and in an ionic solution.   *(4 marks)*

3 A solution of copper sulfate contains Cu^{2+} and $SO_4{}^{2-}$ ions. Sketch a diagram to show how an electric current is carried in a solution of copper sulfate.   *(3 marks)*

4 Compare the direction of conventional current with the movement of the anions and cations in an electrolyte.   *(2 marks)*

5 A solution of magnesium chloride contains Mg^{2+} and Cl^- ions. Calculate the current in the cathode in µA when 6.0×10^{14} cations come into contact with the cathode in 3.0 minutes.   *(3 marks)*

Synoptic link

You will learn more about resistance in Topic 9.4, Resistance.

8.3 Kirchhoff's first law

Specification reference: 4.1.1

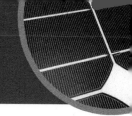

Conservation – it's the law

The large circle in Figure 1 is the outline of the Large Hadron Collider (LHC) near Geneva in Switzerland. The LHC is the most powerful particle accelerator ever built and forms part of the world's largest laboratory. In it, beams of subatomic particles are made to collide in order to replicate the conditions just after the Big Bang.

Physicists at the LHC analyse the data from each collision, applying several conservation laws. Each conservation law states that a particular, measurable physical quantity does not change. These laws are a cornerstone of physics, and include the conservation of energy, of linear momentum, and of charge. Using these laws, physicists are able to demonstrate the existence of exotic particles that vanish after a fraction of a second, including the Higgs boson.

Conservation of charge

In Topic 8.1, Current and charge, we described charge as a physical property (a measure of 'chargedness'). We don't know why some particles have charge and others do not, but if they didn't our universe could not exist in anything like its current form.

Charge is a fundamental physical property and one of only a handful of properties that must be conserved. That is to say in any interaction the total charge before and after must be the same. **Conservation of charge** states that electric charge can neither be created nor destroyed. The total amount of electric charge in the universe is constant.

Kirchhoff's first law

Gustav Kirchhoff was a German physicist born in 1824. He made many valuable contributions to science, including working with Robert Bunsen, who invented the burner, to discovering caesium and rubidium. He also studied electricity.

Kirchhoff's first law deals with electric current. It states that, for any point in an electrical circuit, the sum of currents into that point is equal to the sum of currents out of that point.

▲ **Figure 1** *The Large Hadron Collider is a circular particle accelerator with a circumference of 27 km buried 100 m below the French–Swiss border*

> **Synoptic link**
>
> You will learn more about Kirchhoff's second law in Topic 10.1, Kirchhoff's laws and circuits.

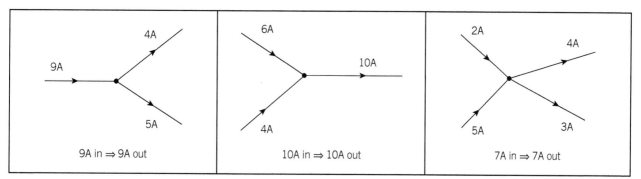

▲ **Figure 2** *Charge must be conserved at all points in a circuit*

9A in ⇒ 9A out   10A in ⇒ 10A out   7A in ⇒ 7A out

> **Learning outcomes**
>
> Demonstrate knowledge, understanding, and application of:
>
> → Kirchhoff's first law and the conservation of charge.

125

The law can be written as

$$\Sigma I_{in} = \Sigma I_{out}$$

where Σ (Greek sigma) denotes 'sum of'. ΣI_{in} is the sum of the current into a point and ΣI_{out} is the sum of the current out of that point.

The law is based on conservation of charge, where the charge (measured in coulombs) is the product of the current (in amperes) and the time (in seconds). Charge cannot be destroyed, so the charge carriers entering a point in a given time must equal the total number of charge carriers leaving that same point during that time.

Summary questions

1 Explain the meaning of conservation of charge. *(2 marks)*

2 Draw a diagram to explain and illustrate an example of Kirchhoff's first law. *(2 marks)*

3 Copy and complete the diagrams below, stating both the magnitude and the direction of the missing electric currents. *(6 marks)*

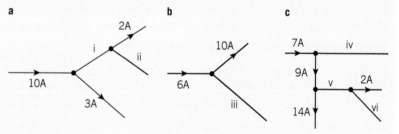

▲ Figure 3

4 A single wire is connected to two other wires, A and B. If the current in the single wire is 15 A, and 1.9×10^{21} electrons pass along wire A in 1.0 minutes, calculate the current in wire B to the nearest ampere. *(3 marks)*

5 Discuss how the idea of conservation of charge relates to the quantisation of charge studied in Topic 8.1, Current and charge. *(3 marks)*

6 Suggest how conservation of charge might be used to determine the existence of a short-lived, negatively charged particle created inside a particle accelerator when two protons collide together. *(2 marks)*

8.4 Mean drift velocity

Specification reference: 4.1.2

A miracle material?

Graphene is pure carbon, arranged in thin sheets just one atom thick. It is incredibly strong and at the same time extremely flexible.

Its unusual composition gives graphene a huge number of free electrons. The number of free charge carriers per unit volume (the **number density**) of graphene is even greater than that of copper, making it an outstanding electrical conductor.

Classification of materials

The number density is the number of free electrons per cubic metre of material. The higher the number density, the greater the number of free electrons per m^3 and so the better the electrical conductor.

We can classify materials into three groups according to their number density. Conductors have a very high number density (of the order of $10^{28} m^{-3}$), insulators have a much lower value with **semiconductors** in between the two, with number densities around $10^{17} m^{-3}$.

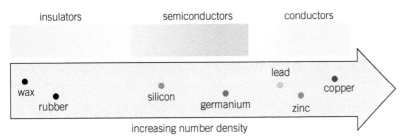

▲ **Figure 2** *The number density of a material determines how easily the material will conduct an electric current*

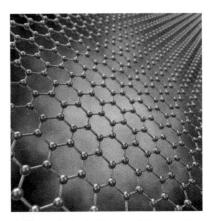

▲ **Figure 1** *Andre Geim and Konstantin Novoselov from the University of Manchester won the 2010 Nobel Prize in Physics for their work on graphene*

▼ **Table 1** *Conduction and number density*

Material	Type	n /m^{-3} (at 300 K)
copper	conductor	8.5×10^{28}
zinc	conductor	6.6×10^{28}
germanium	semiconductor	2.0×10^{18}
silicon	semiconductor	8.7×10^{15}

Semiconductors have a much lower number density than metals, so in order to carry the same current the electrons in semiconductors need to move much faster. This increases the temperature of the semiconductor, a fact relevant in the design of computers, which use processors made of silicon (Figure 3).

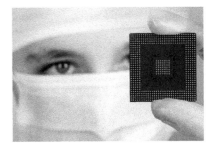

▲ **Figure 3** *Microprocessors are made largely of silicon, and can get very hot when a current passes through them, so computers need carefully designed cooling systems*

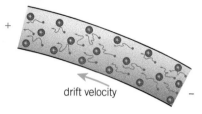

drift velocity

▲ **Figure 4** *Repeated collisions between positive ions and free electrons cause the random motion of the electrons*

How fast do charge carriers move?

When you flick a switch most filament lamps turn on almost instantly. You might imagine that electrons have rushed from the switch to the lamp, but this is not true. In fact, most charge carriers, like electrons, move slowly. Free electrons repeatedly collide with the positive metal ions as they drift through the wire towards the positive terminal. The reason that lights turn on so quickly is that all the free electrons in the wire start moving almost at once.

A new equation for electric current

There is an additional equation for electric current.

$$I = Anev$$

where I is the electric current in the conductor in amperes, A is the cross-sectional area of the conductor in m^2, e is the elementary charge $(1.60 \times 10^{-19}\,C)$, n is the number density, and v is the **mean drift velocity** of the charge carriers in $m\,s^{-1}$.

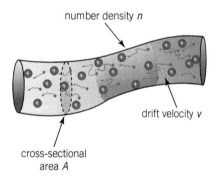

number density n

drift velocity v

cross-sectional area A

▲ **Figure 5** $I = Anev$ *can be derived by considering the properties of the wire and the free electrons*

➕ Derivation of $I = Anev$

We can derive the equation by considering the number density and dimensions of a conductor.

From its definition, electric current (I) is given by $\quad I = \dfrac{\Delta Q}{\Delta t}$

The number of electrons in a given volume V of the conductor is nV, where n is the number density.

The total charge of the electrons in this volume of conductor is neV, where e is the elementary charge. This gives

$$I = \frac{neV}{\Delta t}$$

When there is an electric current in the conductor, a certain volume of charge carriers passes a given point each second. This volume depends on the cross-sectional area A of the conductor and the mean drift velocity v of the charge carriers.

$$\frac{V}{\Delta t} = Av$$

Substituting this into our previous equation for electric current gives

$$I = \frac{neV}{\Delta t} = neAv, \text{ more commonly written as}$$

$$I = Anev$$

1 Show that the equation is homogenous with respect to base units.
2 Describe the effect on the mean drift velocity if all other factors are constant and:
    a current increases;
    b cross-sectional area decreases;
    c number density doubles.

Worked example: Calculating mean drift velocity

A copper wire has a cross-sectional area of $7.85 \times 10^{-7}\,m^2$. The number density of copper is $8.50 \times 10^{28}\,m^{-3}$. Calculate the mean drift velocity of the electrons through the wire when the current is 1.40 A.

Step 1: Identify the equation needed.

$$I = Anev$$

Rearrange to make v the subject. $v = \dfrac{I}{Ane}$

Step 2: Substitute the known values in SI units (including the cross-sectional area in m^2) into the equation and calculate the answer.

$$v = \frac{1.40}{7.85 \times 10^{-7} \times 8.50 \times 10^{28} \times 1.60 \times 10^{-19}}$$

$$v = 1.31 \times 10^{-4}\,m\,s^{-1} \text{ (3 s.f.)}$$

This is very slow: only $0.13\,mm\,s^{-1}$. The electrons have such a low speed because of the large number of random collisions with the fixed positive ions.

The effect of changing cross-sectional area

If the cross-sectional area of a wire changes, so must the drift velocity. The narrower the wire, the greater the drift velocity must be in order for the current to be the same.

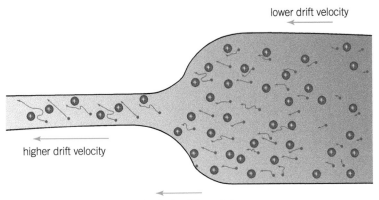

lower drift velocity

higher drift velocity

constant rate of flow of charge

▲ **Figure 6** *In order to maintain the same current (rate of flow of charge), electrons must move faster through narrower wires – the mean drift velocity is inversely proportional to the cross-sectional area of the wire*

 Worked example: Getting thinner

A piece of wire carrying a current narrows such that its radius halves. What effect does this have on the mean drift velocity of the electrons in the wire?

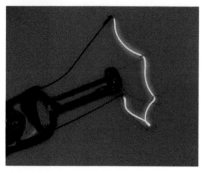

▲ **Figure 7** *The filament of an incandescent bulb is much narrower than the surrounding wire, so electrons move faster through this part of the wire, heating it up so much that it glows white*

Step 1: Identify the equation that relates v and A.

$$I = Anev$$

Rearrange to make v the subject. $v = \dfrac{I}{Ane}$

Step 2: Examine the effect of changing A.

According to Kirchhoff's first law, the current in the wire is the same. The elementary charge e and the number density n are also constants, and n does not change because it is the same material.

Therefore, $v \propto \dfrac{1}{A}$

If the radius halves the cross-sectional area will decrease by a factor of 4.

As a result the mean drift velocity must increase by a factor of 4.

Summary questions

1 Describe the meaning of the terms 'conductor', 'semiconductor', and 'insulator' in terms of their relative number densities. *(2 marks)*

2 Calculate the current through a copper wire with a cross-sectional area of $5.50 \times 10^{-8}\,m^2$, when the mean drift velocity of the charge carriers is $2.0 \times 10^{-3}\,m\,s^{-1}$. Use the values for n and e in the first worked example above. *(3 marks)*

3 A silver wire has a cross-sectional area of $7.10 \times 10^{-6}\,m^2$ and a number density of $5.86 \times 10^{28}\,m^{-3}$. Calculate the mean drift velocity of the electrons through the wire when the current is 500 mA. *(3 marks)*

4 State and explain the effect on the mean drift velocity of electrons if a constant current travels:
   a into the same wire with a larger cross-sectional area; *(2 marks)*
   b from a zinc wire into a copper wire of the same dimensions (see Table 1); *(2 marks)*
   c into the same wire with $\frac{1}{3}$ of the radius. *(3 marks)*

5 A zinc wire has a diameter of 1.0 mm. Calculate the mean drift velocity of the electrons through the wire when the current through it is 3.0 mA. *(3 marks)*

6 A small piece of semiconducting material with a cross-sectional area of $8.2 \times 10^{-6}\,m^2$ is used in an experiment. When the current in the material is 12 mA, the mean drift velocity of the charge carriers is $72\,m\,s^{-1}$. Determine the ratio of the number density in the semiconductor and the number density of copper. *(4 marks)*

Practice questions

1 a State the base units for electrical charge.   *(2 marks)*

   b Figure 1 shows a simple electrical circuit constructed by a student.

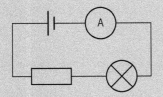

▲ Figure 1

   The ammeter reading is 20 mA.

   (i)   State the direction of conventional current.   *(1 mark)*

   (ii) Calculate the number of electrons entering the resistor per second.   *(3 marks)*

   (iii) State and explain the value of the current in the lamp.   *(2 marks)*

   Jan 2011 G482

2 a A 12 V car battery contains an electrolyte. The battery is connected to an electric motor **M**. There is a current in the motor and the battery. See Figure 2.

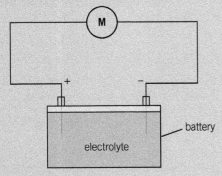

▲ Figure 2

   State

   (i)   the charge carriers in the electrolyte,   *(1 mark)*

   (ii) the charge carriers moving through the electrolyte to the positive terminal of the battery,   *(1 mark)*

   (iii) the charge carriers moving through the wires to the positive terminal of the battery.   *(1 mark)*

   Jan 2011 G482

3 A resistor of length l is connected to a cell, see Figure 3.

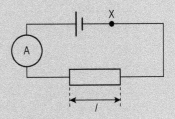

▲ Figure 3

   a State the direction of the flow of electrons in the resistor.   *(1 mark)*

   b In a time of 2.0 minutes, the number of electrons passing through point **X** in the wire is 1.88×10^{20}. Calculate

   (i)   the charge flow through point **X**,   *(2 marks)*

   (ii) the current in the wire.   *(2 marks)*

   c On a copy of the axes of Figure 4, sketch a graph to show the variation of current I with the distance from one end of the resistor.   *(2 marks)*

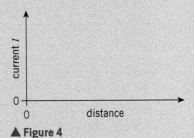

▲ Figure 4

4 Figure 5 shows an electrical circuit with three resistors **A**, **B** and **C**.

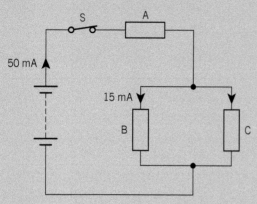

▲ Figure 5

The circuit shows the currents at different points in the circuit when the switch **S** is closed for a time of 30 s. Calculate

a the total charge flow into the resistor **B**, *(2 marks)*

b the number of electrons responsible for the charge in (a), *(2 marks)*

c the current in the resistor **C**. *(1 mark)*

5 a State Kirchhoff's first law and state the quantity conserved according to this law. *(2 marks)*

b Figure 6 shows part of an electrical circuit.

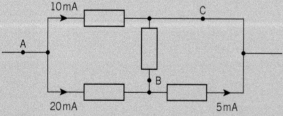

▲ Figure 6

(i) Determine the currents at points **A**, **B** and **C**. *(3 marks)*

(ii) State the direction of the current at **B**. *(1 mark)*

6 a Explain the term *mean drift velocity* of electrons in a metal wire. *(3 marks)*

b Figure 7 shows a resistor made from a conducting material deposited onto a plastic base.

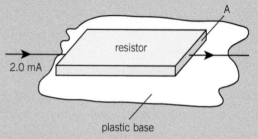

▲ Figure 7

The cross-sectional area A of the resistor is $4.2 \times 10^{-10}\,\text{m}^2$. The material used for the resistor has 8.0×10^{27} free electrons per unit volume. The current in the resistor is 2.0 mA.

(i) Calculate the mean drift velocity of the electrons inside the resistor. *(3 marks)*

(ii) The length of the resistor is 8.0 mm. Calculate the time taken for an electron to drift along the length of this resistor. *(1 mark)*

7 Figure 8 shows a lightning strike between a storm cloud and a tall building.

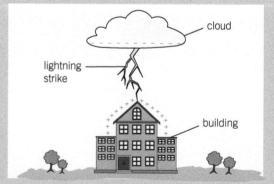

▲ Figure 8

a The bottom of the cloud has a negative charge and the building has a positive charge.

Indicate the direction of the conventional current in the lightning strike in Figure 8.

(1 mark)

b The current in the lightning strike is 8200 A and the strike lasts for 120 ms.

(i) Calculate the total number of electrons transferred between the cloud and the building. *(3 marks)*

(ii) On a copy of the axes below, sketch a graph of charge Q transferred against time t. Explain the shape of the graph. *(3 marks)*

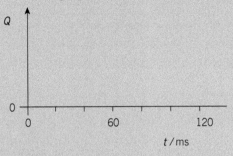

▲ Figure 9

8　a Explain what is meant by *electric current*. *(1 mark)*

b Determine the unit of charge in SI base units. *(1 mark)*

c Figure 10 shows a negatively charged metal sphere attached to a plastic rod.

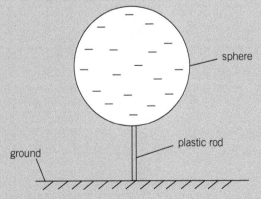

▲ Figure 10

The charge on the sphere is -5.4×10^{-9} C. The charge slowly leaks away through the rod.

This leakage current is 0.20 pA. Estimate the time in hours it would take for the sphere to be completely discharged. *(3 marks)*

9　a Explain what is meant by the *mean drift velocity* of electrons in a current-carrying wire. *(2 marks)*

b Figure 11 shows two wires, A and B, connected in series to a supply.

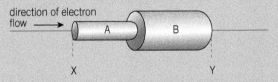

▲ Figure 11

Describe and explain the variation of the mean drift velocity of the electrons from X to Y. *(3 marks)*

c The current in a filament lamp is 0.30 A. The filament has a diameter of 5.0×10^{-5} m. The material of the filament has 3.4×10^{28} free electrons per unit metre.

Calculate the mean drift velocity of the electrons in the filament. *(3 marks)*

9 ENERGY, POWER, AND RESISTANCE

9.1 Circuit symbols

Specification reference: 4.2.1

Learning outcomes

Demonstrate knowledge, understanding, and application of:

→ circuit symbols

→ circuit diagrams using these symbols.

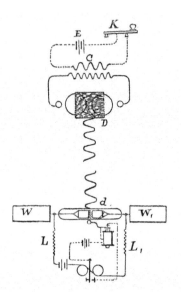

▲ **Figure 1** *This is the sketch Marconi submitted to the Patent Office, with some symbols we still use today*

Cornwall calling

In 1897 the Italian physicist Guglielmo Marconi submitted this circuit diagram to the UK Patent Office (Figure 1). It is a design for a radio transmitter that, a few years later, would send the first radio transmission across the Atlantic Ocean from Cornwall to Canada, and earn him the 1909 Nobel Prize in Physics. Using clearly defined circuit symbols he was able to patent his design and prevent competitors from simply copying his innovations, so his pioneering work made him very wealthy, as well as laying the foundations for all the wireless communications we use today.

Circuit diagrams

We can use a huge number of electrical components to build electrical circuits, from the simple filament lamp to complex components like the **capacitor** or diode.

When drawing circuits it is important to use an internationally recognised set of symbols to represent components, so that scientists and engineers around the world can construct similar circuits to test and verify claims made by anyone working in this field. Every type of component has its own unique circuit symbol – for example, all capacitors have the same symbol, no matter what size or colour they are. Figure 2 contains all the symbols we will require in A Level Physics.

Rules for circuit diagrams

When drawing circuit diagrams you should remember three simple rules.

1   Only use the circuit symbols in Figure 2.

2   Do not leave any gaps in between the wires.

3   When possible use straight lines drawn with a pencil and ruler. However, a carefully drawn free-hand sketch of the circuit will be adequate in assessments.

Cells, batteries, and power supplies

Most circuits will contain a single **cell**, a number of cells, or a mains power supply. A '**battery**' in physics means two or more cells connected end-to-end, or in **series**. In the case of a single cell or a battery, the longer terminal represents the positive terminal. When using a power supply, a small plus sign is often placed next to the positive terminal. Take care to ensure that this **polarity** is represented correctly when you use these symbols, because polarity is very important when using components such as diodes and light-emitting diodes.

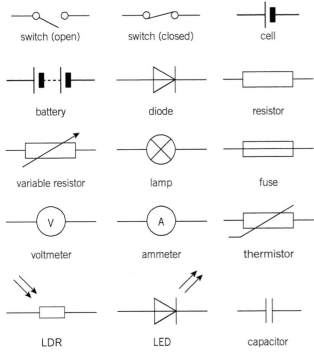

▲ **Figure 2** *Circuit symbols from ASE publication: Signs, symbols, and systematics (The ASE Companion to 16–19 Science, 2000)*

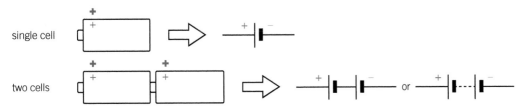

▲ **Figure 3** *You must carefully consider the polarity of a cell when building circuits or drawing circuit diagrams*

Summary questions

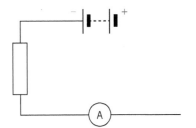

1 Name the components represented by the symbols in Figure 4. (*3 marks*)

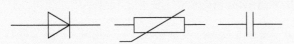

▲ **Figure 4**

▲ **Figure 5**

2 In each case, draw a circuit diagram using the correct symbols.
   a A single cell connected in series with a filament lamp. (*2 marks*)
   b A battery connected to a resistor and an ammeter in series. (*3 marks*)
   c A power supply connected to two resistors connected in series. (*2 marks*)

3 Draw a circuit containing a cell, an open switch, and a filament lamp. Show a voltmeter connected across the lamp (in parallel). (*2 marks*)

4 Identify two errors in the circuit diagram in Figure 5. (*2 marks*)

5 Draw the circuit diagram for the circuit shown in Figure 6. (*3 marks*)

▲ **Figure 6**

9.2 Potential difference and electromotive force

Specification reference: 4.2.2

Potential difference

A Tesla coil in action is a truly impressive sight (Figure 1). It produces artificial lightning in the laboratory by generating a huge **potential difference** (p.d.), also called voltage. Potential difference is a measure of the transfer of energy by charge carriers. The p.d. across a component like a filament lamp is a result of electrical energy being transferred into heat and light as charge carriers move through the lamp. High voltages can be very dangerous because charge carriers can transfer enormous amounts of energy through conductors and, if the voltage is high enough, through insulators like air, or people.

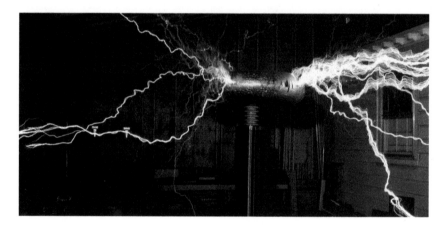

▲ **Figure 1** *A Tesla coil, like this one, can generate p.d.s in excess of 500 000 V. The coil is named after its inventor, the physics genius Nikola Tesla*

The volt

Potential difference is measured in volts, named after Alessandro Volta, an Italian who invented the first battery.

One **volt** is the p.d. across a component when 1 J of energy is transferred per unit charge passing through the component.

$$1\,V = 1\,J\,C^{-1}$$

A p.d. of 1000 V means that 1000 J of energy is transferred per coulomb of charge. The term potential difference is used when charged particles lose energy in a component. Potential difference is defined as the energy transferred from electrical energy to other forms (heat, light, etc.) per unit charge. The equation for potential difference V is.

$$V = \frac{W}{Q}$$

where V is p.d. measured in volts, Q is charge in coulombs, and W is the energy transferred by charge Q.

The voltmeter

A **voltmeter** is used to measure p.d. Like ammeters, voltmeters come in all different shapes and sizes, but they are always connected in parallel across a particular component.

An ideal voltmeter should have an infinite resistance (see Topic 9.4, Resistance), so that when connected, no current passes through the voltmeter itself. Whilst this is not possible in reality, most voltmeters have a resistance of several million ohms.

Electromotive force

At times it is necessary to describe whether the charges in a circuit are losing or gaining energy. Two different terms are used depending on whether the charge carriers are doing work or work is being done on them.

Potential difference is used to describe when work is done *by* the charge carriers. Essentially the charges are losing energy as they pass through the component. The greater the p.d., the more energy per coulomb is transferred from electrical energy into other forms (like light or heat) as the charges move through the component.

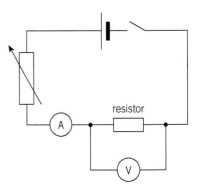

▲ **Figure 2** *Voltmeters are used to measure p.d. and are always connected in parallel*

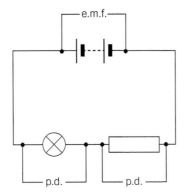

▲ **Figure 3** *There is an e.m.f. across the cell, and p.d.s across the other components*

Electromotive force (e.m.f.) is used to describe when work is done *on* the charge carriers. Essentially the charges are gaining energy as they pass through a component like a cell, battery, or power pack. The greater the e.m.f., the more energy per coulomb has been transferred (often from the form of chemical energy in a cell) into electrical energy. Other sources of e.m.f. include solar cells (from light), dynamos (movement), and thermocouples (heat).

The term electromotive force is used when charged particles gain energy from a source. Electromotive force is defined as the energy transferred from chemical energy (or another form) to electrical energy per unit charge. The equation for electromotive force ε is

$$\varepsilon = \frac{W}{Q}$$

where ε is e.m.f. measured in volts, Q is charge in coulombs, and W is the energy transferred by charge Q.

> **Study tip**
>
> Despite its name, e.m.f. is not a force, which is measured in newtons, whereas e.m.f. is measured in volts.

Calculating energy transfer

Whether it is a p.d. or an e.m.f., the energy transferred from or to the charges can be calculated from the defining equations. The energy transferred depends on the size of the p.d. and the charge passing through the component.

$$W = VQ \quad \text{or} \quad W = \mathcal{E}Q$$

▲ **Figure 4** *A chemical reaction between the acid in the tomato and the different metal electrodes produces an e.m.f.*

▲ **Figure 5** *A small button cell like the ones found in calculators contains a mixture of chemicals that react together to produce an e.m.f., transferring chemical energy into electrical energy*

 Worked example: Energy transferred by an electrical heater

Calculate the energy transferred in 4.0 hours by an electrical heater that draws a current of 15 A when the p.d. across it is 90 V.

Step 1: Identify the correct equation to calculate the charge through the heater in that time.

$$\Delta Q = I\Delta t$$

Step 2: Substitute in the known values in SI units (time in seconds) and calculate the charge in coulombs.

$$\Delta Q = 15 \times 60 \times 60 \times 4.0$$

$$\Delta Q = 216\,000\,\text{C}$$

Step 3: Identify the correct equation to calculate the energy transferred.

$$W = VQ$$

Step 4: Substitute in known values in SI units and calculate the energy transferred in joules.

$$W = 90 \times 216\,000$$

$$W = 19\,\text{MJ (2 s.f.)}$$

Summary questions

1 Sketch a circuit diagram to show how a voltmeter may be used to measure the p.d. across a filament lamp. *(2 marks)*

2 Outline the difference between p.d. and e.m.f. *(2 marks)*

3 Calculate the energy transferred to 4.0 C of charge by a cell with an electromotive force of 80 V. *(2 marks)*

4 Calculate the p.d. across a filament lamp when 168 J of energy is transferred by the lamp by 14 C of charge. *(2 marks)*

5 Define the volt and express it in base units. *(4 marks)*

6 A resistor has a current through it of 500 mA and a p.d. across it of 1.0 kV. Calculate the energy transferred to the resistor in 6.0 hours. *(4 marks)*

Phasers set to stun?

Despite its name, an **electron gun** is not a sci-fi weapon, but an electrical device used to produce a narrow beam of electrons. These electrons can be used to ionise particles by adding or removing electrons from atoms, and they can have very precisely determined kinetic energies. Electron guns are used in scientific instruments such as electron microscopes, mass spectrometers, and oscilloscopes.

How does it work?

All electron guns need a source of electrons. In most cases a small metal filament is heated by an electric current. The electrons in this piece of wire gain kinetic energy. Some of them gain enough kinetic energy to escape from the surface of the metal. This process is called **thermionic emission** – the emission of electrons through the action of heat.

If the heated filament is placed in a vacuum and a high p.d. applied between the filament and an anode, the filament acts as a cathode, and the freed electrons accelerate towards the anode, gaining kinetic energy. If the anode has a small hole in it, then electrons in line with this hole can pass through it, creating a beam of electrons with a specific kinetic energy (Figure 2).

▲ **Figure 1** *Old-style computer monitors and TV screens – cathode-ray tubes – use electron guns to produce images on the screen*

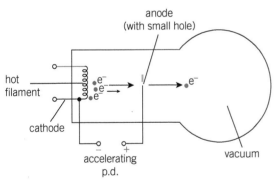

▲ **Figure 2** *An electron gun produces a narrow beam of electrons with a specific kinetic energy depending on the p.d. between the cathode and the anode*

Energy transfers

As the electrons accelerate towards the anode they gain kinetic energy. From the definition for p.d., the work done on a single electron travelling from the cathode to the anode is equal to eV, where e is the elementary charge, the charge on each electron, and V is the accelerating p.d.

By considering the law of conservation of energy we can derive an expression relating the work done on the electron to its increase in kinetic energy.

$$\text{work done on electron} = \text{gain in kinetic energy}$$

$$eV = \frac{1}{2}mv^2$$

This assumes the electrons have negligible kinetic energy at the cathode.

Changing the accelerating p.d. changes the kinetic energy of the electrons within the beam. The greater the p.d. the more energy is transferred to the electrons and so the faster they move.

 Worked example: Calculating the velocity of an electron

Calculate the velocity of an electron from an electron gun with an accelerating p.d. of 4.0 kV.

Step 1: Select the equation relating the velocity of an electron and the p.d.

$$eV = \frac{1}{2}mv^2$$

Rearrange to make v the subject. $\sqrt{\dfrac{2eV}{m}} = v$

Step 2: Substitute in known values (including e and mass of an electron m) in SI units. $v = \sqrt{\dfrac{2 \times 1.60 \times 10^{-19} \times 4000}{9.11 \times 10^{-31}}}$

Calculate the velocity of the electron. $v = 3.7 \times 10^7 \, \mathrm{m\,s^{-1}}$

Summary questions

1 Describe how an electron gun produces a beam of high-speed electrons. *(4 marks)*

2 Calculate the kinetic energy of an electron accelerated through a p.d. of 12 kV. *(3 marks)*

3 Calculate the velocity of an electron that has a kinetic energy of 1.8×10^{-15} J. *(2 marks)*

4 Calculate the magnitude of the accelerating p.d. that produces electrons with a velocity of 9% of the speed of light. (Speed of light $c = 3.00 \times 10^8 \, \mathrm{m\,s^{-1}}$.) *(4 marks)*

5 State and explain how the velocity of an electron and a proton would compare when accelerated through the same p.d. *(3 marks)*

Particle accelerators

A linear particle accelerator (often shortened to LINAC) uses a series of cylindrical electrodes (drift tubes) to accelerate subatomic particles such as electrons. The polarity of the drift tubes is alternated between positive and negative with precise timing so that each time the electrons leave one of the tubes, the polarity changes in order to attract them to the next one. Figure 3 shows a simplified cross-section of a LINAC.

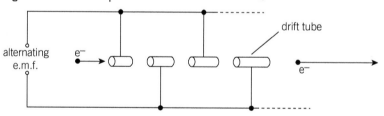

▲ **Figure 3** *A LINAC in cross-section*

Every time an electron moves from one tube to another it gains energy equal to eV, where V is the maximum e.m.f. of the alternating source connected to the tubes. With a large number of electrodes, electrons can be accelerated to extremely high velocities.

LINACs have many applications. They are often used as injectors for even higher energy particle accelerators like the LHC at CERN and they are commonly used in hospitals to produce the X-rays required for radiotherapy.

1 A linear accelerator uses 100 drift tubes, each with a maximum alternating e.m.f. of 50 V. Calculate the maximum speed of the electrons from the LINAC.

2 Suggest why it is necessary for the length of the drift tubes inside a linear accelerator to increase as they get further along the accelerator (Figure 3).

9.4 Resistance

Specification reference: 4.2.3

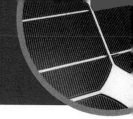

X marks the spot

Archaeologists use a number of different techniques to survey the ground at a dig site. An electrical resistance survey uses specialist detectors to measure the resistance of the Earth. Different features under the soil have different resistances. This property enables archaeologists to build up a map of the features beneath the surface so that they can start digging in a likely place.

Learning outcomes

Demonstrate knowledge, understanding, and application of:

→ resistance; $R = \dfrac{V}{I}$; the unit ohm

→ Ohm's law.

What is resistance?

The electrical components called **resistors** have a known resistance, but in fact all components – including filament lamps, diodes, connecting wires, and even cells – have their own resistances. Each component resists the flow of charge carriers through it. It takes energy to push electrons through a component, and the higher the resistance of that component the more energy it takes.

Determining resistance

The term resistance has a very precise meaning in physics. The resistance of a component in a circuit can be determined by measuring the current I in the component and the p.d. V across the component. The resistance of the component R is defined as the ratio between V and I.

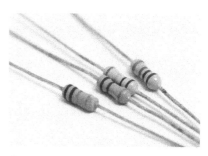

▲ **Figure 1** *Resistors come in all shapes and sizes, like these carbon film resistors – the coloured bands represent the resistance of each one*

$$\text{resistance of component } R = \frac{\text{p.d. across component}}{\text{current in component}}$$

$$R = \frac{V}{I}$$

variable supply

I

ammeter measures current

voltmeter measures potential difference

▲ **Figure 2** *The current in the component and the potential difference across it are used to determine its resistance*

The unit of resistance is the **ohm** (Ω). The ohm was named after the German Georg Ohm, who carried out pioneering work on electrical resistance in the 1800s.

The ohm is defined as the resistance of a component when a p.d. of 1 V is produced per ampere of current.

$$1\,\Omega = 1\,V\,A^{-1}$$

A component with a resistance of $1\,\Omega$ will have a p.d. across it of 1 V per ampere of current in it. A component with a resistance of $500\,\Omega$ would have a p.d. of 500 V when there is a current of 1 A in it.

Ohm's law

Ohm's investigations into the resistances of metallic conductors led him to derive what is now referred to as **Ohm's law**.

For a metallic conductor kept at a constant temperature, the current in the wire is directly proportional to the p.d. across its ends.

In other words, he found that when the p.d. across the wire (kept at constant temperature) doubled, the current in the wire also doubled.

Temperature and resistance

Figure 3 shows a length of insulated metallic wire in the form of a tight bundle connected in a circuit, alongside a graph of the variation of the current I in the circuit with time t. The switch **S** is closed at time $t = 0$. The p.d. across the wire remains constant at 1.5 V, but the current in the wire decreases with time. The shape of the graph can be explained as the resistance of the wire increasing with time. You can confirm this by calculating the resistance at $t = 0$ and at $t = 4.0$ s.

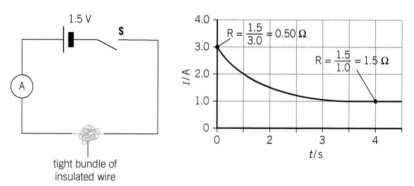

▲ **Figure 3** *A circuit including a bundle of thin wire carries less current as the wire gets hotter*

The current in the circuit changes because the temperature of the wire increases over time as a result of heating caused by the current. As the wire gets hotter, its resistance increases.

A microscopic explanation

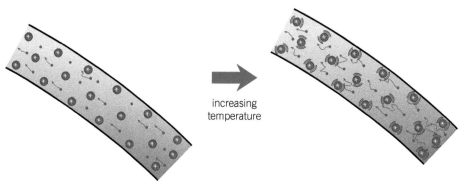

▲ **Figure 4** *As the wire gets hotter the positive ions vibrate more, increasing the resistance of the wire*

When the temperature of the wire increases the positive ions inside the wire have more internal energy and vibrate with greater amplitude about their mean positions. The frequency of the collisions between the charge carriers (free electrons in the metal) and the positive ions increases, and so the charge carriers do more work, in other words, transfer more energy as they travel through the wire.

Other factors as well as temperature affect the resistance of a length of wire. These are discussed in Topic 9.7, Resistance and resistivity.

Summary questions

1 Sketch a graph of I against V for a metallic conductor at constant temperature and use it to describe Ohm's law. **(4 marks)**

2 The current in a resistor is 0.50 A and the p.d. across it is 5.2 V. Calculate its resistance. **(1 mark)**

3 Define the ohm, and express it in base units. **(2 marks)**

4 A length of wire is connected to a cell. The current in the wire is 80 mA and the p.d. across the wire is 2.4 V. Calculate the resistance of the wire. **(3 marks)**

5 A resistor has a resistance of 1.2 kΩ. In 3.0 minutes a charge of 54 C flows through the resistor. Calculate the p.d. across the resistor. **(2 marks)**

6 The current in a filament lamp is 1.5 A. It is operated for a time of 1.0 minutes. The charges flowing through the lamp transfer 500 J of energy to the lamp. Calculate the resistance of the lamp. **(3 marks)**

Tungsten filaments

Tungsten **filament lamps** were invented in Hungary in 1904, and rapidly replaced the carbon filaments used up to that point. Tungsten is a remarkable metal. It is also used in radiation shielding and to make penetrating tips for high-speed military projectiles, because it is incredibly dense (around 1.7 times denser than lead) and very robust. However, it is tungsten's melting point that makes it so useful for filament lamps. It has the highest melting point of all the elements, at over 3000°C.

Filament lamps behave in a complex way when the current in them increases. In order to understand what is going on we need to look carefully at how the current and potential difference are related.

Graphs of *I* against *V*

The current–potential difference characteristic (or simply *I–V* **characteristic**) for any electrical component shows the relationship between the electric current *I* in a component and the potential difference *V* across it.

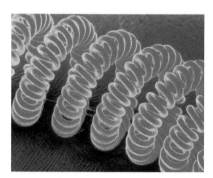

▲ **Figure 1** *The filament is designed so that, when there is a current in it, it gets so hot that it glows brightly*

Collecting data for an *I–V* characteristic

Two methods are commonly used to collect the data needed to plot an *I–V* characteristic (Figure 2). The methods are very similar and both involve a simple way to vary either the current in a component or the potential difference across it.

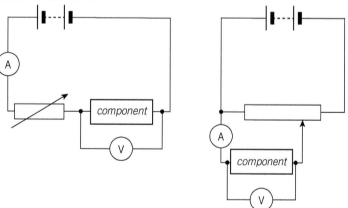

▲ **Figure 2** *Different techniques used to collect data to produce an I–V characteristic: the circuit on the left uses a variable resistor, the one on the right a potentiometer to provide different values of I and V respectively*

When investigating the *I–V* characteristic for a component it is important to look at whether or not the component behaves the same way if the current through it is in the opposite direction. This is normally achieved by reversing the polarity of the battery or the power supply.

Synoptic link

You will learn more about the potentiometer in Topic 10.6, Sensing circuits.

1. Identify the components in the circuit diagrams.
2. Describe how changing the resistance of the variable resistor in the first circuit affects the current in the component being tested.
3. Outline a method for producing an *I–V* characteristic for a filament lamp. You should include a circuit diagram, details of the measurements you would take, and the procedures you would use.

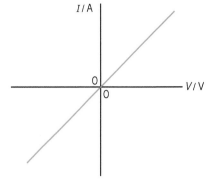

▲ **Figure 3** *The I–V characteristic for a fixed resistor is a straight line through the origin*

Resistors

Using either circuit in Figure 2 we can produce the *I–V* characteristic for a resistor. Fixed resistors are designed to ensure their resistance is constant, regardless of typical changes in temperature as the current in them varies.

Looking at the graph (Figure 3) we can draw a number of conclusions.

- The potential difference across the resistor is directly proportional to the current in the resistor ($V \propto I$). As a result
  - a resistor obeys Ohm's law, and so can be described as an **ohmic conductor**
  - the resistance of the resistor is constant.
- The resistor behaves in the same way regardless of the polarity.

Most wires and other metallic conductors behave in the same way as a resistor; they can be thought of as resistors with very low resistance.

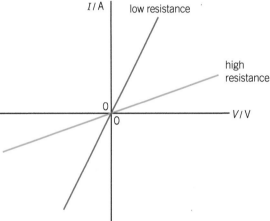

▲ **Figure 4** *Different resistors produce straight lines through the origin with different gradients – the shallower the line, the greater the resistance, which can be determined for each line using $R = \dfrac{V}{I}$ (the inverse of the gradient)*

Filament lamps

Repeating the same experiment for filament lamps produces very different results – see Figure 5.

Looking at the graph we can draw a number of conclusions.

- The potential difference across a filament lamp is not directly proportional to the current through the resistor. In other words
  - a filament lamp does not obey Ohm's law, and so can be described as a **non-ohmic component**
  - the resistance of the filament lamp is not constant.
- The filament lamp behaves in the same way regardless of the polarity.

The resistance of the filament increases as the p.d. across it increases. You can confirm this by determining $\dfrac{V}{I}$ at different points on the graph.

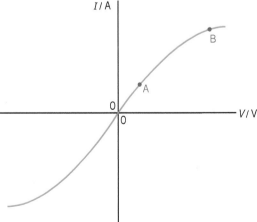

▲ **Figure 5** *The I–V characteristic for a filament lamp is most definitely not a straight line through the origin*

This increase in resistance is caused by the wire getting so hot that it glows. As the current increases so does the rate of flow of charge through the filament – more electrons per second pass through it, so more collisions occur between the electrons and the positive metal ions per second. When the electrons collide with the ions they transfer energy to the ions, causing the ions to vibrate more, or in other words to increase in temperature, and to collide with still more electrons.

Summary questions

1 The graph in Figure 6 shows the *I−V* characteristic for two different components.
   a State and explain which component obeys Ohm's law. *(2 marks)*
   b Calculate the resistance of each component at 4.0 V. *(2 marks)*

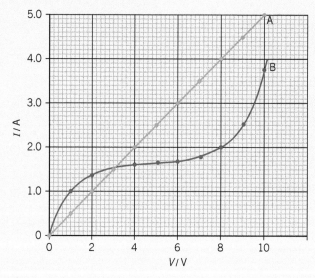

▲ **Figure 6**

2 Plot an *I−V* characteristic from the data in the table below. *(3 marks)*

I / A	−1.65	−1.63	−1.52	−1.34	−1.14	−0.82	0.00	0.81	1.12	1.36	1.53	1.62	1.64
V / V	−9.0	−7.5	−6.0	−4.5	−3.0	−1.5	0.0	1.5	3.0	4.5	6.0	7.5	9.0

3 Use your graph to calculate the resistance of the component at:
   a 0.50 A;   b 1.00 A;   c 1.60 A. *(3 marks)*

4 Identify the component used to produce the *I−V* characteristic in question 2 and explain the shape of the graph. *(4 marks)*

5 On the same set of axes, sketch a labelled *I−V* characteristic for a metallic conductor:
   a at room temperature; *(2 marks)*
   b at a much higher temperature. Explain the shape of the graphs. *(2 marks)*

9.6 Diodes

Specification reference: 4.2.3

Lighting the way

If you look closely at rear lights on modern cars you will see they are very different from the lights on older cars (Figure 1). Instead of a single filament lamp, the lights are made up of arrays of **light-emitting diodes** (LEDs). They emit light by a process very different from that in a filament lamp – electrical energy is transferred directly into light, and LEDs do not get hot, so they are much more efficient and draw much less power.

The diode

Diodes are everywhere, but you very rarely see them (Figure 2). They are a vital part of nearly every modern electronic circuit, from mobile phones to washing machines.

A diode only allows a current in one particular direction. This is the unique property that makes diodes so important for modern electronics. All the components you have seen so far are unaffected by the direction of the current – a filament lamp works equally well regardless of which way round you connect its terminals. Not so for a diode.

The light-emitting diode

Some diodes are made of a material that emits light when they conduct. However, unlike most light sources, these light-emitting diodes (LEDs) emit light of a single specific wavelength (see Topic 11.2, Wave properties).

As LEDs are so efficient and take very little energy to run, they are sometimes used as simple indicators to show the direction of current through a particular part of a circuit. They light up when there is current in them, showing that that part of the circuit is live (Figure 5).

Learning outcomes

Demonstrate knowledge, understanding, and application of:

→ *I–V* characteristics of diodes and light-emitting diodes.

▲ **Figure 1** *Some modern cars have arrays of LEDs in place of filament lamps – LEDs are much more efficient and last much longer*

▲ **Figure 2** *Diodes don't appear to be anything special, but they are: unlike the components you have seen so far, a diode is made from a semiconductor*

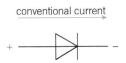

conventional current

▲ **Figure 3** *The circuit symbol for a diode hints at its key feature: it allows current in one direction only*

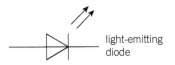

light-emitting diode

▲ **Figure 4** *Circuit symbol for a light-emitting diode*

I–V characteristic for a diode

Repeating the experiment described in Topic 9.5 to collect data for a diode produces a very different *I–V* characteristic from those for a resistor or a filament lamp (Figure 6).

Looking at the graph for a semiconducting diode we can draw a number of conclusions.

Synoptic link

Wavelengths of light are covered in Topic 11.2, Wave properties.

- The potential difference across a diode (or LED) is not directly proportional to the current through it. This means
  - a diode does not obey Ohm's law, and so can be described as a non-Ohmic component
  - the resistance of the diode is not constant.
- The diode's behaviour depends on the polarity.

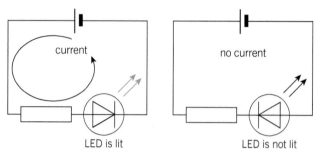

▲ **Figure 5** *An LED will only light up if the polarity allows a current to pass through it*

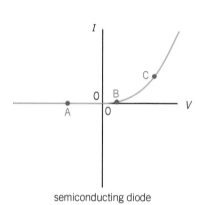

semiconducting diode

▲ **Figure 6** *The I–V characteristic for a diode*

At **A** in Figure 6 the resistance of the diode is very high – infinite for practical purposes. With the p.d. in this reverse direction, the diode does not conduct. At **B**, as the p.d. increases, the resistance gradually starts to drop. For a silicon diode this happens at around 0.7 V (the **threshold p.d.**). Above this value, the resistance drops sharply for every small increase in p.d. (at **C**). Above this point the diode has very little resistance.

Different LEDs have different values for their threshold p.d., related to the colour of the light they emit.

Summary questions

1 The circuit in Figure 7 was used to collect data in order to produce an *I–V* characteristic for a diode. Identify two errors in the circuit. *(2 marks)*

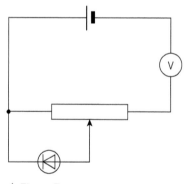

▲ **Figure 7**

2 State and explain which of the lamps in the circuit in Figure 8 would be lit. *(3 marks)*

3 Plot the *I–V* characteristic for a diode using the data in the table below. *(4 marks)*

I/mA	0.00	0.00	0.00	0.00	0.00	0.02	0.35	1.50	29.20	124.10
V/mV	−750	−500	−250	0	200	400	600	650	700	750

4 Use your graph to determine the resistance of the diode at:
   a −300 mV;   b 625 mV;   c 0.72 V. *(3 marks)*

5 A diode and a fixed resistor are connected together in a circuit. Sketch the *I–V* characteristic when the diode and resistor are connected:
   a in series;   b in parallel. *(5 marks)*

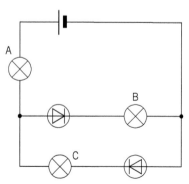

▲ **Figure 8**

9.7 Resistance and resistivity

Specification reference: 4.2.4

Not all wires are the same

Gold is an excellent conductor. It has a number density almost as high as copper and is much less reactive, so it corrodes very little over time. It is used in high-energy electrical applications, and for the fine wires that connect the pins on computer chips to integrated circuit boards. Some expensive electrical cables use gold contacts, which have a much lower resistance than the tin in cheaper cables, but many audiovisual experts consider this simply a marketing ploy and a pointless expense.

A gold contact and a similar tin contact have different resistances. You already know that the temperature of a wire affects its resistance (Topic 9.4). Material is one of three factors beside temperature that affects the resistance.

- The material of the wire.
- The length of the wire L.
- The cross-sectional area of the wire A.

Resistance and material: resistivity

The term 'resistance' refers only to a specific component. The term **'resistivity'** is used to describe the electrical property of a material. For example, different components made from copper may have different resistances: copper wires may have different resistances as their lengths or cross-sectional areas differ, but copper has a unique resistivity.

Resistance and length

For any given current, increasing the length of the wire will increase the p.d. across it. Doubling the length doubles the p.d., so the resistance must have doubled.

The resistance R of a wire is directly proportional to its length L.

$$R \propto L$$

Resistance and cross-sectional area

When the cross-sectional area of the wire increases, the opposite happens – the resistance drops. Wires with a greater cross-sectional area have a lower resistance.

For any given p.d., doubling the cross-sectional area will double the current in the wire, so the resistance must have halved.

The resistance R of a wire is inversely proportional to its cross-sectional area A.

$$R \propto \frac{1}{A}$$

Learning outcomes

Demonstrate knowledge, understanding, and application of:

→ resistivity of a material; the equation $R = \frac{\rho L}{A}$

→ the variation of resistivity of metals and semiconductors with temperature.

▲ **Figure 1** *The resistance of a material depends on a number of factors*

Calculating resistance from resistivity

We can combine the two relationships above to give

$$R \propto \frac{L}{A}$$

The resistivity ρ of a particular material at a given temperature is the constant of proportionality in this equation.

$$R = \frac{\rho L}{A}$$

Resistivity has the unit ohm metre ($\Omega\,m$). This equation may be used for a known constant temperature. As you will see later, the resistivity of a material is affected by its temperature too.

 Worked example: Resistance of a nichrome wire

Nichrome has a resistivity of $1.50 \times 10^{-6}\,\Omega\,m$. Calculate the resistance of a wire made from nichrome with a length of 80 cm and a radius of $2.0 \times 10^{-4}\,m$.

Step 1: Identify the correct equation to calculate the resistance.

$$R = \frac{\rho L}{A}$$

Step 2: Determine the cross-sectional area of the wire in m^2 using $A = \pi r^2$.

$$A = \pi \times (2.0 \times 10^{-4})^2 = 1.26 \times 10^{-7}\,m^2$$

Step 3: Substitute known values in SI units into the equation for resistance.

$$R = \frac{\rho L}{A} = \frac{1.50 \times 10^{-6} \times 0.80}{1.26 \times 10^{-7}}$$

Calculating the answer to an appropriate number of significant figures gives

$$R = 9.5\,\Omega \text{ (2 s.f.)}$$

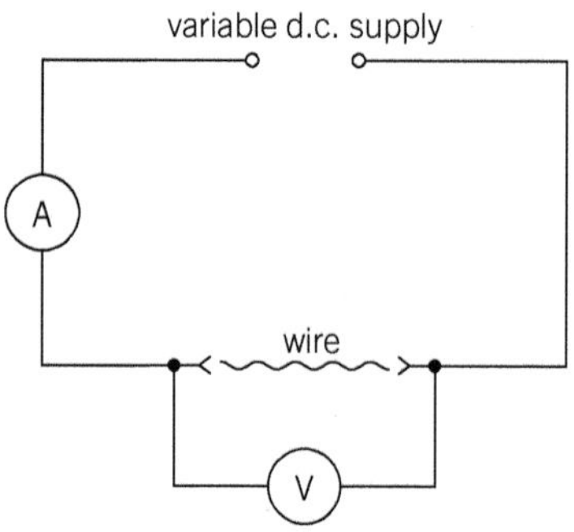

variable d.c. supply

A

wire

V

▲ **Figure 2** *Circuit for determining resistivity*

Defining resistivity

We can define the resistivity of a material by rearranging the equation above to make the resistivity the subject.

$$\rho = \frac{RA}{L}$$

The resistivity of a material at a given temperature is the product of the resistance of a component made of the material and its cross-sectional area, divided by its length.

The resistivity of a material varies with temperature in the same way as the resistance of most components varies with temperature. As the material gets hotter, its resistivity increases (see Topic 9.4).

Determining ρ

A simple experiment can be carried out to determine the resistivity of a material, in this case a piece of wire. Amongst the different wires that might be used are copper and some common alloys including constantan and nichrome.

A typical way to determine the resistivity is to investigate how the resistance of a wire varies with its length. Using the circuit illustrated in Figure 2 we can obtain values for the p.d. across different lengths of wire. If the current in each wire is measured too, we can use $R = \dfrac{V}{I}$ to calculate the resistance for each length.

As $R \propto L$, a graph of R against L is a straight line through the origin. By considering the general equation for a straight line $(y = mx + c)$ and the equation $R = \dfrac{\rho L}{A}$ we see that the gradient of this graph is $\dfrac{\rho}{A}$.

The resistivity can be determined by multiplying the gradient of the graph by the cross-sectional area of the wire.

Conductors, semiconductors, and insulators

Different materials have widely different values for resistivity. Good conductors like metals have a resistivity of the order of $10^{-8}\,\Omega\,\text{m}$, insulators have a value of the order of $10^{16}\,\Omega\,\text{m}$, and semiconductors have values in between these extremes. Table 1 shows the resistivity values for some materials at 20 °C.

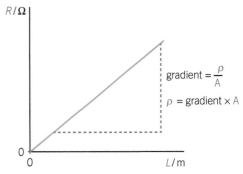

gradient $= \dfrac{\rho}{A}$

$\rho = $ gradient $\times A$

▲ **Figure 3** *A graph of the resistance of a wire against its length produces a straight line through the origin, with a gradient equal to $\dfrac{\rho}{A}$*

Superconductivity

A strange quantum effect occurs when some materials are cooled. As they get colder their resistivity drops, as expected, but then at a critical temperature the resistivity suddenly falls to zero. Not just very low, but truly zero [Figure 4].

This is **superconductivity**. Any components made from a superconducting material have no electrical resistance. No energy is lost when there is a current in the material, and so huge amounts of charge can pass through a superconductor without it even getting warm. The LHC at CERN uses superconducting wires carrying up to 20 000 A to produce exceptionally strong magnetic fields. The only problem is that the wires become superconducting at around 4 K (−269°C, even colder than deep space). This is a typical temperature for superconductors. A few high-temperature superconductors have been developed that become superconducting around 80 K, but this is still around −190°C. The race is on to develop room-temperature superconductors.

1 Compare how the resistance of a normal metallic wire and a superconductor changes with temperature.
2 Outline some potential advantages of room-temperature superconductors.

Synoptic link

You can find more information about the gradient of straight-line graphs in Appendix A2, Recording results.

▼ **Table 1** *Some materials and their resistivity values*

Material	Type	ρ at 20°C / Ω m
silver	conductor	1.6×10^{-8}
copper	conductor	1.7×10^{-8}
iron	conductor	1.0×10^{-7}
carbon	semiconductor	3.5×10^{-5}
silicon	semiconductor	640
glass	insulator	$10^{10} - 10^{14}$
quartz	insulator	7.0×10^{17}

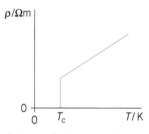

▲ **Figure 4** *Below a critical temperature the resistivity of some materials falls to zero*

▲ **Figure 5** *Superconductivity has another strange effect: a superconducting material does not allow magnetic fields to pass through it (the Meissner effect), which results in the permanent magnet hovering perfectly over the superconductor*

▼ **Table 2** *I–V data for a nichrome wire*

Length / m	p.d. / V	Current / A
0.100	0.31	0.50
0.200	0.61	0.50
0.300	0.92	0.50
0.400	1.23	0.50
0.500	1.53	0.50
0.600	1.84	0.50
0.700	2.14	0.50
0.800	2.45	0.50
0.900	2.76	0.50
1.000	3.06	0.50

Summary questions

1 Describe the difference between resistance and resistivity. *(2 marks)*

2 Calculate the resistance of a copper wire of length 1.0 m and cross-sectional area of $3.32 \times 10^{-6}\,\text{m}^2$. *(3 marks)*

3 Explain the effect of changing the temperature on the resistivity of a metal. *(3 marks)*

4 Calculate the resistivity of a piece of wire that has a resistance of 170 Ω when it has a length of 12 m and a radius of 3.0 mm. *(5 marks)*

5 A wire has resistance of 8.0 W. State and explain the effect on the resistance of a piece of wire if:
  a the length of the wire doubles (all other factors remain the same); *(1 mark)*
  b the cross-sectional area of the wire doubles (all other factors remain the same); *(2 marks)*
  c the radius of the wire halves (the length remains constant); *(2 marks)*
  d the volume remains constant but the length of the wire doubles. *(3 marks)*

6 Table 2 lists data collected for a nichrome wire. The diameter of the wire was measured in three places using a micrometer and an average value of 0.46 mm was calculated from these results.
  a Explain why the diameter of the wire should be measured in several places along the wire. *(2 marks)*
  b i Use the data in the table to plot a graph of the resistance of the wire against length. *(4 marks)*
     ii Use the graph to determine the resistivity of the wire. *(4 marks)*
  c Describe how the current is kept constant and why it is important to do so in this experiment. *(3 marks)*

7 Outline an experiment you could do using several different thicknesses of wire of set length to determine the resistivity of the wire. *(4 marks)*

9.8 The thermistor

Specification reference: 4.2.3, 4.2.4

Not too hot, not too cold

Babies born prematurely need extra care. They are kept inside an incubator (Figure 1) where everything from humidity to CO_2 concentration is precisely monitored to help the baby develop. Temperature-sensing components called **thermistors** are used to ensure the temperature inside the incubator is perfect, automatically alerting medical staff if the temperature falls outside of a predetermined range.

Temperature and number density

We have already seen the effect of increasing temperature on the resistance of a metal wire. The hotter the wire becomes, the more the positive ions vibrate, and so the greater the resistance becomes. But a few materials behave differently. Some semiconductor components have a **negative temperature coefficient**, meaning that their resistance drops as the temperature increases. This sounds odd, but it is nothing new: it was first observed by Michael Faraday in 1833.

The effect can be explained in terms of the number density of the charge carriers within the material from which the component is made. In some semiconductors, as the temperature increases, the number density of the charge carriers also increases.

The thermistor

A thermistor (Figures 2 and 3) is an electrical component made from a semiconductor with a negative temperature coefficient. As the temperature of the thermistor increases, its resistance drops.

The change in resistance is often dramatic (Figure 4). This makes thermistors particularly useful in temperature-sensing circuits. A small change in temperature can be detected by monitoring the resistance of the thermistor.

Thermistors are used:

- in simple thermometers
- in thermostats to control heating and air-conditioning units
- to monitor the temperature of components inside electrical devices like computers and smartphones so that they can power down before overheating damages them
- to measure temperature in a wide variety of electrical devices like toasters, kettles, fridges, freezers, and hair dryers
- to monitor engine temperatures to ensure the engine does not overheat.

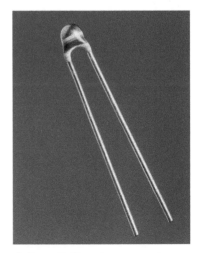

▲ **Figure 1** *Thermistors are sensitive enough to precisely monitor the temperature of incubators for premature babies*

thermistor

▲ **Figure 2** *The circuit symbol for a thermistor (the term 'thermistor' comes from combining 'thermal' and 'resistor')*

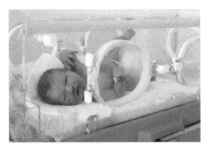

▲ **Figure 3** *Thermistors come in different shapes and sizes, but can be small enough to fit inside most electronic devices*

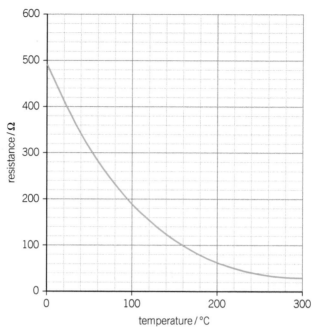

▲ **Figure 4** *The drop in resistance of a thermistor as the temperature increases is not linear – there is often a significant decrease, followed by a more gradual drop as it gets hotter and hotter*

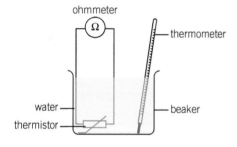

▲ **Figure 5** *Use a water bath to control the temperature of a thermistor and an ohmmeter for a quick and simple recording of the resistance in an experiment to investigate the effect of temperature on resistance*

Thermistor experiment

A simple investigation into how the resistance of a thermistor changes with temperature change be carried out using an ohmmeter and a water bath (Figure 5). Alternatively, an ammeter and a voltmeter can be used to measure current in the thermistor and p.d. across it at different temperatures. The resistance can then be calculated with $R = \dfrac{V}{I}$.

The results from this experiment may be used in the choice of a thermistor for a particular application. A thermistor is selected to ensure it provides the greatest change in resistance over the range of temperatures in which it will operate. In Figure 4, this change takes place at 20–50°C, making that particular thermistor perfect for incubators, but less useful for monitoring the temperature inside a car engine, typically around 100°C.

I–V characteristics of thermistors

Like most semiconducting components, thermistors are non-ohmic. The *I–V* characteristic has some features similar to that of a filament lamp, and one crucial difference (Figure 6). With a filament lamp, as the current increases, electrons transfer energy to the positive ions, which raises the temperature. This causes an increase in resistance (see Topic 9.4).

With a thermistor, like a lamp, as the current increases the temperature increases. But unlike the lamp, this temperature increase leads to a drop in resistance because the number density of charge carriers increases. This may be confirmed by comparing $R = \dfrac{V}{I}$ at various points on the graph.

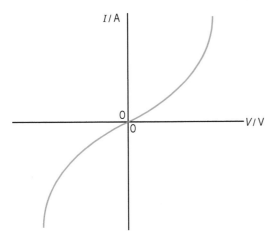

▲ **Figure 6** *The I–V characteristic for a thermistor is not a straight line, so its resistance must be changing*

An increase in temperature leads to an increase in the number density of the free electrons. This means that the resistance of the thermistor decreases as its temperature increases.

Summary questions

1 A thermistor is used in a thermostat in order to maintain a constant temperature inside a lorry delivering frozen goods.
   a Describe how any changes in the temperature inside the lorry would affect the resistance of the thermistor. *(1 mark)*
   b Suggest how the thermistor might be used to ensure the produce transported by the lorry is kept at a constant temperature. *(2 marks)*

2 Sketch a labelled graph to show how the resistance of a typical thermistor used inside an oven varies with temperature. You should include typical values for the temperature and an explanation of why you selected these values. *(3 marks)*

3 Table 1 contains data collected using a thermistor in a water bath. The thermistor was connected to a 1.5 V cell. Plot a graph of resistance against temperature for the thermistor. *(5 marks)*

   ▼ **Table 1** *Variation of current with changing temperature*

Temperature /°C	−20.0	−10.0	0.0	10.0	20.0	30.0
Current in the thermistor /mA	10.7	17.3	32.5	50.2	82.5	121.2

4 Use your graph from question 3 to determine the resistance of the thermistor at: a −5.0°C;   b 25°C. *(2 marks)*

5 The thermistor used in question 3 is now used to measure the temperature in a water heater. The current in the thermistor was 45.2 mA and the p.d. across it was 1.03 V. Determine the temperature of the water heater. *(3 marks)*

9.9 The LDR

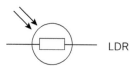

▲ **Figure 1** *The circuit symbol for a light-dependent resistor*

▲ **Figure 2** *LDRs are often made of cadmium sulfide formed into a flat disc on the surface of the LDR*

Definitely not, just, cricket

A **light-dependent resistor** (LDR) is an essential part of many modern sports, including cricket and tennis. LDRs are small electrical components which change their resistance depending on the light intensity. If it gets too dark, play is postponed until the light intensity increases, maybe even to the following day.

As well as sports, LDRs are used in automatic street lights, the brightness meters in smartphones and laptops, and even in some space telescopes.

The light-dependent resistor

With the thermistor we saw how the resistance of some semiconductors varies in an unusual way. An LDR is another component that makes use of the unusual properties of a different type of semiconductor.

A typical LDR is made from a semiconductor in which the number density of charge carriers changes depending on the intensity of the incident light. In dark conditions the LDR has a very high resistance. The number density of the free electrons inside the semiconductor is very low, so the resistance is very high (often into MΩ). When light shines onto an LDR, the number density of the charge carriers increases dramatically, leading to a rapid decrease in the resistance of this component.

Investigating LDRs

Figure 4 shows an experiment to investigate how the resistance of an LDR varies with distance from a constant light source (like a simple filament lamp). A narrow tube made of black cardboard placed around the LDR will greatly reduce the effect of other background sources of light.

The results give a calibration graph that relates the resistance of the LDR to the light intensity. This graph can then be used to determine light intensity from different sources.

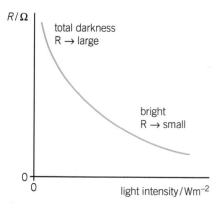

▲ **Figure 3** *The resistance of an LDR can vary from millions of ohms to just a few, depending on the light intensity*

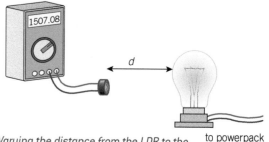

▲ **Figure 4** *Varying the distance from the LDR to the lamp has the effect of changing the light intensity received by the LDR. The further it is away from the filament lamp, the higher its resistance becomes.*

156

Infrared astronomy and LDRs

Some LDRs are particularly sensitive to infrared radiation (see Topic 11.6, Electromagnetic waves), so are useful as sensors for the very dim infrared received from space.

Our eyes cannot detect infrared wavelengths, so using infrared telescopes enables us to discover more about the universe than we ever could with our naked eyes. Astrophysicists can even see *inside* dense clouds of gas and dust with infrared, which passes through these clouds, whereas visible light does not. This allows us to see objects like new stars forming in stellar nurseries and to peer into the very centre of our galaxy.

But infrared astronomy has limitations. Infrared is absorbed by water in the atmosphere, so the European Space Agency has recently launched an infrared telescope into space, the Herschel Space Observatory.

1 Explain why most infrared telescopes are positioned on mountain tops.
2 Suggest how the changing resistance of an LDR might be used to detect extrasolar planets as they pass in front of distant stars.

▲ **Figure 5** *This image of the Helix Nebula, made with data collected from infrared telescopes using LDRs, shows the fine detail in the clouds of hydrogen and helium around the centre*

Synoptic link

You can read more about intensity in Topic 11.5, Intensity.

Summary questions

1 LDRs are used in smartphones to measure the ambient light level and automatically adjust the screen brightness when necessary. Describe how the resistance of the LDR changes in different conditions. *(2 marks)*

2 The data in Table 1 was collected using an LDR and an ohmmeter in different conditions. Identify two likely errors in the table. *(2 marks)*

▼ **Table 1**

Conditions	complete darkness	dim light	normal daylight	very bright
Resistance of LED	400	2000	18 000	1 000 000

3 Compare a resistor, a thermistor, and an LDR. *(4 marks)*

4 The graph in Figure 6 is a logarithmic plot of the resistance of an LDR against light intensity. Use the graph to estimate the relative light intensity when the resistance is: a   1000 Ω;   b   50 kΩ. *(2 marks)*

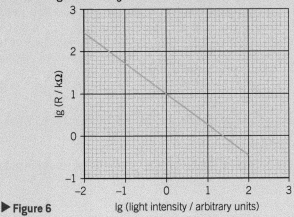

▶ **Figure 6**

9.10 Electrical energy and power

Specification reference: 4.2.5

▲ **Figure 1** *The Three Gorges Dam*

Ten million kettles

The Three Gorges Dam, the world's largest power plant, is a hydroelectric power station that spans the Yangtze river in eastern China. It has an output of 22.5 GW, in other words 22 500 000 000 J s^{-1}, or enough to boil around 10 million average kettles at once.

The electricity generated by the dam is distributed to consumers through the Chinese grid. Electrical energy is relatively easy to transfer from one place to another and from one form to another, and this is one of the reasons it has become essential to us.

Transferring energy

Electric circuits are often used to transfer energy from one place to another, often from a power source like a cell or power supply to a component that transfers this energy into another form. A filament lamp transfers electrical energy to heat and light, whereas a loudspeaker transfers electrical energy to sound and heat.

Whenever there is a current in a component, energy is transferred from the power source to that component. Open a switch, or disconnect the power source, and the current falls to zero. No energy is transferred and the circuit is off.

Electrical power

The rate of energy transfer by each electrical component is called the electrical power (or just power). This depends on the current I in the component, measured in amperes (A), and the p.d. V across it, in volts (V).

The equation for electrical power P (in watts, W) is

$$\text{electrical power} = \text{p.d.} \times \text{current}$$

$$P = VI$$

 Worked example: Charging a tablet

A typical tablet charger has a power of 12 W compared with 6.0 W for a mobile phone charger. Calculate the current drawn by a 12 W charger when the p.d. is 5.0 V.

Step 1: Identify the correct equation.

$$P = VI$$

Rearrange the equation to make I the subject.

$$I = \frac{P}{V}$$

Step 2: Substitute in known values and calculate the value for the current.

$$I = \frac{12}{5.0}$$

$$I = 2.4 \, \text{A}$$

Additional equations for power

We can combine the equation above with $V = IR$ to give two additional equations for power.

$$P = VI$$
$$= (IR) \times I$$
$$= I^2R$$

Rearranging $V = IR$ to make I the subject gives $I = \dfrac{V}{R}$

Substituting this into $P = VI$ gives

$$P = V \times \frac{V}{R}$$
$$= \frac{V^2}{R}$$

Each version of the power equation is missing one of V, I, or R.

$$P = VI \qquad P = I^2R \qquad P = \frac{V^2}{R}$$

▲ **Figure 2** *Heating element of a kettle*

 ## Worked example: Kettle power I

The heating element in a kettle has a small resistance, typically $20.0\,\Omega$, in order to draw the large current needed to heat the element. For a current of $12\,A$, calculate the rate of energy transfer.

Step 1: Identify the correct equation to calculate the rate of energy transfer (the power).

$$P = I^2R$$

Step 2: Substitute the known values in SI units into the equation and calculate the power.

$$P = 12^2 \times 20.0 = 2880$$

Express the answer to an appropriate number of significant figures.

$$P = 2900\,W \ (2\ \text{s.f.})$$

Two significant figures are appropriate in this case as the value of the current ($12\,A$) is to two significant figures.

Deriving $P = VI$

Our equation for electrical power can be derived from the generic equation

$$P = \frac{W}{t}$$

Using our defining equation for p.d., rearrange to make W the subject.

$$V = \frac{W}{Q}$$

$$W = VQ$$

Substituting work done into our generic equation for power gives

$$P = \frac{VQ}{t}$$

Synoptic link

If you need to review the equation for power, look at Topic 5.4, Power and efficiency.

However, $\frac{Q}{t}$ is equal to the current I. Therefore

$$P = VI$$

Calculating energy transferred

The energy transferred in a given time can be determined by combining our general equation for power and our equation for electrical power.

$$P = \frac{W}{t} \text{ is rearranged to give } W = Pt$$

Substituting in $P = VI$ gives

$$W = VIt$$

where W is the energy transferred in joules, V is the p.d. in volts, I is the current in amperes and t is the time in seconds.

 Worked example: Kettle power II

Calculate the energy transferred into heat in 2.0 minutes by the kettle in the first worked example.

Step 1: Identify the correct equation to calculate energy transfer.

$P = \frac{W}{t}$ Rearrange to make the energy transferred the subject. $W = Pt$

Step 2: Substitute the values in SI units into the equation and calculate the value for the energy transferred.

$$W = 2900 \times 120 = 3.48 \times 10^5 \text{ J}$$

$$W = 3.5 \times 10^5 \text{ J (2 s.f.)}$$

Summary questions

1 Using the appropriate circuit symbols outline a simple experiment that could be used to determine the power of a filament lamp (you should include a circuit diagram in your answer). *(2 marks)*

2 Calculate the rate of energy transfer from a resistor with a current of 5.0 A and a p.d. of 8.0 V. *(2 marks)*

3 A 1.2 kW heater has a p.d. of 20 V when working normally. Calculate:
   a the current in the heater; *(2 marks)*
   b the energy transferred in one hour. *(2 marks)*

4 State and explain the effect on the rate of energy transfer for a component when:
   a its resistance doubles (the current in the component remains unchanged);
   b the current in the component doubles (the resistance is unchanged). *(4 marks)*

5 Using base units, show that one watt is equivalent to one volt amp (1 W = 1 V A). *(4 marks)*

Electricity in the home

What would our homes be like without electricity? Most of the devices we use at home require electricity. They transfer electrical energy into other useful forms for everything from heating and cooking to entertaining ourselves. Think about all the electrical devices used in your home in the last week, and what you would have to do without them.

However, there is a financial cost, and we need to pay to keep all of our devices running.

Paying for energy

We pay electricity companies for the total amount of electrical energy they transfer to our homes. By law, each home contains an **electricity meter** that accurately records the transfer of energy from the National Grid to the house (Figure 1). All the electricity supplied to the house passes through the meter.

The energy transferred to each individual electrical device, and so how much it costs to run, depends on two factors (Figure 1):

- the power of the device
- how long the device is used for.

These two factors must be considered together when calculating the energy transferred by the device. For example, a powerful microwave oven may be cheaper to run than a less powerful one, despite transferring more energy per second, as it may take less time to cook food.

The defining equation for power is

$$\text{power} = \frac{\text{energy transferred}}{\text{time taken}} \qquad P = \frac{W}{t}$$

Rearranging this equation to determine the energy transferred by an electrical device gives

$$W = Pt$$

The kilowatt-hour

The SI unit for energy is the joule, but one joule is a tiny amount on the scale of the energy transferred to our homes. It takes at least 100 kJ to boil just one cup of water. Electricity bills therefore use a derived unit, the **kilowatt-hour** (kWh), defined as the energy transferred by a device with a power of 1 kW operating for a time of 1 hour.

From its definition 1 kWh is equivalent to 3.6 MJ.

Our equation for energy transferred therefore has two versions, depending on the units used.

▲ **Figure 1** *A typical electricity meter, which is calibrated to ensure it accurately measures the electrical energy transferred to the home*

▲ **Figure 2** *A label on the base normally indicates the power rating of a device, meaning the energy transferred to the device each second*

Worked example: Calculating energy transferred in kW h

A 1450 W dishwasher is used for 15.0 minutes. Calculate the energy transferred in kW h.

Step 1: Identify the correct equation to calculate energy transferred.

$$W = Pt$$

Step 2: Substitute in the values, paying close attention to the units.

$$W = 1.450 \times 0.250$$

$$= 0.363 \, \text{kW h (3 s.f.)}$$

SI units: energy transferred (J) = power of device (W) × time for which the device is used (s)

kWh: energy transferred (kW h) = power of device (kW) × time for which the device is used (h)

Electricity Statement

Electricity Readings					
Meter Serial no.	Read Date	Read Type	Read	Last Read	Units Used
		Removal	28619	28170	449
		Smart	1749	0	1749
Total units					**2198 kWh**

Electricity Charges 05 Jan 2010 - 30 Jun 2010				
Electricity supply standing charge	177 days	22.0p per day	£	38.94
Electricity total unit charge	2198.0 kWh	8.085p per kWh	£	177.71
Total supply charges			**£**	**216.65**
VAT @5.00%			£	10.83
Total cost of electricity			**£**	**227.48**

▲ **Figure 3** *An electricity bill shows how much energy has been transferred to each home in a certain time (usually each quarter) in kWh*

The cost of each kW h of electrical energy (sometimes simply referred to as a 'unit') varies between operating companies. It depends on the particular tariff from the electricity company, and even in some cases on the time of day the energy is transferred. Typical costs are around 6–15 p per kW h.

Study tip

Make sure that you are confident about the meaning of 'power'. Phrases like 'the rate of energy transfer', 'the energy transferred per second', 'the rate of work done' are all the same thing, just written differently.

Study tip

It's always important to look carefully at the units used in calculations. It's even more important where there might be different units used depending on the context, as with joules and kilowatt-hours.

Summary questions

1. Calculate the energy transferred in joules by a 60 W television in 2.0 hours. *(2 marks)*

2. Two readings are taken from an electricity meter. Feb: 0034512 – Jun: 0035387. If each unit costs 12 p calculate the cost of the electrical energy transferred to the home during this time. *(2 marks)*

3. Show that one kilowatt-hour is equal to 3 600 000 J. *(3 marks)*

4. Calculate the energy transferred in kW h by a 9000 W power shower used for 15 minutes:
   a in J;   b in kWh. *(4 marks)*

5. If one kWh costs 11.2 p, calculate the cost of leaving a 60 W lamp on continuously for 5.0 weeks. *(3 marks)*

6. A household decides to swap all their filament lamps for energy-efficient LEDs. They have a total of 18 lamps each with a power of 100 W. The replacement LEDs have a power of 15 W. If each lamp is used for an average of 2.0 hours during the day, calculate the difference in energy transferred over one year. *(5 marks)*

Practice questions

1 a With the help of a sketch graph, describe and explain how the resistance of a negative temperature coefficient (NTC) thermistor is affected by its temperature.
(*3 marks*)

b Figure 1 shows a circuit for a simple light-meter designed by a student.

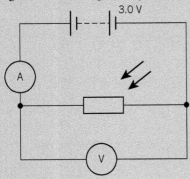

▲ **Figure 1**

The battery and the ammeter both have negligible internal resistances. Discuss how each meter responds as the intensity of light incident on the LDR is increased.
(*4 marks*)

2 a State one similarity and one difference between potential difference and electromotive force (e.m.f.). (*2 marks*)

b The e.m.f of a cell is 1.5 V. The energy transformed by the cell is 100 J. Calculate the total charge flowing through the cell.
(*2 marks*)

c According to a student, an electron accelerated from rest can be made to travel at a speed greater than $10 \, \text{km s}^{-1}$ by using the cell from (b) connected across two electrodes. With the help of a calculation, show whether or not this suggestion is correct. (*4 marks*)

3 a Show that the SI unit for resistivity is $\Omega \, \text{m}$. (*2 marks*)

b Figure 2 shows a cube made from a material of resistivity $5.0 \times 10^{-2} \, \Omega \, \text{m}$.

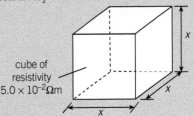

cube of resistivity $5.0 \times 10^{-2} \, \Omega \text{m}$

▲ **Figure 2**

The resistance of the cube across any of its two opposite faces is $5.0 \, \Omega$. The length of each edge of the cube is x. Calculate the length x. (*3 marks*)

c A student conducts an experiment to determine the resistivity of a metal in the form of a wire. Figure 2 shows a graph of resistance R against its length L.

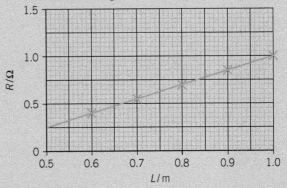

▲ **Figure 3**

(i) Explain how the student may have determined the resistance of each length of the wire using meters. Assume an ohmmeter is not available. (*3 marks*)

(ii) The wire has cross-sectional area of $7.8 \times 10^{-7} \, \text{m}^2$. Use Figure 3 to determine the resistivity of the metal.
(*4 marks*)

4 The power of a 230 V mains filament lamp is 40 W.

a Define *power*. (*1 mark*)

b The lamp is connected to the 230 V supply. Calculate

(i) the current I in the filament,
(*2 marks*)

(ii) the resistance R of the filament.
(*1 mark*)

c The cross-sectional area of the filament is $3.0 \times 10^{-8} \text{m}^2$. The resistivity of the filament when the lamp is lit is $7.0 \times 10^{-5} \Omega \text{m}$. Use your answer to **(b)(ii)** to calculate the length L of the filament wire. *(3 marks)*

May 2012 G482

5 Figure 4 shows the I–V characteristic of a slice of semiconducting material.

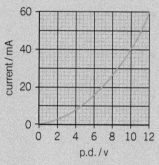

▲ **Figure 4**

a **(i)** Define *resistance*. *(1 mark)*

(ii) Show that the resistance of the slice is about 250Ω when there is a current of 40 mA in it. *(2 marks)*

b The dimensions of the slice are shown in Figure 5.

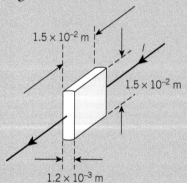

▲ **Figure 5**

Calculate the resistivity ρ of the semiconducting material when there is a current I of 40 mA in the slice. *(3 marks)*

c Explain how the I–V characteristic shows that the resistivity of the semiconducting material decreases with increasing temperature. *(4 marks)*

June 2013 G482

6 A student connects a component across a battery of negligible internal resistance. Figure 6 shows the variation of the current I in this component with time t from the

moment the component is connected to the battery.

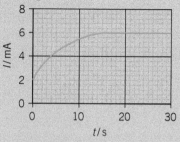

▲ **Figure 6**

The student suggests that the component must be a negative temperature coefficient (NTC) thermistor.

a Calculate the ratio
$$\frac{\text{power dissipated at } t = 30 \, \text{s}}{\text{power dissipated at } t = 0}.$$
(3 marks)

b Explain why the current changes as shown in the graph of Figure 6 *(4 marks)*

7 This question is about the rigid copper bars which carry the very large currents generated in a power station to the transformers. Figure 7 shows such a copper bar.

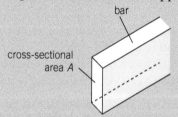

▲ **Figure 7**

a Write down a suitable word equation to define the *resistivity* of a material.
(1 mark)

b **(i)** The cross-sectional area A of the bar is $6.4 \times 10^{-3} \text{m}^2$. Calculate the resistance of a 1.0 m length of the bar. The resistivity of copper is $1.7 \times 10^{-8} \Omega \text{m}$. *(2 marks)*

(ii) The bar carries a constant current of 8000 A. Calculate the power dissipated as heat along a 1.0 m length of it. *(3 marks)*

(iii) The bar is 9.0 m long. Estimate the total energy in kW h lost from the bar in one day. *(2 marks)*

(iv) Calculate the cost per day of operating the copper bar. The cost of 1 kW h is 15 p. *(1 mark)*

Jan 2012 G482

8 a Define electrical resistivity. (*1 mark*)

b Figure 8 shows the 'lead' of a pencil.

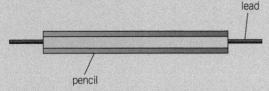

▲ **Figure 8**

Describe how you can determine the resistivity of the material of the 'lead' in the laboratory.

In your description pay particular attention to

- how the circuit is connected
- what measurements are taken
- how the data is analysed. (*5 marks*)

c Figure 9 shows a glass tube with some conducting paint.

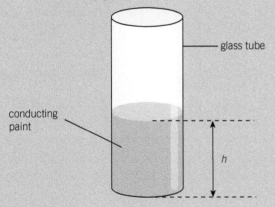

▲ **Figure 9**

The volume of the paint is $8.0 \times 10^{-6}\,m^3$. The internal cross-sectional area of the base of the tube is $1.2\,cm^2$.

  (i) Calculate the height h of the paint column. (*2 marks*)

  (ii) The resistivity of the paint is $5.2 \times 10^{-4}\,\Omega\,m$.

     Calculate the resistance of the paint column. (*3 marks*)

  (iii) State and explain how your answer to **(c)(ii)** would change if the same volume of paint is poured into another glass tube with double the internal diameter. (*3 marks*)

9 Figure 10 shows an electrical cable consisting of bare copper wires encased in plastic insulation.

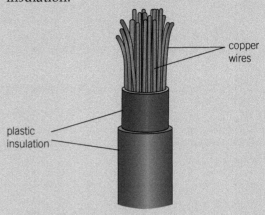

▲ **Figure 10**

a A particular cable contains 26 copper wires and is 12.0 m long. The radius of each copper wire is $3.50 \times 10^{-5}\,m$. The resistivity of copper is $1.70 \times 10^{-8}\,\Omega\,m$.

  (i) Show that the resistance of a single copper wire is about $53\,\Omega$. (*3 marks*)

  (ii) Explain why the resistance of the electrical cable is about $2\,\Omega$. (*1 mark*)

b Figure 11 shows two electrical cables used to connect a power supply to a lamp. Each cable has length 12.0 m and is identical to that described in **(a)**.

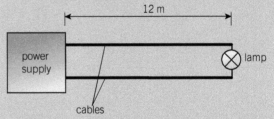

▲ **Figure 11**

The lamp is rated at 24 W, 6.0 V. The power supply has negligible internal resistance and its output is adjusted so that the potential difference across the lamp is 6.0 V.

  (i) Calculate the resistance of the lamp when operating at 6.0 V (*2 marks*)

  (ii) Explain why the e.m.f of the power supply is greater than 6.0 V. (*1 mark*)

  (iii) Calculate the e.m.f. of the power supply. (*2 marks*)

May 2008 2822

ELECTRICAL CIRCUITS
10.1 Kirchhoff's laws and circuits
Specification reference: 4.3.1

Learning outcomes

Demonstrate knowledge, understanding, and application of:

→ Kirchhoff's second law

→ conservation of energy

→ Kirchhoff's first and second laws applied to electrical circuits.

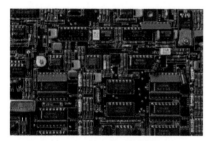

▲ **Figure 1** *Circuits found in modern electronic devices can be incredibly complex, but they rely on just a few simple laws and relationships*

Synoptic link

You will recall Kirchhoff's first law from Topic 8.3, Kirchhoff's first law.

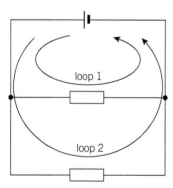

▲ **Figure 2** *A closed loop is one possible path for the current – in a series circuit there is only one loop, but in parallel there are often several possible loops*

Complex circuitry

Circuits found in modern electronic devices can look very daunting. They are rarely simple series or parallel circuits. Instead hundreds of components are connected together in complex ways. No matter how complex the circuit, though, it follows a few simple rules governing the current and p.d. through the circuit.

Kirchhoff's laws

We have already met Kirchhoff's first law, $\Sigma I_{in} = \Sigma I_{out}$. It is essentially the law of conservation of charge applied to electric circuits.

Kirchhoff's second law takes the law of conservation of energy and applies it to electrical circuits. It states: In any circuit, the sum of the electromotive forces is equal to the sum of the p.d.s around a closed loop.

$$\Sigma\mathcal{E} = \Sigma V \text{ around a closed loop}$$

where $\Sigma\mathcal{E}$ is the sum of the e.m.f.s and ΣV is the sum of all the p.d.s. A closed loop can be thought of as a single possible path for the current (Figure 2).

Essentially Kirchhoff's second law says that the total energy transferred to the charges in a circuit ($\Sigma\mathcal{E}$) is always equal to the total energy transferred from the charges (ΣV) as they move around the circuit.

Series circuits

A **series circuit** has only one path for the current, a single loop from one terminal of the source of e.m.f. (e.g., a cell) back to the other terminal (Figure 3).

In a series circuit the current is the same in every position.

From Kirchhoff's first law you know that the rate of flow of charge is the same at all points in the circuit, the charge is not used up; it just flows around the circuit.

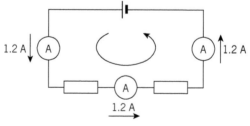

▲ **Figure 3** *In a series circuit there is one possible path, or one loop for the current, and the current is the same at all places*

Since a series circuit has only one closed loop, then from Kirchhoff's second law the e.m.f. is shared between the components. The sum of the p.d.s across the components is always equal to the e.m.f. (Figure 4). If the circuit contains two components with the same resistance, the e.m.f. is shared equally between them. If the components have different resistances the component with the greater resistance will take a greater proportion of the e.m.f.

In circuits with more than one source of e.m.f., the same rule applies, but we need to add the e.m.f. from each source, before sharing it between the components (Figure 5). The sources of e.m.f. are connected with opposing polarities. The sum of the e.m.f.s is equal to $9.0\,V - 6.0\,V = 3.0\,V$, and not $15\,V$.

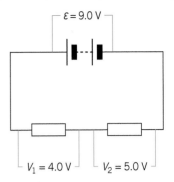

▲ **Figure 4** *In a series circuit the e.m.f. from the power source is shared between components: the total p.d. always adds up to the e.m.f.*

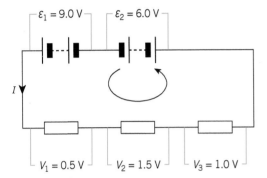

▲ **Figure 5** *Where there are multiple sources of e.m.f., the p.d. across each component must add up to the total e.m.f. (in this case 3 V)*

Parallel circuits

A **parallel circuit** provides more than one possible path for the charges. How much charge flows down each path depends on the resistance of the path. Kirchhoff's first law tells us that the current into each junction must be equal to the current out of that junction (Figure 6).

The greater the resistance of the branch, the lower the current that passes through it. If one branch has half the resistance of the other, it will have twice the current through it, so two-thirds of the total current will go through the branch with the lower resistance.

Each parallel branch can be thought of as a separate circuit. If changes are made to one branch, the other branches are not affected (Figure 7).

In a parallel circuit there are several different loops. Each branch forms its own loop. Kirchhoff's second law tells us that around each loop the e.m.f. must equal the p.d., so in other words the total p.d. across each branch is equal to the e.m.f. from the power supply (Figure 8). If one branch contains several components then the sum of the p.d.s across these components must equal the e.m.f. (Figure 9).

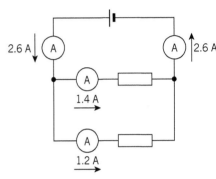

▲ **Figure 6** *In a parallel circuit the current in each branch might be different, but the current into each junction must always equal the total current leaving it (in this case $2.6\,A = 1.4\,A + 1.2\,A$)*

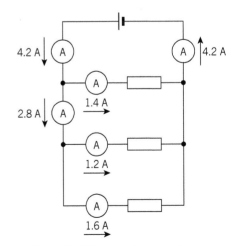

▲ **Figure 7** *Adding another branch has no effect on the current through the first two branches, but an additional 1.6 A is drawn from the cell for the final branch, increasing the current through the cell to 4.2 A*

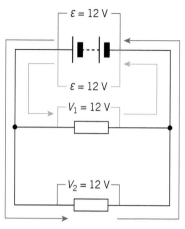

▲ **Figure 8** *In parallel circuits, the p.d. across each branch is the same as the e.m.f., no matter how many branches there are*

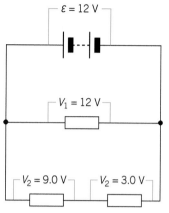

▲ **Figure 9** *If there is more than one component on a branch, then within that branch the sum of the p.d.s across the components must equal the e.m.f., as in a series circuit*

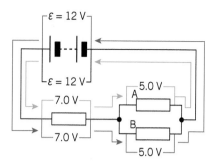

▲ **Figure 10** *It's often helpful to draw separate loops in more complex circuits to determine any unknown values: here resistors A and B are in parallel, so the p.d. across them must be the same; they don't 'share' the 5.0 V between them*

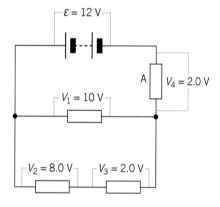

▲ **Figure 11** *In this example resistor A has a p.d. 2.0 V across it and so 10 V remains from the supplied 12 V for each branch – the p.d.s around each loop must add up to 12 V*

Multiple loops and adding components

In more complex circuits it is useful to consider each loop separately, paying particular attention to components in parallel (Figure 10).

Adding additional components in series reduces the e.m.f. that is shared between the original components (Figure 11). In this case the addition of a resistor means there is a lower p.d. across the original components as 2.0 V of the 12 V supplied has been taken up by the new resistor.

Summary questions

1 Fill in the six missing values A–F for the current in the circuits in Figure 12. *(6 marks)*

▲ **Figure 12**

Study tip

When solving circuit problems, it is really helpful to label all the values you know for current, e.m.f., and p.d. on the circuit diagram. This often makes it easier to determine any missing values.

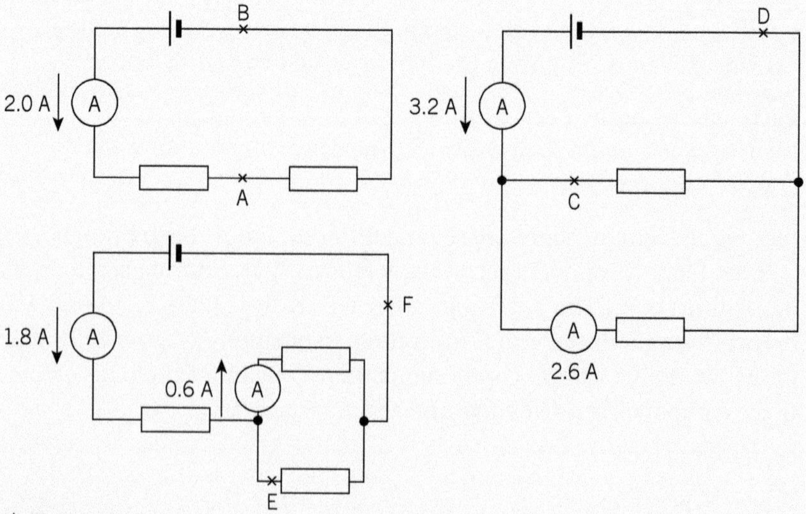

2 Fill in the seven missing values **A–G** for the p.d.s in the
   circuits in Figure 13.   (*7 marks*)

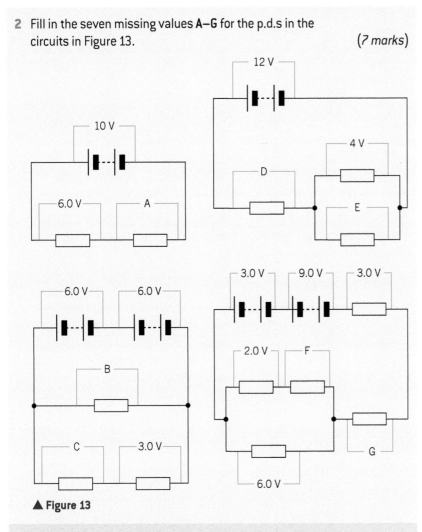

▲ **Figure 13**

3 Two identical resistors are connected to a 10 V power supply.
   Draw two circuit diagrams and label the p.d. across each
   resistor when connected:
   a in series;   b in parallel.   (*4 marks*)

4 One of the resistors in question 3 is replaced with one of twice
   the resistance. Redraw the circuit diagrams and label the
   p.d. across each resistor.   (*4 marks*)

5 Describe what happens to the current in a parallel circuit
   when additional branches are added.   (*3 marks*)

6 A parallel circuit contains three identical lamps connected to a
   battery with an e.m.f. of 12 V. Two lamps are on the first branch
   and one is on the second.
   a Sketch a circuit diagram for this arrangement.   (*2 marks*)
   b If a current of 6.0 A is drawn from the cell, determine the
      current through each lamp in the circuit.   (*4 marks*)
   c Calculate the power dissipated by each lamp.   (*4 marks*)
   d Over time, the current drawn from the battery decreases as the
      battery begins to go flat. If a current of 5.0 A is drawn from
      the battery determine the current through each lamp.   (*3 marks*)

10.2 Combining resistors

Specification reference: 4.3.1

Learning outcomes

Demonstrate knowledge, understanding, and application of:

→ total resistance of resistors in series $R = R_1 + R_2 + ...$

→ total resistance of resistors in parallel $\frac{1}{R} = \frac{1}{R_1} + \frac{1}{R_2} + ...$

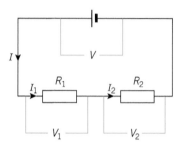

▲ **Figure 1** *In series, the total resistance is just the sum of the resistances of the resistors*

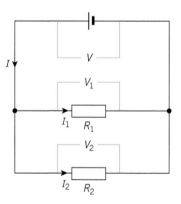

▲ **Figure 2** *The total resistance decreases as more resistors are added in parallel*

Added resistance

Resistors come in many shapes and sizes, but if a specific resistance is needed for a task, sometimes it is necessary to combine standard resistors to get a specific non-standard resistance. For example, a mobile phone charger is designed to provide the correct output p.d. in order to charge the phone most efficiently, and so chargers for different phone models are built with different resistances.

Adding a resistor to a circuit can increase its resistance, or, perhaps counterintuitively, decrease its resistance – it depends on how the resistors are connected together.

Resistors in series

When resistors are connected in series, each additional resistor effectively increases the length of the path taken by the charges, and so the more resistors you add, the greater the resistance becomes. Therefore, the total resistance R of a number of resistors connected in series is equal to the sum of the individual resistances.

$$R = R_1 + R_2 + ...$$

The relationship can be derived by considering Kirchhoff's two laws on electric circuits.

From Kirchhoff's second law, the total p.d. is equal to the sum of the p.d.s across each resistor: $V = V_1 + V_2 + ...$

Because $V = IR$, this can be rewritten as $IR = IR_1 + IR_2 + ...$

According to Kirchhoff's first law the current through each resistor must be the same, so I is a constant. Giving $R = R_1 + R_2 + ...$

Resistors in parallel

When resistors are connected in parallel the outcome is very different: the total resistance actually drops. The additional resistor provides another path for the current, effectively increasing the cross-sectional area and so lowering the resistance.

We can use the same approach as we did with series circuits to derive an expression for the total resistance of resistors connected in parallel. From Kirchhoff's first law, the total current is equal to the sum of the current in each resistor, giving $I = I_1 + I_2 + ...$

In this case, from Kirchhoff's second law, the p.d. across each resistor is constant and must be equal to V. Dividing our first equation by V gives

$$\frac{I}{V} = \frac{I_1}{V} + \frac{I_2}{V} + ...$$

$V = IR$, so $\frac{I}{V} = \frac{1}{R}$, so the equation for the total resistance R of resistors connected in parallel is

$$\frac{1}{R} = \frac{1}{R_1} + \frac{1}{R_2} + \ldots$$

Study tip

When using the equation for the total resistance of resistors connected in parallel, don't forget that your initial calculation gives you $\frac{1}{R}$. To find R you must find the reciprocal of your value (use the x^{-1} button on your calculator).

 Worked example: Three resistors in parallel

Resistors of $6.0\,\Omega$, $4.0\,\Omega$, and $3.0\,\Omega$ are connected in parallel. Calculate the total resistance of this combination.

Step 1: Identify the correct equation to calculate the resistance for resistors connected in parallel.

$$\frac{1}{R} = \frac{1}{R_1} + \frac{1}{R_2} + \frac{1}{R_3}$$

Step 2: Substitute in known values and calculate a value for $\frac{1}{R}$.

$$\frac{1}{R} = \frac{1}{6.0} + \frac{1}{4.0} + \frac{1}{3.0} = 0.75$$

Step 3: Invert $\frac{1}{R}$ to calculate a value for R.

$$\frac{1}{R} = 0.75 \text{ therefore } R = \frac{1}{0.75}$$

Calculate R and express it to an appropriate number of significant figures.

$$R = 1.3\,\Omega \text{ (2 s.f.)}$$

The total resistance is always lower than the lowest resistance of any resistor in the combination. This is a quick way to check whether your answer is correct.

Resistor circuits

Different combinations of resistors in series and parallel can be used to build more complex **resistor circuits**. Using the relationships above we can determine the total resistance of each circuit (Figure 3).

Study tip

The equations for total resistance of series and parallel circuits can also be applied to other circuit components.

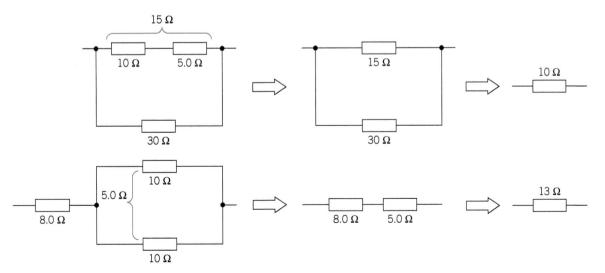

▲ **Figure 3** *It is possible to simplify the parts of a complex resistor circuit before calculating the total resistance*

Summary questions

1 State what happens to the resistance when resistors are connected:
   a in series; *(1 mark)*
   b in parallel. *(1 mark)*

2 Calculate the total resistance of the resistor circuits in Figure 4. *(5 marks)*

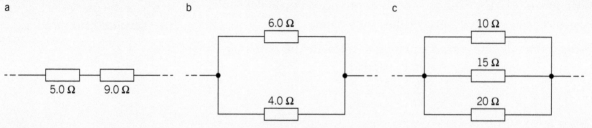

a

b

c

▲ **Figure 4**

3 Calculate the total resistance of the resistor circuits in Figure 5. *(6 marks)*

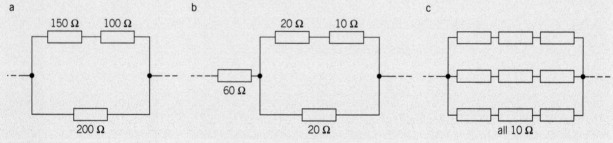

a

b

c

▲ **Figure 5**

4 Determine all the possible combinations of one or more resistors from a selection of three resistors with resistance 100 Ω, 50 Ω, and 200 Ω. For each combination draw a diagram of the resistor circuit and calculate the total resistance. *(15 marks)*

5 The total resistance of three resistors in parallel is 1030 Ω. Two of the resistors are known to have resistances 2.2 kΩ and 4.7 kΩ. Calculate the resistance of the third resistor. *(3 marks)*

10.3 Analysing circuits

Specification reference: 4.3.1

Call an electrician

The Gaia space probe (Figure 1) launched in 2013 is now around 1.6 million kilometres from the Earth. Its five-year mission is to search for extrasolar planets, asteroids within the solar system, and to collect further evidence to test Einstein's theory of general relativity. All the complex electrical circuitry in Gaia had to be carefully designed and thoroughly tested using the rules you have been studying. If scientists got their calculations wrong, there would be no way to repair the probe so far from the Earth.

Learning outcomes

Demonstrate knowledge, understanding, and application of:

→ analysis of circuits with components in both series and parallel

→ analysis of circuits with more than one source of e.m.f.

An electrical mathematical toolkit

When tackling any complex circuit problem we should start with Kirchhoff's circuit laws and these four key electrical relationships.

$$I = \frac{\Delta Q}{\Delta t} \qquad V = \frac{W}{Q} \qquad P = VI \qquad \text{and most importantly} \qquad V = IR$$

In the following three examples we shall determine any missing values for p.d. or current. Resistors are used for simplicity, but circuits could contain a variety of components, including diodes, filament lamps, and so on.

▲ **Figure 1** *The circuits found in cutting-edge technologies still rely on Kirchhoff's laws and the other relationships you have studied*

🖩 Worked example: Circuit 1

For the circuit in Figure 2, calculate the current in, p.d. across, and power rating for each resistor and in total.

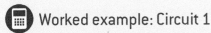

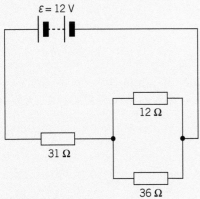

▲ **Figure 2** *Circuit 1*

Step 1: We can use our formulae for resistor networks to determine the total resistance of the network.

Using $\frac{1}{R} = \frac{1}{R_1} + \frac{1}{R_2} + \dots$ we first find the resistance of the two resistors in parallel (9.0 Ω) and add this to the single resistor, giving a total resistance of $R = 40\,\Omega$. →

Step 2: We can then use $V = IR$ to determine the current in the circuit, $I = 0.30\,A$. This must also be the current in the $31\,\Omega$ resistor, as it is in series.

Step 3: We can use $V = IR$ to determine the p.d. across this resistor, $V_{31\Omega} = 9.3\,V$.

Step 4: Using Kirchhoff's second law we now know that the p.d. across both the $12\,\Omega$ and $36\,\Omega$ resistors must be 2.7 V. Using $V = IR$ we can then calculate the current in each resistor and using $P = VI$ the power of each component.

The answers are summarised in Table 1. All values are expressed to two significant figures.

▼ **Table 1**

Component	Current / A	p.d. / V	Resistance / Ω	Power / W
31 Ω resistor	0.30	9.3	31	2.8
12 Ω resistor	0.23	2.7	12	0.62
36 Ω resistor	0.075	2.7	36	0.20
total circuit	0.30	12	40	3.6

The total values can be used to check for any calculation errors (using $V = IR$ and $P = VI$).

If we are given a known time we could also determine the energy transferred by each component and the charge passing through each component in that time using $I = \dfrac{\Delta Q}{\Delta t}$ and $V = \dfrac{W}{Q}$.

 Worked example: Circuit 2

In our second example circuit the orientation of the circuit appears different and an assortment of values of R, I, and V are available to us.

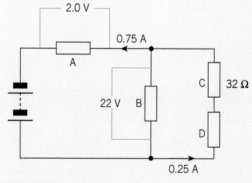

▲ **Figure 3** *Circuit 2*

Step 1: By considering the loop including the battery, resistor B, and resistor A, we can determine that the e.m.f. must be 24 V (2 s.f.). We can also use Kirchhoff's first law to find the current in resistor B, $I_B = 0.50$ A.

Step 2: Using $V = IR$ we can calculate the p.d. across resistor C ($R_c = 8.0$ V) and from Kirchhoff's second law the p.d. across resistor D must be 14 V.

Step 3: From this point we can use $V = IR$ and $P = VI$ to determine all the other values for V, I, R, and P, summarised in Table 2. All values are expressed to two significant figures.

▼ **Table 2**

Component	Current / A	p.d. / V	Resistance / Ω	Power / W
resistor A	0.75	2.0	2.7	1.5
resistor B	0.50	22	44	11
resistor C	0.25	8.0	32	2.0
resistor D	0.25	14	56	3.5
total circuit	0.75	24	32	18

 Worked example: Circuit 3

In our final example circuit we have two sources of e.m.f., several resistors of unknown resistance (although we know the power of resistor A) and a known p.d. across resistor C.

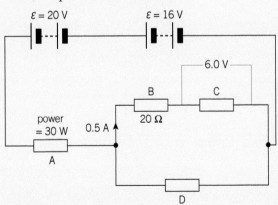

▲ **Figure 4** *Circuit 3*

Step 1: We know that the total e.m.f. in this circuit must be 36 V. We can use $V = IR$ to determine the p.d. across resistor B

(10 V), giving 16 V for the branch. From Kirchhoff's second law it follows that the p.d. across resistor A must be 20 V, and the p.d. across resistor D must be 16 V.

Step 2: We can also determine the resistance of resistor C using $V = IR$ ($R_C = 12\,\Omega$).

Step 3: Using $P = VI$ we can determine the current through resistor A as 1.5 A. From Kirchhoff's first law it follows that the current in resistor D must be 1.0 A.

Step 4: We can then determine the resistance of resistors A and D using $V = IR$, making sure to use the appropriate values for the p.d. across each resistor and the current in the resistor.

$$R_A = \frac{V_A}{I_A} = \frac{20}{1.5} = 13\,\Omega$$

$$R_D = \frac{V_D}{I_D} = \frac{16}{1.0} = 16\,\Omega$$

Step 5: We now have all values for current, p.d., and resistance, and can calculate the power of each component.

The total power in this case is 54 W, with 30 W from the 20 V battery and 24 W from the 16 V battery. These values should be checked carefully using $P = IV$.

Component	Current / A	p.d. / V	Resistance / Ω	Power / W
resistor A	1.5	20	13	30
resistor B	0.50	10	20	5.0
resistor C	0.50	6.0	12	3.0
resistor D	1.0	16	16	16
total circuit	1.5	36	24	54

Summary questions

1 Name all the quantities represented in the following equations:

$$I = \frac{\Delta Q}{\Delta t} \qquad V = \frac{W}{Q} \qquad P = VI \qquad V = IR$$ *(4 marks)*

2 Using circuit 1 calculate:
   a   the charge passing through the 31 Ω resistor in 45 seconds;   *(2 marks)*
   b   the energy transferred by the 36 Ω resistor in 2.0 minutes.   *(2 marks)*

3 For the circuit in Figure 5:
   a   determine the current through resistor B;   *(1 mark)*
   b   calculate the p.d. across resistors A and D;   *(4 marks)*
   c   show that the resistance of resistor D = 2.0 Ω.   *(2 marks)*

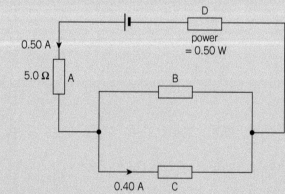

▲ **Figure 5**

4 If resistor B in Figure 5 has a resistance of 8.0 Ω, calculate the power of each resistor and the e.m.f. of the cell.   *(6 marks)*

10.4 Internal resistance

Specification reference: 4.3.2

Not all batteries are created equal

Different power sources have different internal resistances, often by design, depending on the job they are made for. The key consideration is the current in the circuits they power. If a large current is needed, a power source with a small **internal resistance** is required.

Car batteries have a very low internal resistance so that they can provide the large current needed (often hundreds of amperes) to turn the starter motor in the car. Even if you connected enough AA batteries together to give an e.m.f. of 12.6 V, the same as a car battery, they would not provide the necessary current because of their internal resistance.

Internal resistance and lost volts

Whenever there is a current in a power source, work has to be done by the charges as they move through the power source. In a chemical cell this work is due to reactions between the chemicals. In a solar cell it is due to the resistance of the materials of the cell.

As a result some energy is 'lost' (transferred into heat) when there is a current in the power source, and not all the energy transferred to the charge is available for the circuit. The p.d. measured at the terminals of the power source (the **terminal p.d.**) is less than the actual e.m.f. We call this difference **lost volts**.

▲ **Figure 1** Car batteries have a negligible internal resistance and so they can supply very large currents

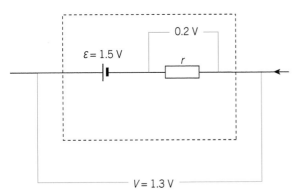

▲ **Figure 2** This cell has an e.m.f. ε of 1.5 V, but the terminal p.d. V across it is only 1.3 V, because 0.2 V has been 'lost' to the internal resistance of the cell r

From Kirchhoff's second law, the relationship between the e.m.f., the terminal p.d., and the lost volts is

$$\text{electromotive force} = \text{terminal p.d.} + \text{lost volts}$$

In normal use the e.m.f. does not change. However, changing the current affects the lost volts and the terminal p.d. Increasing the current means that more charges travel through the cell each second and so more work is done by the charges, increasing the lost volts. This lowers the terminal p.d.

If we apply the equation $V = IR$ to the internal resistance we can see that

$$\text{lost volts} = I \times r$$

where r is the internal resistance in ohms. If r remains fixed then the current in the power source is directly proportional to the lost volts.

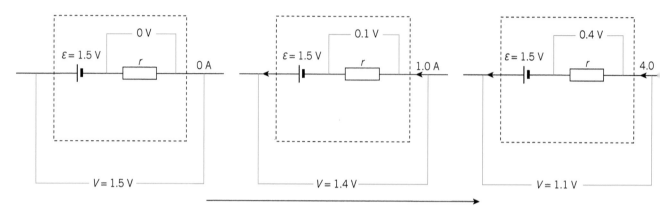

increasing current (terminal p.d. drops)

▲ **Figure 3** *As the current through the cell increases, the e.m.f. does not change, but the terminal p.d. drops as the lost volts increase*

The e.m.f. is always more than the terminal p.d. (unless there is no current). When the current is very small $\varepsilon \approx V$. This is why a high-resistance digital voltmeter connected directly across a cell gives a reading that approximates the e.m.f.

Combining the relationship electromotive force = terminal p.d. + lost volts and the equation above, we can derive an equation for the e.m.f. from a power source.

$$\varepsilon = V + \text{lost volts}$$

$$\varepsilon = V + Ir$$

where V is the terminal p.d. in volts, I is the current though the power supply in amperes, and r is the internal resistance of the power supply in ohms.

The terminal p.d. V is also equal to IR, where R is the resistance of the circuit, and so

$$\varepsilon = IR + Ir$$

As the current through the circuit and through the power supply must be the same, I is a common factor.

$$\varepsilon = I(R + r)$$

This relationship is essentially a version of $V = IR$ that takes into account the internal resistance of the power source.

Connecting cells

We can connect cells together to produce either a higher e.m.f. or a higher current. Depending on the desired effect we can connect them in series or in parallel.

Connecting cells in series increases the available e.m.f., but also increases the internal resistance. This limits the current that the combination can produce (Figure 4).

The same two cells connected in parallel produce the same e.m.f. as one cell, but have a much smaller internal resistance, so provide a greater current.

1 Two identical 1.5 V cells, each with an internal resistance of 0.75 Ω, are connected in series. Calculate the terminal p.d. when:
   a the cells supply a current of 0.80 A to an external circuit;
   b the cells are connected to a resistor of 10 Ω.
2 Repeat the above calculations for the cells connected in parallel.

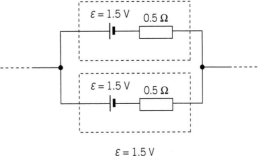

$\varepsilon = 3.0$ V
$r = 1.0$ Ω

$\varepsilon = 1.5$ V
$r = 0.25$ Ω

▲ **Figure 4** *The same two 1.5 V cells produce very different outcomes when connected in series and when connected in parallel*

Investigating internal resistance

Using the circuit in Figure 5, we can record values for terminal p.d. for different values of current. We use the variable resistor to change the resistance of the circuit, drawing different currents from the power source. Alternatively, several different resistors can be connected in various combinations.

By considering the general equation for a straight line $(y = mx + c)$ and rearranging the equation $\varepsilon = V + Ir$ to give $V = -rI + \varepsilon$, we can see that if we plot a graph of terminal p.d. (V) against current (I) we can find both the e.m.f. ε (constant) and the internal resistance of a power source r. The graph of V against I will have a gradient equal to $-r$ and a y-intercept equal to the e.m.f. ε of the power source.

Sketch the graph and you will see that, as the current through the cell increases, the terminal p.d. drops and the lost volts increase. When the current is zero the terminal p.d. is equal to the e.m.f. As the current increases so do the lost volts. The lost volts and the terminal p.d. always add up to the e.m.f. The current should not be allowed to get too high or it will raise the temperature of the cell, increasing its internal resistance.

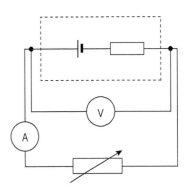

▲ **Figure 5** *Using a variable resistor we can take measurements of the terminal p.d. for various values of current in the cell*

Synoptic link

You will find more information about gradients of straight-line graphs in Appendix A2, Recording results.

▼ Table 1

Terminal p.d. /V	Current /A
1.52	0.15
1.50	0.20
1.47	0.25
1.44	0.30
1.43	0.35
1.40	0.40
1.37	0.45
1.34	0.50
1.32	0.55

Table 1 contains data collected using a typical AA cell.

1 Use the data in Table 1 to plot a graph of the terminal p.d. against current.
2 Explain how the graph suggests that the internal resistance is not affected by increasing the current through the cell.
3 Use the graph to determine the e.m.f. and the internal resistance of the cell.
4 Explain why the cell should be disconnected between each reading.

Graphs of *V* against *I*

Different power sources (including different cells) have different e.m.f.s and internal resistances. The graphs for different power sources will follow the same general trend but have different values for the e.m.f. and *r*. Three examples can be seen in Figure 6.

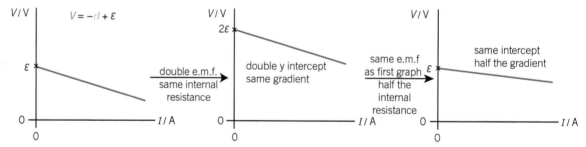

▲ **Figure 6** *Changing the e.m.f. ε or the internal resistance r of the power source changes the graph*

High or low internal resistance?

Like car batteries, many rechargeable batteries, including those in mobile phones and laptops as well as the more traditional types, have a small internal resistance. This allows them to be recharged using higher currents without overheating or wasting a lot of energy, so that recharging is fast. Some batteries in mobile phones charge to 80% capacity from flat in under an hour.

In contrast, the high-voltage power supplies used in classrooms have a very high internal resistance (often millions of ohms). This acts as a safety feature, preventing the power supply from delivering a fatal electric current.

Summary questions

1 With examples, outline why it is necessary for different
power sources to have different internal resistances. *(3 marks)*

2 Describe what happens to the terminal p.d. from a power
source as the current through the source increases. *(2 marks)*

3 A 9.0 V battery with an internal resistance of 2.0 Ω is connected in series
with a filament lamp. The lamp draws a current of 1.5 A. Calculate:
  a the lost volts; *(1 mark)*
  b the terminal p.d.; *(1 mark)*
  c the energy lost per second by the battery; *(1 mark)*
  d the resistance of the lamp. *(1 mark)*

4 A 12 V battery with an unknown internal resistance is connected with
three resistors as shown in Figure 7. If the current through the cell is
0.10 A, calculate:
  a the total resistance of the resistor circuit; *(2 marks)*
  b the terminal p.d. and the lost volts; *(2 marks)*
  c the internal resistance of the battery. *(1 mark)*

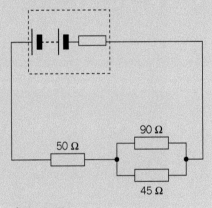

▲ Figure 7

5 Sketch a graph of V against I and label the internal resistance r and
the e.m.f. ε. Sketch a second graph to show how the graph
changes if two identical cells are connected in series. *(4 marks)*

10.5 Potential divider circuits

Specification reference: 4.3.3

Learning outcomes

Demonstrate knowledge, understanding, and application of:

→ potential divider circuit with components

→ potential divider equations, for example,

$$V_{out} = \left(\frac{R_2}{R_1 + R_2}\right) \times V_{in}$$

and $\dfrac{V_1}{V_2} = \dfrac{R_1}{R_2}$

▲ **Figure 1** *A mechanical volume control uses a type of potential divider to change the p.d. across a speaker, increasing or decreasing the intensity of sound*

Turn it up to 11

Mains-powered speakers are always connected to the same power source with a fixed p.d. So why do they not all always produce the same volume?

Speakers are amongst hundreds of electrical devices that make use of **potential divider** circuits. These circuits can vary the p.d. across an output (like a speaker) when connected to a fixed input.

Potential dividers

You may find you need a p.d. of 4.0 V for a specific task, but the power source is a 9.0 V battery. You can use potential dividers to divide the p.d. to give any value you require up to the maximum supplied from the power source (Figure 2).

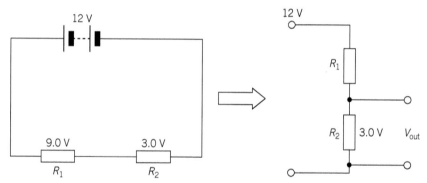

▲ **Figure 2** *A pair of resistors in series will 'share' or divide the p.d. across them, depending on the ratio of their resistances, a fact used in simple potential divider circuits, which are normally drawn like the diagram on the right*

A circuit can be connected across one of the resistors in parallel. The p.d. supplied to this circuit (V_{out}) can be varied to any value from zero to the maximum supplied from the power source, depending on the resistances of R_1 and R_2. From Kirchhoff's second law the p.d. across each resistor must always add up to the p.d. from the power source.

Ratio of resistances

The p.d. across each resistor in a potential divider depends on their resistances. If they have the same resistance then the p.d. is shared equally between them. If one has twice the resistance of the other, then this one will receive two-thirds of the total p.d.

Mathematically, this can be expressed as

$$\frac{V_1}{V_2} = \frac{R_1}{R_2}$$

where V_1 is the p.d. (in volts) across the resistor with resistance R_1 (in ohms) and V_2 is the p.d. (in volts) across the resistor with resistance R_2 (in ohms).

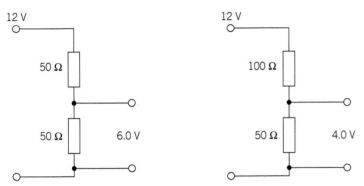

▲ **Figure 3** *When the two resistors have the same resistance the p.d. is split evenly between them (left), but when they are different the p.d. is shared according to the ratio of the resistances, so here (right) the upper resistor makes up two-thirds of the total resistance and so receives two-thirds of the p.d., leaving one-third or 4.0 V for the second resistor*

The potential divider equation

By considering the total p.d. V_{in} and the fraction of the total resistance provided by R_2 we can determine the value of V_{out}.

$$V_{out} = \left(\frac{R_2}{R_1 + R_2}\right) \times V_{in}$$

This relationship is often simply called the **potential divider equation**.

 Worked example: Calculating V_{out}

A $270\,\Omega$ resistor and a $170\,\Omega$ resistor are connected as part of a potential divider circuit to a $36\,V$ supply. The output is connected across the $270\,\Omega$ resistor. Calculate V_{out}.

Step 1: Identify the correct equation to calculate V_{out}.

$$V_{out} = \left(\frac{R_2}{R_1 + R_2}\right) \times V_{in}$$

Step 2: Substitute in known values (taking care to ensure that R_2 is the correct resistor – in this case $270\,\Omega$) and calculate V_{out}.

$$V_{out} = \left(\frac{270}{170 + 270}\right) \times 36 = 22\,V$$

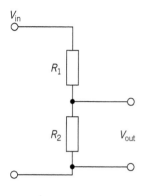

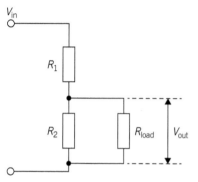

▲ **Figure 4** *When loaded, the resistance of the loaded part of the circuit is less than R_2 and so V_{out} drops*

 Loading a potential divider

Loading refers to connecting a component or circuit to V_{out}, that is, placing a component in parallel with R_2. This lowers the resistance of this part of the potential divider circuit, which lowers the fraction of the total p.d. across this part of the circuit, and so lowers V_{out}.

Adding a large load (high resistance) to the circuit has little effect on V_{out}, but if the load has a small resistance, V_{out} is significantly reduced.

1   Outline why V_{out} drops when a component is connected across it.
2   A 2.2 kΩ resistor and a 4.7 kΩ resistor are connected as part of a potential divider circuit to a 12 V supply. V_{out} is connected across the 4.7 kΩ resistor. Calculate V_{out} when
 a   the potential divider is not loaded
 b   the potential divider is loaded with a resistor of resistance 10 kΩ
 c   the potential divider is loaded with a resistor of resistance 100 Ω.

Summary questions

1   Outline how a pair of identical resistors can be used in a potential divider to produce an output of 10 V from a 20 V supply. Include a labelled diagram in your answer.   (*4 marks*)

2   Calculate V_{out} from the two potential dividers in Figure 5.   (*4 marks*)

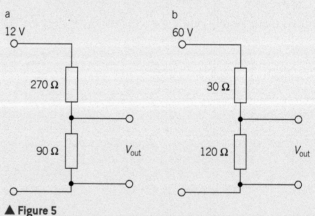

▲ **Figure 5**

3   Using a 24 V battery with negligible internal resistance, a 30 Ω resistor, and a 90 Ω resistor, draw a labelled diagram of a potential divider that would produce a V_{out} of:
 a   6.0 V;   **b**   18 V.   (*2 marks*)

4   A potential divider is connected to a 360 V power source with negligible internal resistance. The resistance of R_1 is 110 Ω. Calculate the resistance of R_2 if V_{out} is 3.0 V.   (*3 marks*)

10.6 Sensing circuits

Specification reference: 4.3.3

Stay in lane

A line-follower robot is a mobile robot that can follow a coloured line painted on the ground. Early designs used simple sensor circuits made up of potential dividers with an LDR in place of one of the resistors. The robot can be programmed to respond to a changing V_{out} caused by a change in the resistance of the LDR as it moves, ensuring that it always follows the line.

This technology has come a long way. Some cars now include lane departure warning systems. These warn the driver if the car begins to move out of its lane unless it is indicating to do so. The system is designed to reduce accidents caused by drivers drifting out of their lane when tired or distracted.

Producing a varying V_{out}

Using a pair of fixed resistors in series in a potential divider has the effect of splitting the p.d., but what if you need V_{out} to vary?

The simplest way to vary V_{out} is to replace one of the fixed resistors with a variable resistor (Figure 1). In the configuration shown, increasing the resistance of the variable resistor will increase V_{out} and vice versa.

Learning outcomes

Demonstrate knowledge, understanding, and application of:

→ potential divider circuits with variable components.

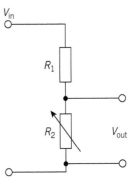

▲ **Figure 1** A variable resistor can be used to change the output voltage

 Temperature-sensing circuits – using a thermistor

Replacing the variable resistor with a thermistor allows V_{out} to vary automatically depending on the temperature of the surroundings.

As the temperature increases, the resistance of the thermistor decreases and so V_{out} drops (Figure 2).

Light-sensing circuits – using an LDR

An LDR can be used in the same way as a thermistor, producing a potential divider that gives an output that depends on the light intensity.

In this configuration (Figure 3), as the light intensity increases, the resistance of the LDR falls and so the p.d. across it decreases. R_2 receives a greater proportion of the p.d. and so V_{out} increases.

1 Design a potential divider which increases V_{out} as the temperature increases.
2 An LDR and a resistor with a resistance of 1000 Ω are connected to a 9.0 V battery with negligible internal resistance as part of a potential divider. The resistance of the LDR varies from 500 Ω in bright light to 50 MΩ in darkness. Calculate the minimum and maximum values for V_{out}.

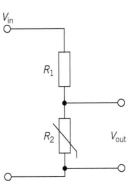

▲ **Figure 2** The output voltage will depend on the temperature of the thermistor

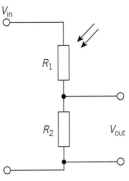

▲ **Figure 3** The output voltage will depend on the intensity of light

▲ **Figure 4** *Potentiometers with sliding contacts*

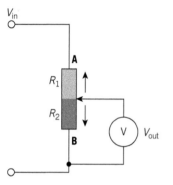

▲ **Figure 5** *The sliding contact found inside every potentiometer is a simple, cheap, and often compact way to produce a variable V_{out}*

The potentiometer

Many low-voltage electrical circuits that need a varying p.d. use a **potentiometer** rather than a potential divider.

A potentiometer is a variable resistor with three terminals and a sliding contact. Adjusting this contact varies the p.d. between two of the terminals, giving a variable V_{out} (Figures 4 and 5). Potentiometers can be made very compact, making them useful for portable electronic devices.

When the contact is moved towards **A**, V_{out} increases, until at **A** it is equal to V_{in}. When the contact is moved towards **B**, V_{out} decreases until at **B** it is zero.

Sometimes a dial is used rather than a slider, making the potentiometer even more compact. The potentiometer can also be constructed so that the change in resistance is either linear or logarithmic.

Summary questions

1 State two advantages of using a potentiometer over a potential divider. *(2 marks)*

2 Design a potential divider in which V_{out} increases as it gets darker. *(3 marks)*

3 A thermistor and a resistor with a resistance of 2.2 kΩ are connected to a 12 V battery with negligible internal resistance as part of a potential divider. V_{out} is connected across the thermistor. Calculate the resistance of the thermistor when V_{out} is:
   a 6.0 V;   b 10 V;   c 1.0 V.   *(6 marks)*

4 The potential divider used in question 3 is modified so that V_{out} is connected across the 2.2 kΩ resistor. Describe how changes in temperature would affect the value of V_{out} from this new potential divider circuit. *(2 marks)*

5 A thermistor and a resistor of resistance 220 Ω are connected in series as part of a potential divider to a 12 V battery with negligible internal resistance. The output is connected across the resistor. The thermistor has a resistance of 200 Ω at 0°C and 50 Ω at 100°C.
   a Sketch a graph of resistance against temperature for the thermistor. *(2 marks)*
   b Use your graph to estimate the temperature that would give an output of 4.0 V. *(3 marks)*

6 Suggest a reason for connecting a variable resistor, rather than a fixed resistor, with a thermistor as part of a potential divider. *(2 marks)*

Practice questions

1 a A student is given three resistors of resistances $10\,\Omega$, $20\,\Omega$ and $40\,\Omega$. These resistors are connected in different combinations. Calculate

    (i) the minimum possible resistance,

                                          (3 marks)

    (ii) the maximum possible resistance.

                                          (2 marks)

  b A filament lamp connected to a power supply glows at its maximum brightness.

    The output voltage from the power supply is halved. Explain why the current in the lamp is not halved.   *(2 marks)*

2 a Figure 1 shows combination of resistors connected to a power supply of e.m.f. $\mathcal{E}$.

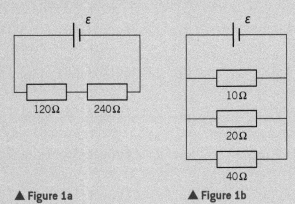

▲ Figure 1a   ▲ Figure 1b

    (i) For the circuit of Figure 1a

        1 calculate the total resistance R_s,

                            (1 mark)

        2 state one electrical quantity which is the same for both resistors.

                            (1 mark)

    (ii) For the circuit of Figure 1b

        1 calculate the total resistance R_p,

                            (2 marks)

        2 state one electrical quantity which is the same for all the resistors.   *(1 mark)*

                            Jan 2012 G482

3 Figure 2 shows a circuit connected to a d.c. supply.

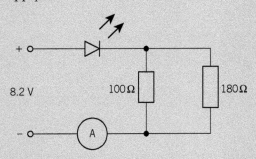

▲ Figure 2

The supply has e.m.f. 8.2 V and negligible internal resistance.

  a Calculate the total resistance of the resistors.   *(2 marks)*

  b The current measured by the ammeter is 100 mA. Calculate

    (i) the potential difference across the LED,   *(3 marks)*

    (ii) the total power dissipated in the resistors.   *(2 marks)*

4 a Kirchhoff's laws can be used to analyse any electrical circuit. State each of Kirchhoff's laws and the physical quantity associated with each law that is conserved in the circuit.

    (i) Kirchhoff's first law   *(2 marks)*

    (ii) Kirchhoff's second law   *(2 marks)*

  b The circuit in Figure 3 consists of a battery of e.m.f. 45 V and negligible internal resistance, and three resistors.

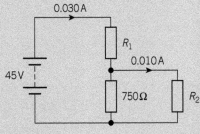

▲ Figure 3

The resistors have resistances R_1, R_2 and $750\,\Omega$. The current in the resistor of resistance R_1 is 0.030 A. The current in the resistor of resistance R_2 is 0.010 A.

Calculate

(i) the current I in the 750 Ω resistor,

(1 mark)

(ii) the p.d. V across the 750 Ω resistor,

(1 mark)

(iii) the resistances R_1 and R_2.

(2 marks)

May 2013 G482

5 Figure 4 shows how the resistance of a thermistor varies with temperature.

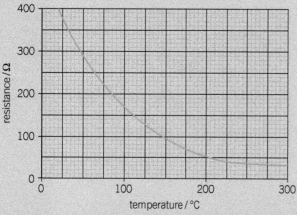

▲ **Figure 4**

The thermistor is used in the potential divider circuit of Figure 5 to monitor the temperature of an oven. The 6.0 V d.c. supply has zero internal resistance and the voltmeter has infinite resistance.

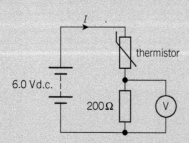

▲ **Figure 5**

a State and explain how the current I in the circuit changes as the thermistor is heated.

(3 marks)

b Use Figure 4 to calculate the voltmeter reading when the temperature of the oven is 240 °C.

(4 marks)

c A light-dependent resistor (LDR) is another component used in sensing circuits.

(i) Copy and complete Figure 6 with an LDR between **X** and **Y**.

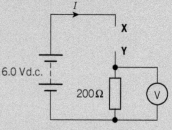

▲ **Figure 6**

(1 mark)

(ii) State with a reason how the voltmeter reading varies as the intensity of the light incident on the LDR increases.

(2 marks)

May 2012 G482

6 This question is about possible heating circuits used to demist the rear window of a car. The heater is made of 8 thin strips of a metal conductor fused onto the glass surface. Figure 7 shows the 8 strips connected in parallel to the car battery of e.m.f. ε and internal resistance r.

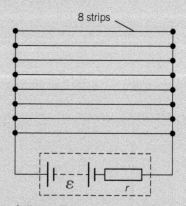

▲ **Figure 7**

▲ **Figure 7**

a The potential difference across each strip is 12 V when a current of 2.0 A passes through it.

(i) Calculate the resistance r_P of one strip of the heater.

(1 mark)

(ii) Calculate the total resistance R_P of the heater. (*3 marks*)

(iii) Show that the power P dissipated by the heater is about 200 W. (*2 marks*)

b Each strip is 0.90 m long, 2.4×10^{-4} m thick and 2.0×10^{-3} m wide.

Calculate the resistivity ρ of the metal of the strip. Give the unit with your answer.

(*4 marks*)

June 2011 G482

7 Figure 8 shows an electrical circuit.

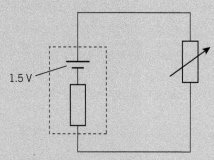

▲ **Figure 8**

The cell has e.m.f. 1.5 V. The resistance of the variable resistor is set to $1.0\,\Omega$. The current in the cell is 0.50 A.

a Calculate the internal resistance of the cell. (*3 marks*)

b The resistance R of the variable resistor is changed from $1.0\,\Omega$ to $4.0\,\Omega$.

(i) Copy and complete Table 1 to show the current I in the circuit and the power P dissipated in the variable resistor. (*2 marks*)

▼ **Table 1**

R/Ω	I/A	P/W
1.0		
2.0		
4.0		

(ii) Use your answer to (i) to suggest how the power P dissipated in the variable resistor is linked to the value of the internal resistance of the cell. (*1 mark*)

8 Figure 9 shows a circuit used to monitor the level of water in a container.

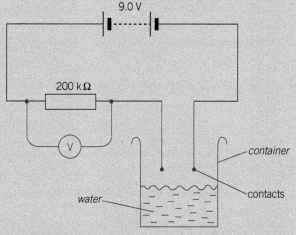

▲ **Figure 9**

The battery has electromotive force (e.m.f.) of 9.0 V and negligible internal resistance. The digital voltmeter shows a reading of 0.0 V when the contacts are dry and 6.5 V when the contacts are in water.

a Explain why the voltmeter reading is 0.0 V when the contacts are dry. (*2 marks*)

b Calculate the resistance of the water between the contacts when they are placed in the water. (*3 marks*)

c Without doing any calculations, explain how the voltmeter reading would change when the contacts are in water and the resistance of the resistor is much smaller than 200 kΩ. (*3 marks*)

WAVES 1
11.1 Progressive waves
Specification reference: 4.4.1

▲ **Figure 1** *By studying how seismic waves travel through the Earth, scientists may one day be able to predict earthquakes, potentially saving lives*

Earth-shaking

The effects of an earthquake can be devastating. An earthquake produces two main types of seismic wave, primary or **P-waves** and secondary or **S-waves**. Both are types of **progressive wave** that travel rapidly through parts of the Earth's interior. They can cause significant damage to structures on the surface.

The P-waves are **longitudinal waves** and S-waves are **transverse waves**. Differences in the way these types of wave travel through the ground allow scientists to study the interior structure of the Earth. The wave paths are calculated from the time delays between monitoring stations, and this information enables scientists to determine the densities and thicknesses of the layers inside the Earth.

Progressive waves

A progressive wave is an oscillation that travels through matter (or in some cases, through a vacuum). All progressive waves transfer energy from one place to another, but not matter. In other words, although the particles in the matter vibrate, they do not move along the wave.

Sound is an example of a progressive wave. When you hear someone talking to you, vibrations travel to your ears, but the air particles do not. Instead they vibrate in a plane parallel to the direction of energy transfer as the wave passes through the air.

When a progressive wave travels through a medium, like air or water, the particles in the medium move from their original **equilibrium position** to a new position. The particles in the medium exert forces on each other. A displaced particle experiences a **restoring force** from its neighbours and it is pulled back to its original position.

For example, waves on the surface of water propagate via water molecules interacting with their neighbours. As one molecule moves up it attracts its neighbours, which in turn pull the original molecule back down towards its equilibrium position, whilst at the same pulling up the neighbouring particles. No single water molecule moves along the wave. Instead they oscillate at right angles to the energy transfer.

▲ **Figure 2** *A ripple on water is another example of a progressive wave: it transfers energy from the centre of the ripple to the edges, but the water molecules vibrate perpendicular to the direction of energy transfer without moving out from the centre*

Transverse waves

When you think of waves, the first image in your mind is likely to be a transverse wave like a water wave. In a transverse wave the oscillations or vibrations are perpendicular to the direction of energy transfer (Figure 3).

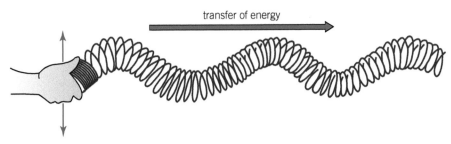

▲ **Figure 3** *A transverse wave travelling along a horizontal slinky spring*

As the wave moves from left to right, the oscillations are at 90° to the direction of the wave's movement – up and down, or side to side, or at any orientation as long as it is in a plane that is perpendicular to the direction of energy transfer.

Transverse waves have **peaks** and **troughs** where the oscillating particles are at a maximum displacement from their equilibrium position (Figure 4).

Examples of transverse waves include:

● waves on the surface of water

● any electromagnetic wave – radio waves, microwaves, infrared, visible light, ultraviolet, X-rays, and gamma rays

● waves on stretched strings

● S-waves produced in earthquakes.

You will learn more about electromagnetic waves in Topic 11.6, Electromagnetic waves.

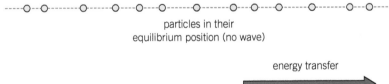

particles in their
equilibrium position (no wave)

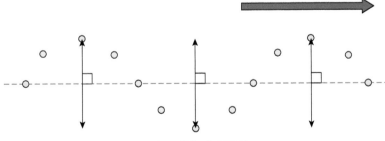

each particle vibrates at
90° to the direction of energy transfer

▲ **Figure 4** *For a transverse wave, the particles of the medium vibrate in a plane at 90° to the direction of energy transfer*

Longitudinal waves

In longitudinal waves the oscillations are parallel to the direction of energy transfer.

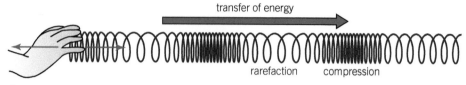

▲ **Figure 5** *A longitudinal wave looks very different to a transverse wave*

particles in their equilibrium positions (no wave)

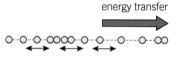

each particle vibrates parallel to the direction of energy transfer

▲ **Figure 6** *In a longitudinal wave the particles vibrate parallel to the direction of energy transfer*

Examples of longitudinal waves include:

- sound waves
- P-waves produced in earthquakes.

Longitudinal waves are often called compression waves. When they travel through a medium they create a series of **compressions** and **rarefactions**.

When sound waves travel through air, air particles are displaced and bounce off their neighbours. These collisions provide the restoring force. As the wave moves, regions of higher pressure (where the particles are close together) and regions of lower pressure (where the particles are more spread out) travel through the air, but no single air particle travels along the wave. Instead they oscillate about their equilibrium positions.

Study tip

Do not be tempted to use language like 'side to side' and 'up and down' to describe progressive waves. Instead use 'perpendicular' or 'parallel' to describe the oscillations.

Summary questions

1   Describe the similarities and the differences between transverse and longitudinal waves. *(3 marks)*

2   Describe how you can produce the two types of wave using a slinky spring. *(2 marks)*

3   Copy Figure 7 of a transverse wave on a rope and label the direction of motion of the particles labelled A, B, and C. *(3 marks)*

4   Suggest why the speed of sound is faster through a medium with a higher density. *(2 marks)*

5   Sketch a series of diagrams to show how the particles in a sound wave vibrate about their equilibrium positions during one complete oscillation. *(4 marks)*

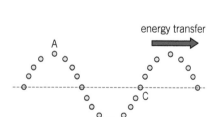

▲ **Figure 7**

11.2 Wave properties

Specification reference: 4.4.1

Notes on terminology

Humans have played flutes and similar instruments for 40 000 years. A modern flute can produce notes with frequencies ranging from around 250 Hz to over 2 kHz. The characteristics of these notes can be described with musical terms like pitch and volume, but they can also be described with scientific terms like frequency and wavelength.

Table 1 contains a list of key terms used to describe waves.

▼ **Table 1** *Wave terminology*

Term	Symbol	Unit	Definition
displacement	s	m	distance from the equilibrium position in a particular direction; a vector, so it can have either a positive or a negative value
amplitude	A	m	maximum displacement from the equilibrium position (can be positive or negative)
wavelength	λ	m	minimum distance between two points in phase on adjacent waves, for example, the distance from one peak to the next or from one compression to the next
period of oscillation	T	s	the time taken for one oscillation or time taken for wave to move one whole wavelength past a given point
frequency	f	Hz	the number of wavelengths passing a given point per unit time
wave speed	v (or c)	m s^{-1}	the distance travelled by the wave per unit time

Learning outcomes
Demonstrate knowledge, understanding, and application of:

→ displacement, amplitude, wavelength, period, phase difference, frequency, and speed of a wave

→ the equation $f = \dfrac{1}{T}$

→ the wave equation $v = f\lambda$

→ graphical representations of transverse and longitudinal waves.

▲ **Figure 1** *Playing the flute*

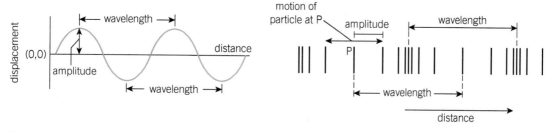

▲ **Figure 2** *Wavelength and amplitude of a transverse and a longitudinal wave*

The wave equation

From the definitions above we can derive the **wave equation**. This relates the frequency f in hertz, the wavelength λ in metres, and the wave speed v in m s^{-1}.

$$v = f\lambda$$

If a wave has a frequency of 5.0 Hz, each second there are 5 complete oscillations. If each wave has a wavelength of 2.0 m, then the wave has travelled 10 m from the source in that time. Therefore, its speed must be 10 m s^{-1}. So, for a frequency f, the wave would have travelled a distance of $f \times \lambda$ per second, that is, the wave speed v.

We can also see from the definitions that the period of oscillation and the frequency of a wave are reciprocals of each other. If a wave has a frequency of 2.0 Hz, there are two complete wave cycles each second; therefore, the period for each wave must be 0.50 s or $\frac{1}{2.0}$. Therefore, we have a second important equation relating the frequency f of a wave in Hz to its period T in s.

$$f = \frac{1}{T}$$

 Worked example: Finding the wavelength of a musical note

A flute produces a high-pitched note that has a time period of 0.45 ms. The speed of sound through air is 330 m s^{-1}. Calculate the wavelength of the note produced.

Step 1: Identify the correct equation to calculate the speed of the wave.

$$v = f\lambda$$

As $f = \frac{1}{T}$ substituting this into the first equation gives $v = \frac{\lambda}{T}$
Rearranging gives $\lambda = vT$.

Step 2: Substitute in known values in SI units (including the time in seconds) and calculate the wavelength of the note.

$$\lambda = 330 \times 0.45 \times 10^{-3} = 0.15 \, \text{m} \ (2 \, \text{s.f.})$$

Wave profile: displacement–distance graphs

A graph showing the displacement of the particles in the wave against the distance along the wave is sometimes called the **wave profile** (Figure 3). It may be helpful to think of such a graph as a 'snapshot' of the wave.

The wave profile can be used to determine the wavelength and amplitude of both types of wave. As the displacement of the particles in the wave is continuously changing, the wave profile changes shape over time.

Figure 4 shows how the wave profile for a progressive wave changes shape for four consecutive quarters of the period T, starting at $t = 0$ and increasing by $\frac{T}{4}$ each time. After one complete period the particles are back in their original positions.

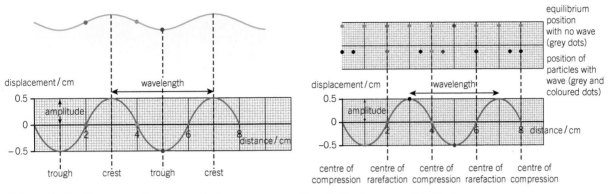

▲ **Figure 3** *Displacement–distance graphs for transverse and longitudinal waves are identical*

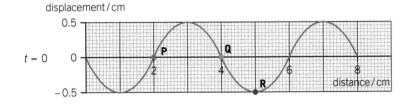

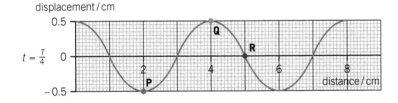

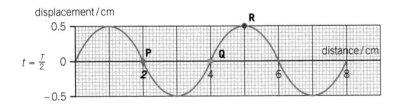

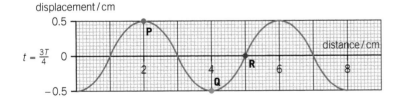

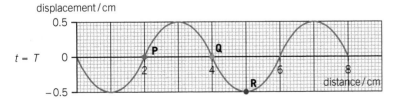

▲ **Figure 4** *Particles at P, Q, and R oscillate from their equilibrium position to a maximum positive displacement, back through their equilibrium position to a maximum negative displacement, and back again*

Phase difference

Phase difference describes the difference between the displacements of particles along a wave, or the difference between the displacements of particles on different waves. It is most often measured in degrees or radians, with each complete cycle or wave representing 360° or 2π radians.

If particles are oscillating perfectly in step with each other (they both reach their maximum positive displacement at the same time) then they are described as **in phase**. They have a phase difference of zero.

If two particles are separated by a distance of one whole wavelength (Figure 5), we say their phase difference is 360°, or 2π radians (angles can also be measured in radians). If they are two complete cycles out of step their phase difference is 720° or 4π radians, and so on.

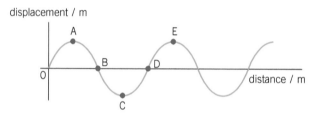

▲ **Figure 5** *Wave profiles are helpful when determining phase differences: particles at A and E have a phase difference of 360°, particles at C and D have a phase difference of 90°, whereas those at C and E are in antiphase and have a have a phase difference of 180°*

If particles are oscillating completely out of step with each other (one reaches its maximum positive displacement at the same time as the other reaches its maximum negative displacement) then they are described as being in **antiphase**. They have a phase difference of 180°, or π radians.

Two particles can have any phase difference as phase difference depends on the separation of particles in terms of the wavelength.

> ### ✚ Relating position and phase difference
>
> The phase difference ϕ between two points on a wave of wavelength λ separated by a distance x is given by
>
> $$\phi = \frac{x}{\lambda} \times 360°$$
>
> From this relationship we can see that if the distance between two points is equal to one wavelength the phase difference will be 360°.
>
> > **1** Calculate the phase difference in degrees between two points on a wave of wavelength 40 cm separated by:
> >   **a** 20 cm;   **b** 40 cm;   **c** 80 cm.

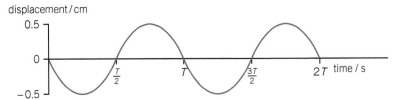

2　Calculate the distance between two points on a wave of wavelength 1.60 m when they have a phase difference of:
a　90°;　　　b　540°;　　　c　5π radians.

Displacement–time graphs

A second type of graph can be used to show how the displacement of a given particle of the medium varies with time as the wave passes through the medium. This graph looks the same for both transverse and longitudinal waves (Figure 6).

displacement / cm

$\frac{T}{2}$　　T　　$\frac{3T}{2}$　　$2T$　time / s

▲ **Figure 6** *A graph of displacement against time for a wave looks similar to a wave profile*

A graph of displacement against time can easily be used to determine the period T and amplitude of both types of wave. From the graph we can see that at time $t = 0$ the displacement of the particle is zero. After one-quarter of the period ($t = \frac{T}{4}$) the particle is at its maximum possible negative displacement -0.5 m. At time $t = \frac{T}{2}$ the particle is back in its equilibrium position (displacement = 0), before moving to its maximum positive displacement at $t = \frac{3T}{4}$ and returning again to its equilibrium position after one complete cycle ($t = T$). The amplitude of the progressive wave is 0.5 cm.

Using an oscilloscope to determine wave frequency

An **oscilloscope** can be used to determine the frequency of a wave. For example, using a microphone we can produce a trace on the screen (Figure 7). The oscilloscope screen shows a graph of p.d. against time for any signal fed into it.

Each horizontal square on the oscilloscope screen represents a certain time interval. This is called the **timebase**. If this is set to 1.0 ms cm^{-1}, then each square represents a time interval of 1.0 ms (the squares are normally 1 cm across). The height of the trace on the screen can be changed by adjusting the y sensitivity, measured in V cm^{-1}. For example, a setting

of 10 V cm^{-1} would result in each square representing a p.d. of 10 V.

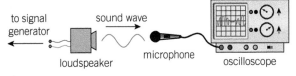

▲ **Figure 7**

From the timebase of the oscilloscope we can determine the time period T of the wave. Using this we can calculate the frequency with

$$f = \frac{1}{T}$$

In Figure 8 the timebase is set to $0.50\,\text{ms cm}^{-1}$. The horizontal distance from one peak to the next is exactly two squares on the screen, giving a period of 1.0 ms and therefore a frequency of 1.0 kHz.

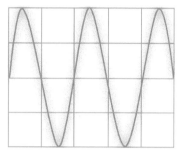

▲ Figure 8

As the frequency of the varying p.d. from the microphone is the same as the sound frequency, we can use the oscilloscope to determine the sound frequency.

1 Describe and explain the effect on the trace produced on the oscilloscope screen from a source of constant frequency if the timebase is changed.
2 Sketch the trace on an oscilloscope screen with a timebase set to $0.02\,\text{s cm}^{-1}$ when a loudspeaker producing a sound at 50 Hz is directed towards a microphone connected to the oscilloscope.

Summary questions

1 Calculate the speed of a water wave with a wavelength of 50 cm and a frequency of 2.0 Hz. (2 marks)

2 Copy Figure 5 and label the direction of the velocity of particles at A and D. (2 marks)

3 Plan an experiment to determine the frequency of sound emitted from a whistle. (4 marks)

4 A flute produces a note with a time period of 2.0 ms. The speed of sound through air is $340\,\text{m s}^{-1}$. Calculate the frequency of the note and wavelength of the sound produced. (4 marks)

5 The wave profile in Figure 8 shows several particles at $t = 0$. The wave has a period of 4.0 seconds.

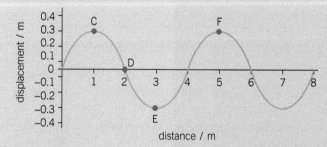

▲ Figure 9

a Sketch a wave profile showing the positions of the particles after:
 i 1.0 s; ii 2.0 s. (4 marks)
b Determine the displacement of particle C after:
 i 4.0 s; ii 11 s; iii 1 minute. (4 marks)

6 Determine the phase difference in degrees and radians between the following particles in Figure 8 ($360° = 2\pi$ radians)
 a CD; b CE; c DF. (3 marks)

11.3 Reflection and refraction

Specification reference: 4.4.1

Getting the right shot

There are certain features of some photographs that make them enthralling. One key aspect of a good shot is what the light is doing in the photo: how it reflects off and highlights different objects, and sometimes unusual effects caused by the light changing direction as it passes from one medium to another.

All waves can be reflected and refracted. Not just light but sound, radio waves, and even X-rays can be reflected and refracted under the right conditions.

Reflection

Reflection occurs when a wave changes direction at a boundary between two different media, remaining in the original medium.

A simple example is light reflecting off a mirrored surface (Figure 1). The light waves remain in the original medium (the air). We often represent the direction taken by a wave as a **ray** (like those in Figure 1). The ray shows the direction of energy transfer and so the path taken by the wave.

The **law of reflection** applies whenever waves are reflected. It states that the **angle of incidence** is equal to the **angle of reflection**.

When waves are reflected their wavelength and frequency do not change. This can be seen by reflecting water waves using a ripple tank. In Figure 2 we have represented the wave as a series of **wavefronts**. Each wavefront is a line joining points of the wave which are in phase. They can be thought of as the peak of each ripple. By definition (see Topic 11.2, Wave properties) the distance between wavefronts is equal to the wavelength of the wave.

Like plane (straight) waves, circular waves – like ripples from dropping a stone into a pond – can be reflected too (Figure 3).

Refraction

Refraction occurs when a wave changes direction as it changes speed when it passes from one medium to another. You will look at refraction of light in more detail in Topic 11.8, Refractive index.

Whenever a wave refracts there is always some reflection off the surface (partial reflection).

If the wave slows down it will refract towards the normal, if it speeds up it refracts away from the normal. Sound waves normally speed up when they enter a denser medium, whereas electromagnetic waves, like light, normally slow down. This results in the waves refracting in different directions.

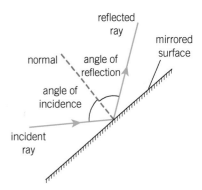

▲ **Figure 1** *When waves are reflected they always obey the law of reflection (note that all angles are measured to the normal)*

▲ **Figure 2** *Plane waves can be made by bobbing a ruler up and down in a ripple tank – when they reflect off a surface their frequency and wavelength remain unchanged*

Study tip

When drawing ray diagrams, always measure angles to the normal.

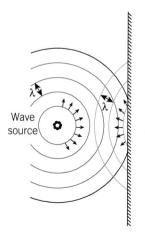

▲ **Figure 3** *When circular waves reflect off surfaces their wavelength and frequency remain the same, just like plane waves*

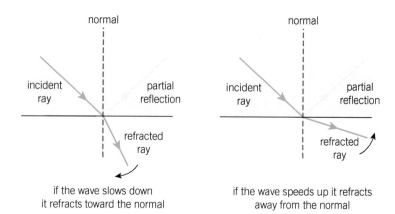

▲ **Figure 4** *A change in wave speed causes the wave to refract towards or away from the normal*

Unlike reflection, refraction does have an effect on the wavelength of the wave, but not its frequency. If the wave slows down its wavelength decreases and the frequency remains unchanged, and vice versa.

Refraction of water waves

The speed of water waves is affected by changes in the depth of the water, which gives us an easy way to investigate refraction of water waves. When a water wave enters shallower water, it slows down and the wavelength gets shorter.

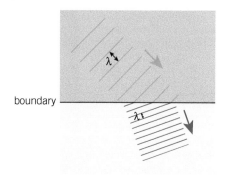

▲ **Figure 5** *When waves refract their wavelength changes: if the wave slows down, as shown here, the wavelength decreases so the wavefronts get closer together*

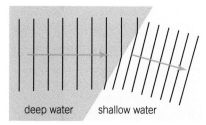

deep water   shallow water

▲ **Figure 6** *Water waves are refracted when there is a change in depth*

Summary questions

1 Outline the similarities and differences between reflection and refraction.
   (3 marks)

2 Complete the diagrams in Figure 9 to show the reflection of light off various mirrored surfaces.
   (4 marks)

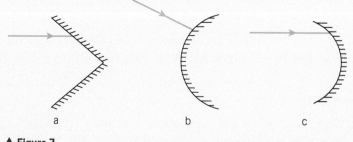

a   b   c

▲ **Figure 7**

3 Sketch a diagram to show a ray of light refracting when it travels from water into air.
   (2 marks)

4 Use the wave equation to explain why the wavelength of a refracted wave changes when it enters a different medium.
   (2 marks)

5 A swimming pool changes in depth from the shallow end to the deep end. Draw a wavefront diagram to show how circular ripples from a source in the centre of the pool travel to the edges.
   (3 marks)

11.4 Diffraction and polarisation

Specification reference: 4.4.1

Seeing atoms

Optical microscopes have enabled us to make huge advances in the fields of science and medicine. However, they have a limitation. We cannot keep on magnifying an object, seeing ever more detail with light. There is a limit.

This limit is around a few hundred nanometres. It results from a particular property of all waves. At high magnifications the image gets blurry because of the spreading of light, diffraction, as it passes through the apertures in the microscope. Diffraction cannot be avoided, and so if we want further magnification we must use different microscopes that do not rely on light. A scanning tunnelling microscope (STM) uses electrons to form images with much greater magnification. Objects down to individual atoms can be detected using this technology.

Diffraction

Diffraction is a property unique to waves. When waves pass through a gap or travel around an obstacle, they spread out.

All waves can be diffracted. The speed, wavelength, and frequency of a wave do not change when diffraction occurs.

How much a wave diffracts depends on the relative sizes of the wavelength and the gap or obstacle. Diffraction effects are most significant when the size of the gap or obstacle is about the same as the wavelength of the wave. This is why sound diffracts when it passes through a doorway, allowing you to hear conversations around the corner. The wavelength of the sound is similar to the size of the gap. However, light has a much smaller wavelength, so it does not diffract through such a large gap. In order to observe the diffraction of light we need a much smaller gap.

Polarisation

It is also possible to polarise some waves. **Polarisation** means that the particles oscillate along one direction only (e.g., up and down in the vertical direction), which means that the wave is confined to a single plane. This 'plane of oscillation' contains the oscillation of the particles and the direction of travel of the wave. The wave is said to be **plane polarised**. You will learn more about the polarisation of electromagnetic waves in Topic 11.7.

Light from an **unpolarised** source, like a filament lamp, is made up of oscillations in many possible planes. As light is a transverse wave, these oscillations are always at 90° to the direction of energy transfer. If you could observe the wave travelling towards you, you might see these oscillations as up and down, side to side, or at any angle.

▲ **Figure 1** *This image of a small number of gold atoms (around 5 nm across at their base) on top of a graphite layer (with individual carbon atoms seen as green dots) could not possibly be seen using a light microscope – it was recorded with a scanning tunnelling microscope*

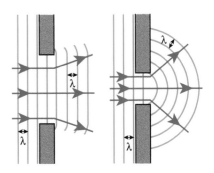

▲ **Figure 2** *The size of the gap compared with the wavelength of the wave affects how much diffraction takes place, and the spacing between the wavefronts shows that there is no change in wavelength*

201

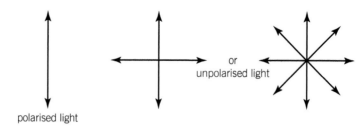

▲ **Figure 3** *A plane polarised wave is a wave in which the oscillations are in one direction only (left), whereas an unpolarised wave has oscillations in several directions*

In longitudinal waves, the oscillations are always parallel to the direction of energy transfer, so longitudinal waves cannot be plane polarised. Their oscillations are already limited to only one plane (the direction of energy transfer).

Partial polarisation

When transverse waves reflect off a surface they become **partially polarised**. This means there are more waves oscillating in one particular plane, but the wave is not completely plane polarised. For example, light reflected off the surface of water is partially polarised. Most of the light reflected off the surface becomes horizontally polarised. Some sunglasses contain polarising filters. These only allow light oscillating in one plane to pass through them, reducing the glare reflected off flat surfaces like lakes.

> **Study tip**
>
> When drawing diagrams to show diffraction, make sure the wavelength does not change.

▲ **Figure 4** *This composite image shows the effect of a polarising filter (right) on the reflections off the surface of a rock pool – when the polarised reflections are screened out by the filter it is much easier to see the seaweed growing in the pool*

Summary questions

1. Explain why it is not possible to polarise a sound wave. *(2 marks)*

2. Give two examples of a wave that can be plane polarised. *(2 marks)*

3. Explain why the diffraction of sound is regularly observed, but the diffraction of light is observed less frequently. *(2 marks)*

4. Two different waves pass through a 3.0 m gap. The first wave has a wavelength of 3.0 cm, the second wave 3.0 m. Describe the effect of the gap on each wave. *(3 marks)*

5. Explain why it is possible to receive long-wavelength radio signals at the bottom of some valleys in which the higher-frequency TV signal cannot be received. *(3 marks)*

6. Sound waves are directed towards a slit of width 0.30 m. The speed of sound in air is 340 m s^{-1}. State and explain whether or not each of the following frequency sound waves will be diffracted significantly at this slit:
   a. 1200 Hz; *(2 marks)*
   b. 1.0 MHz. *(2 marks)*

11.5 Intensity

Specification reference: 4.4.1

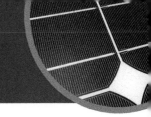

Can you hear me at the back?

The Andromeda galaxy (Figure 1) is one of the most distant objects you can detect with the naked eye. It is around 24 000 000 000 000 000 000 000 m (2.4×10^{22} m) from Earth. Like any progressive wave, the light from the galaxy spreads out as it travels further away from the source. The energy and power transferred becomes less concentrated. In the case of light, this means the further it travels from a source, the dimmer it becomes.

If the power of the wave source is known, this drop in brightness can be used to calculate how far the source is from the receiver. Certain types of supernovae always reach the same maximum brightness. Astronomers can use this information to determine our distance from the Andromeda galaxy and other astronomical objects.

This decrease in intensity with distance from the source occurs for all waves. Sound gets quieter and water waves decrease in amplitude.

Intensity

The **intensity** of a progressive wave is defined as the radiant power passing through a surface per unit area. Intensity has units watts per square metre (W m^{-2}) and can be calculated using the equation

$$I = \frac{P}{A}$$

where I is the intensity of the wave at a surface, P is the radiant power passing through the surface, and A is the cross-sectional area of the surface.

Intensity and distance – an inverse square relationship

When the wave travels out from a source the radiant power spreads out, reducing the intensity. For a point source of a wave, the energy and power spread uniformly in all directions, that is, over the surface of a sphere (Figure 2).

The total radiant power P at a distance r from the source is spread out over an area equal to the surface area of the sphere ($A = 4\pi r^2$). Substituting this area into our equation for intensity gives

$$I = \frac{P}{A} = \frac{P}{4\pi r^2}$$

 Worked example: Finding the power of the Sun

The average distance from the Earth to the Sun is 150 million km. The intensity of the radiation received by the upper atmosphere is 1400 W m^{-2}. Calculate the total power output of the Sun.

Learning outcomes

Demonstrate knowledge, understanding, and application of:

→ intensity of a progressive wave $I = \frac{P}{A}$

→ intensity $\propto$ (amplitude)^2.

▲ **Figure 1** *The Andromeda galaxy is around two million light years away, so its light has taken two million years to reach us*

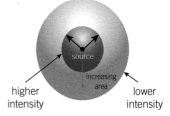

▲ **Figure 2** *The radiant power from a point source spreads out in a sphere, so the intensity of a wave depends on the area over which the power is spread out – the greater the area, the lower the intensity*

Step 1: Identify the correct equation to calculate the power from the Sun.

$$I = \frac{P}{A} = \frac{P}{4\pi r^2}$$

Step 2: Rearrange to make the power the subject.

$$P = I \times 4\pi r^2$$

Substitute in known values in SI units (including the distance in metres.

$$P = 1400 \times 4\pi \times (150 \times 10^9)^2 = 4.0 \times 10^{26}\,\text{W (2 s.f.)}$$

We can see from the equation that the intensity has an inverse square relationship with the distance from the source ($I \propto 1/r^2$). If the distance doubles, the intensity decreases by a factor of 4 (2^2), and if the distance increases by a factor of 100 the intensity will be 100^2 times smaller (Figure 2).

Synoptic link

LDRs (light-dependent resistors) were introduced in Topic 9.9, The LDR.

Intensity and LDRs

We can use an LDR to investigate how the intensity varies with distance from a constant power source (like a simple filament lamp).

In order to determine the intensity a **calibration curve** is used. Each LDR has its own calibration curve that allows the user to convert the resistance of the LDR into intensity.

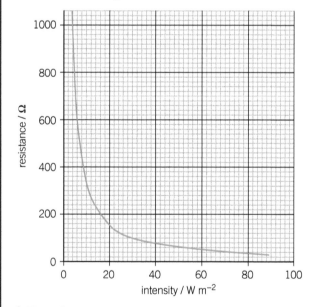

▲ **Figure 3** *Calibration curve for an LDR*

1 Use the data collected in Table 1 and the calibration curve to plot a graph of intensity against distance.

▼ Table 1

Distance from the source / cm	10	20	30	40	50	60
Resistance of the LDR / Ω	30	140	300	580	880	1400

2 Use the graph to test whether the intensity follows an inverse square relationship with distance from source.

Intensity and amplitude

When ripples travel out across the surface of a pond the intensity drops as the energy becomes more spread out. This causes a drop in amplitude. That is, the ripple height decreases the further the wave is from the source.

Decreased amplitude means a reduced average speed of the oscillating particles. Halving the amplitude results in particles oscillating with half the speed, and a quarter of the kinetic energy ($E_K = \frac{1}{2}mv^2$). So for any wave the intensity is directly proportional to the square of the amplitude. Double the amplitude of a wave and the intensity will quadruple.

$$\text{intensity} \propto (\text{amplitude})^2$$

Summary questions

1 State what happens to the intensity of a wave when the amplitude:
  a increases by a factor of 3;   b decreases by a factor of 4.   (2 marks)

2 Calculate the intensity when a power of 400 W is received over a cross-sectional area of 20 m². (2 marks)

3 Calculate the intensity 20 m from a source of light with a power of 60 W. (3 marks)

4 Figure 4 shows the cone of light created when light passes through a converging lens. Describe and explain how the intensity of light changes from A to B. (4 marks)

5 A satellite in orbit around the Earth uses two solar panels for power. The intensity of sunlight received at the height of the satellite is 1.4 kW m⁻². The surface area of each solar panel is 8.0 m². Calculate the total energy transferred to the panel in a period of 2.0 hours. (4 marks)

6 At a distance of 15 m from a point source the intensity of a sound wave is 1.0×10^{-4} W m⁻².
  a Show that the intensity 120 m from the source is approximately 1.6×10^{-6} W m⁻². (3 marks)
  b Discuss how the amplitude of the wave has changed. (2 marks)

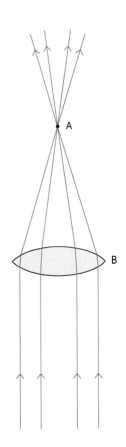

▲ Figure 4

11.6 Electromagnetic waves

Specification reference: 4.4.2

Travel through nothing

Electromagnetic waves (EM waves) do not need a medium. Unlike all other waves, they can travel through a vacuum. Without this ability to travel through space there could be no life on Earth, because this is how energy is transferred to our planet from the Sun.

What is an electromagnetic wave?

An EM wave is an example of a transverse wave, but it is a little more complex than a ripple on a pond. EM waves can be thought of as electric and magnetic fields oscillating at right angles to each other (Figure 2).

▲ **Figure 1** *These radio telescopes are designed to detect radio waves from objects deep in space, allowing physicists to learn more about our universe and ultimately about how life on Earth was possible*

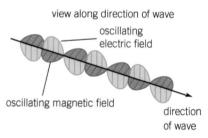

▲ **Figure 2** *An electromagnetic wave does not need a medium to be able to transfer energy*

The electromagnetic spectrum

The different types of EM wave are classified by wavelength. The full range of EM waves is called the **electromagnetic spectrum** and ranges from **radio waves**, with the longest wavelength, to **gamma rays**, with the shortest. Some radio waves have wavelengths longer than a million metres, whilst high-frequency gamma waves have wavelengths of just 10^{-16} m (less than the diameter of an atomic nucleus).

As you can see from Figure 3, the wavelength ranges of **X-rays** and gamma rays overlap. Unlike other parts of the spectrum, these EM

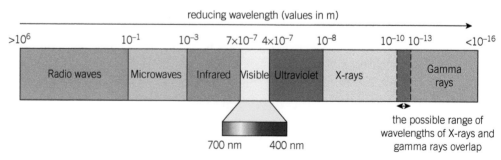

▲ **Figure 3** *The electromagnetic spectrum ranges from radio waves down to gamma rays*

waves are not classified by their wavelength, but by their origin. X-rays are emitted by fast-moving electrons, whereas gamma rays come from the unstable atomic nuclei.

Properties of EM waves

Like all waves, EM waves can be reflected, refracted, and diffracted. As EM waves are transverse waves they can also be plane polarised (see Topic 11.7, Polarisation of electromagnetic waves).

All EM waves travel at the same speed through a vacuum (c), $3.00 \times 10^8 \, \text{m s}^{-1}$. This is a very close approximation to their speed through air. Therefore, we can modify the wave equation for EM waves to

$$c = f\lambda$$

 Using c to find wavelength

A radio station transmits at a frequency of 107.3 MHz. Calculate the wavelength of the radio wave to an appropriate number of significant figures.

Step 1: Identify the correct equation to calculate the speed of the wave.

$$c = f\lambda$$

Rearranging this equation for λ gives

$$\lambda = \frac{c}{f}$$

Step 2: Substitute in the known values in SI units and calculate the wavelength.

$$\lambda = \frac{3.00 \times 10^8}{107.3 \times 10^6} = 2.80 \, \text{m} \ (3 \text{ s.f.})$$

 Studying the universe

Only certain EM waves are transmitted through our atmosphere: some radio waves (wavelengths between 10 cm and 10 m), visible wavelengths, and longer wavelengths of UV make it to the surface, but most other frequencies are either reflected or absorbed by the atmosphere.

EM radiation from distant stars and galaxies is studied in great detail. Often this is only achievable by placing telescopes high on mountain tops or sending them into space.

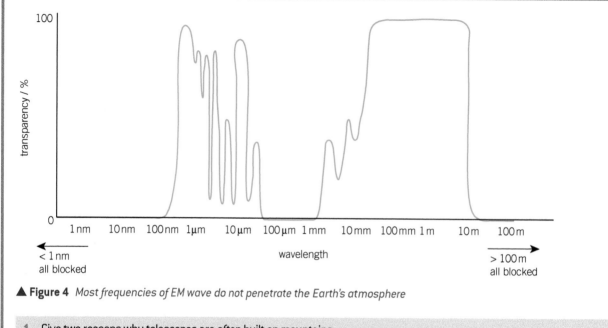

▲ **Figure 4** *Most frequencies of EM wave do not penetrate the Earth's atmosphere*

1   Give two reasons why telescopes are often built on mountains.
2   Outline the benefits and drawbacks of sending a telescope into space.
3   Look at Figure 4.
   a   Determine the parts of the EM spectrum for which the radiation received at ground level is:
     i   less than 10%;   ii   more than 80% of the radiation level above the atmosphere.
   b   Calculate the highest and lowest frequencies of EM radiation that reach the Earth's surface.

Summary questions

1   List the EM spectrum in order of frequency from highest to lowest.   *(2 marks)*

2   State the property of EM waves that confirms they are a type of transverse wave.   *(1 mark)*

3   Calculate the wavelength of the following EM waves:
   a   a radio wave with a frequency of 88.0 MHz;   *(1 mark)*
   b   microwaves with a frequency of 2.4 GHz;   *(1 mark)*
   c   X-rays with a frequency of 9.0×10^{16} Hz.   *(1 mark)*

4   The human eye can detect EM waves with wavelengths from around 400 to 700 nm. Calculate the minimum and maximum frequencies of light that the eye can detect.   *(2 marks)*

5   The Earth is on average 150 million km from the Sun. Calculate the time taken for light to travel from the Sun to the Earth.   *(2 marks)*

6   A radar system uses microwave pulses that reflect off incoming aircraft. If the time delay from transmitting a pulse to receiving the reflection is 0.56 μs, calculate the distance to the aircraft.   *(3 marks)*

11.7 Polarisation of electromagnetic waves

Specification reference: 4.4.1, 4.4.2

An echo from inflation?

We have seen how electromagnetic waves, like all transverse waves, can be plane polarised. In 2014 a pattern in the polarisation of the cosmic microwave background radiation was observed that may provide the first glimpse of evidence for gravitational waves and for inflation, a key part of the Big Bang theory (when the very early universe rapidly expanded from smaller than a proton to the size of a grapefruit).

Learning outcomes

Demonstrate knowledge, understanding, and application of:

→ plane polarised waves

→ the polarisation of electromagnetic waves.

Using polarising filters

Most naturally occurring electromagnetic waves are unpolarised. The electric field oscillates in random planes, all at 90° to the direction of energy transfer.

Unpolarised electromagnetic waves can be polarised using filters called polarisers. The nature of the polariser depends on the part of the electromagnetic spectrum to be polarised, but each filter only allows waves with a particular orientation through (Figure 1).

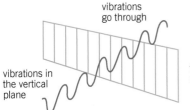

▲ **Figure 1** *A polarising filter acts like a slatted fence, only allowing electromagnetic waves polarised in the same direction as the filter to pass through*

Polarisation of light

Polaroid filters are plastic films that contain very long thin crystals and polarise light. They are used in sunglasses and over liquid crystal displays such as watches.

If you take two pieces of Polaroid filter, place them together, and rotate them, you can observe the effect of the plane polarisation of light passing through the filters (Figure 2).

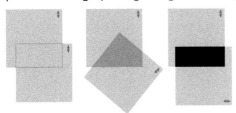

▲ **Figure 2** *As one filter is rotated with respect to the other, the intensity of the light passing through the filters varies*

Unpolarised light passing through the first filter becomes plane polarised. If the second filter (sometimes called the analyser) is in the same plane as the first, then the light passes through it unaffected. However, if the second Polaroid is slowly rotated, the intensity of the light transmitted through it drops. When the second filter has turned through 90°, no light is transmitted and the intensity falls to zero (Figure 3).

◀ **Figure 3** *Change in the intensity of the light transmitted through a pair of Polaroid filters as their relative orientation is rotated through 360°*

Polarisation of microwaves

Microwaves produced artificially tend to be plane polarised. Any unpolarised microwaves can be polarised like light, but in place of a Polaroid filter, a metal grille is used (Figure 4).

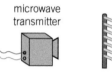

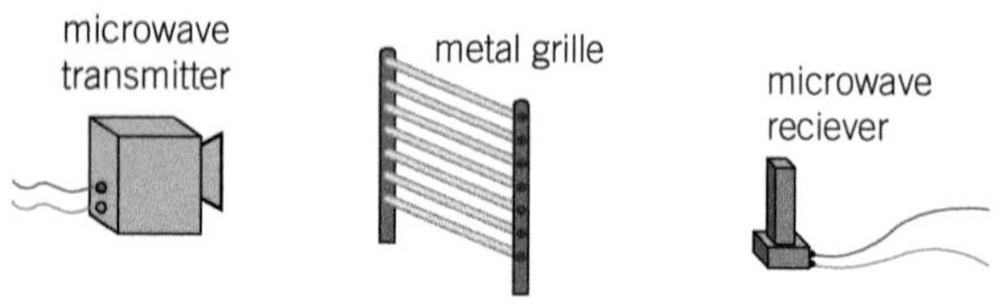

microwave transmitter

metal grille

microwave reciever

▲ **Figure 4** *A metal grille is able to polarise microwaves*

Inside the door of each microwave oven is a metal sheet with many holes in it. This allows light to pass through, enabling the user to see what is cooking, but at the same time preventing microwaves from escaping.

1 Outline why the intensity varies from a maximum value to zero.
2 A metal grille is placed between a source of plane polarised microwaves and a receiver. Describe and explain the effect of rotating the grille through 180° around the axis of the beam on the intensity recorded by the receiver.
3 Suggest why the metal sheet in the door of microwave ovens contains little holes, rather than a series of slits.

Aligning aerials

One use for the polarisation of electromagnetic waves is in communications transmitters. In order to reduce interference between different transmitters, some transmit vertically plane polarised waves and others nearby transmit horizontally plane polarised waves. An aerial aligned to detect vertically polarised radio waves will suffer less interference from horizontally polarised waves and vice versa.

Summary questions

1 State why the polarisation of light supports the view that light is a transverse wave. *(1 mark)*

2 Look at Figure 3. Explain why the maximum intensity occurs at 0°, 180°, and 360° and the minimum at 90° and 270°. *(2 marks)*

3 A student holds a polarising filter in front of a laptop screen and then rotates it. At a particular angle, the laptop screen appears to go dark.
   a Suggest what you can deduce about the nature of light emitted from the laptop screen from the student's observation. *(1 mark)*
   b Explain how the laptop screen can be viewed once again though the filter. *(3 marks)*

4 A beam of polarised light is directed normally at a polarising filter of cross-sectional area $9.0 \times 10^{-4}\,\text{m}^2$. The polarising filter is slowly rotated in a plane at right angles to the beam. The transmitted intensity I plotted against the angle θ resembles Figure 3, with a maximum intensity of $20\,\text{W m}^{-2}$.
   a Calculate the power of light transmitted through the filter at $\theta = 0°$.
   b Use the graph to calculate the ratio:
   $$\frac{\text{amplitude of light at } 0°}{\text{amplitude of light at } 60°}.$$
   (2 marks)

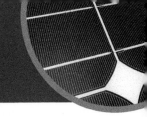

Seeing the light

Opticians use refractometers, which measure precisely how beams of light are refracted as they pass through the lens in a patient's eye. The angle at which the light refracts depends on a property of the material from which the lens is made called the **refractive index**. By accurately determining the refractive index of the lenses in the patient's eyes an appropriate prescription of glasses or contact lenses can be recommended.

Refractive index

Different materials refract light by different amounts. The angle at which the light is bent depends on the relative speeds of light through the two materials. Each material therefore has a refractive index, calculated using the equation

$$n = \frac{c}{v}$$

where n is the refractive index of the material (it has no units), c is the speed of light through a vacuum ($3.00 \times 10^8 \, \text{m s}^{-1}$), and v is the speed of light through the material in m s^{-1}.

If $n = 1$ then the speed of light through the material is the same as the speed of light through a vacuum.

▼ **Table 1** *Refractive indices for some common substances*

Material	n
vacuum	1.00
air	1.00 (actually 1.000293)
water	1.33
olive oil	1.47
crown glass	1.52
diamond	2.42

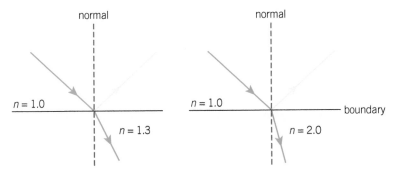

▲ **Figure 1** *The greater the refractive index, the more the light entering the material is refracted towards the normal*

Study tip

When drawing ray diagrams to show refraction, remember that angles are measured between the ray and the normal.

 ## Worked example: The speed of light through olive oil

Use the data in Table 1 to determine the speed of light through olive oil.

Step 1: Identify the correct equation to use.

$$n = \frac{c}{v} \quad \text{which can be rearranged for } v \text{ to give}$$

$$v = \frac{c}{n}$$

Step 2: Substitute in known values in SI units and calculate the speed of light through the olive oil.

$$v = \frac{3.00 \times 10^8}{1.47} = 2.04 \times 10^8 \, \text{m s}^{-1}$$

The speed of light through the material will always be less than the speed of light through a vacuum.

Summary questions

1 Describe the relationship between refractive index and the speed of light through a material. *(3 marks)*

2 The speed of light in ethanol is measured as $220 \times 10^6 \, \text{m s}^{-1}$. Calculate the refractive index of ethanol. *(2 marks)*

3 Using the data in Table 1, calculate the speed of light through water. *(3 marks)*

4 Use the information in Figure 2 to determine the refractive index of material B. *(2 marks)*

5 A ray of light travels from olive oil to crown glass. It strikes the boundary between the media at an angle of 45°. Carefully draw a diagram showing the path of the refracted ray, labelling the angles with their correct values. *(4 marks)*

6 A ray of light travels from diamond into water. The light strikes the boundary between the diamond and the water at 20°. Calculate the angle of refraction. *(3 marks)*

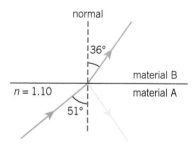

▲ Figure 2

Refraction law

The angles made by a ray of light at a boundary between two media was first investigated in 984 by the Persian physicist Ibn Sahl, then much later by Willebrord Snellius in 1621. Their findings can be simplified and expressed mathematically as

$$n \sin \theta = k$$

where n is the refractive index of the material, θ is the angle between the normal and the incident ray, and k is a constant.

We can apply this equation to describe what happens when light travels from one medium to another.

$$n_1 \sin \theta_1 = n_2 \sin \theta_2$$

 Worked example: Calculating the angle of refraction

A ray of light travels from water to crown glass (a glass often used to make lenses). The light strikes the boundary between the two at an angle of 40.0° to the normal. Calculate the angle of refraction.

Step 1: Identify the correct equation to use.

$$n \sin \theta = k$$

Apply this equation to a ray of light travelling from water into glass.

$$n_{\text{water}} \sin \theta_{\text{water}} = n_{\text{glass}} \sin \theta_{\text{glass}}$$

Rearrange for $\sin \theta_{\text{glass}}$.

$$\sin \theta_{\text{glass}} = \frac{n_{\text{water}} \sin \theta_{\text{water}}}{n_{\text{glass}}}$$

Step 2: Substitute in known values and calculate $\sin \theta_{\text{glass}}$ (make sure your calculator is in degree mode).

$$\sin \theta_{\text{glass}} = \frac{1.33 \times \sin 40.0°}{1.52} = 0.562\ldots$$

$$\theta_{\text{glass}} = 34.2°$$

This angle is less than 40.0°, as the light has slowed down when it entered the glass from the water, bending towards the normal.

Giving diamonds their sparkle

Diamonds have a number of unusual physical properties. They are extremely hard (the term diamond comes from the ancient Greek word for unbreakable) and are superb thermal conductors. But it is perhaps their sparkle that makes diamonds so attractive.

At $n = 2.42$, the refractive index of diamond is one of the highest of all natural materials. Once light enters a diamond it is usually reflected off the inner surfaces several times before it leaves the diamond. A skilled diamond cutter exploits this property to give diamonds their characteristic sparkle.

Conditions for total internal reflection

The **total internal reflection** (TIR) of light occurs at the boundary between two different media. When the light strikes the boundary at a large angle to the normal, it is totally internally reflected. All the light is reflected back into the original medium. There is no light energy refracted out of the original medium.

Two conditions are required for TIR.

1   The light must be travelling through a medium with a higher refractive index as it strikes the boundary with a medium with a lower refractive index. For example, TIR is possible when light in glass meets air, but not the other way around.

2   The angle at which the light strikes the boundary must be above the **critical angle**. This angle depends on the refractive index of the medium.

Learning outcomes

Demonstrate knowledge, understanding, and application of:

→   critical angle and $\sin C = \dfrac{1}{n}$

→   total internal reflection for light.

▲ **Figure 1** *The high refractive index of diamond means light travels at a slow speed of 124 000 000 m s^{-1} through it*

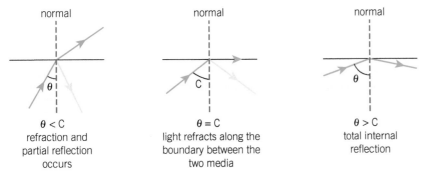

$\theta < C$	$\theta = C$	$\theta > C$
refraction and partial reflection occurs	light refracts along the boundary between the two media	total internal reflection

▲ **Figure 2** *Light meeting the boundary at exactly the critical angle travels along the boundary between the two media*

By using $n_1 \sin \theta_1 = n_2 \sin \theta_2$, you can determine the relationship between the refractive index of the medium and the critical angle when light travels from the medium into air.

At the critical angle C, θ_{air} is 90° (see Figure 2).

$$n \sin C = n_{air} \sin 90°$$

Both the refractive index of air and $\sin 90°$ are equal to 1, so this becomes

$$n \sin C = 1 \times 1$$

$$\sin C = \frac{1}{n}$$

From this we can see that the greater the refractive index the lower the critical angle.

 Worked example: Crown glass

Crown glass has a refractive index of 1.52. Determine the critical angle between crown glass and air.

Step 1: Identify the correct equation to use.

$$\sin C = \frac{1}{n}$$

Rearranging for C gives

$$C = \sin^{-1} \frac{1}{n}$$

Step 2: Substitute in known values and calculate the critical angle (making sure your calculator is in degree mode).

$$C = \sin^{-1} \frac{1}{1.52} = 41.1°$$

If light strikes the internal surface of crown glass at above 41.1° then it will be totally internally reflected.

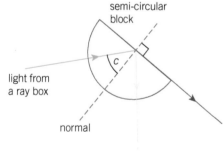

▲ **Figure 3** *Using a semi-circular block to measure the critical angle and calculate the refractive index of the block*

Determining refractive index from the critical angle

A simple experiment to determine the refractive index of a material can be carried out by carefully measuring the critical angle of a semi-circular block (Figure 3).

Directing the ray of light towards the centre of the semi-circular block ensures that light enters the block at 90° to the boundary and does not change direction, so the critical angle can be measured accurately.

▲ **Figure 4** *Optical fibres are usually made from flexible glass that has a high refractive index*

 Optical fibres

Optical fibres are designed to totally internally reflect pulses of visible light (or occasionally infrared) travelling through them. They have many uses, including transmitting data for fast broadband connections and images from inside patients during keyhole surgery.

A simple optical fibre has a fine glass core surrounded by a glass cladding with a lower refractive index (Figure 5, left). Light travelling through the fibre is contained within the core because of total internal reflection at the core/cladding boundary.

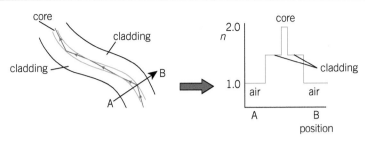

▲ **Figure 5** *Simple optical fibres are step indexed to ensure that the light remains inside the core of each fibre*

Simple optical fibres are step indexed. The refractive index changes suddenly between the core and cladding, and between the cladding and the air. When plotted on a graph of refractive index against distance across the fibre, the refractive index changes as a series of steps (Figure 5).

1   Explain why the cladding of an optical fibre must have a lower refractive index than the core.

2   Two pulses of light are sent along an optical fibre. One travels along the central axis of the fibre, the other undergoes many total internal reflections off the core–cladding boundary. Outline the differences in the received pulses at the end of the fibre.

3   In some optical fibres, called graded index fibres, the refractive index changes gradually across the fibre. It is lowest at the edge, increasing towards a maximum in the centre. Draw a diagram to show how a ray of light travels through a graded index optical fibre.

Summary questions

1   Describe what would happen to the critical angle if a semi-circular block with a lower refractive index was used in an investigation similar to the one described above.   *(2 marks)*

2   Calculate the critical angle for diamond (refractive index for diamond is 2.42).   *(2 marks)*

3   The critical angle for a Perspex (polyacrylate) block was measured as 42.8°. Calculate the refractive index of Perspex.   *(2 marks)*

4   Use Figure 6 to calculate the refractive index of material A.   *(3 marks)*

5   Crown glass has a critical angle of 41.1°. Draw diagrams to show the path of a ray of light that strikes the boundary between crown glass and air at:
   a   41.1°;   b   30.0°;   c   65.0°.   *(6 marks)*

6   The speed of light through flint glass is measured as 185 Mm s^{-1}. Calculate the critical angle for flint glass with air.   *(3 marks)*

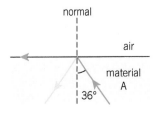

▲ **Figure 6**

Practice questions

1 a   Explain what is meant by a *progressive wave*. (*2 marks*)

   b   Describe how a *transverse wave* differs from a *longitudinal wave*. (*2 marks*)

   c   (i) Explain what is meant by diffraction of a wave. (*1 mark*)

       (ii) Describe how you could demonstrate that a sound wave of wavelength 0.10 m emitted from a loudspeaker can be diffracted. (*4 marks*)

Jan 2012 G482

2 a   State two main properties of electromagnetic waves. (*2 marks*)

   b   A scientist is investigating a device emitting electromagnetic waves of frequency 15 GHz. Calculate the wavelength and identify the type of electromagnetic waves being investigated by the scientist. (*4 marks*)

3 a   Define the refractive index of a transparent material such as glass. (*1 mark*)

   b   A ray of light is incident normally at the surface of a triangular shaped block of glass, see Figure 1.

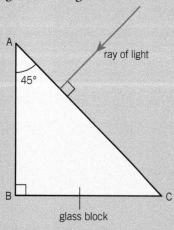

▲ Figure 1

The refractive index of the glass is 1.48.

   (i) Calculate the speed of light in the glass block. (*2 marks*)

   (ii) Calculate the critical angle for the glass–air interface. (*2 marks*)

   (iii) Describe and explain the path of the ray of light **inside** the block. (*4 marks*)

4 a   Define the following terms as applied to wave motion.

       (i) *displacement and amplitude* (*2 marks*)

       (ii) *frequency and phase difference* (*2 marks*)

   b   Figure 2 shows a transverse pulse on a *slinky*, a wound spring, at time $t = 0$. The pulse is travelling at a speed of $0.50\,\text{m s}^{-1}$ from left to right. The front of the pulse is at point **X**, 0.25 m from the point **P**.

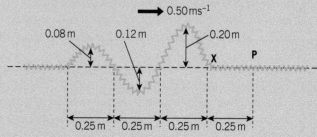

▲ Figure 2 (*4 marks*)

On a copy of Figure 3, draw a displacement y against time t graph of the motion of point P on the slinky from $t = 0$ to $t = 2.5\,\text{s}$.

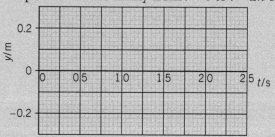

▲ Figure 3

May 2012 G482

5 A transverse wave travels on a stretched cord from left to right. Figure 4 shows, at a given instant, the shape of the cord.

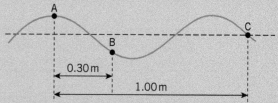

▲ Figure 4

The frequency of the wave is 2.4 Hz. **A**, **B** and **C** are particles on the rubber cord.

a Use Figure 4 to determine the speed of the wave. *(4 marks)*

b Calculate the phase difference, in degrees, between the points **A** and **B**. *(2 marks)*

c Describe and explain the direction in which **A** would move as the wave travels a very short distance. *(2 marks)*

June 2011 G482

6 a Describe a *plane polarised wave*. *(2 marks)*

b Light reflected from the surface of water is partially plane polarised in the horizontal direction. The reflected light is totally plane polarised when the angle of reflection is about 53°.

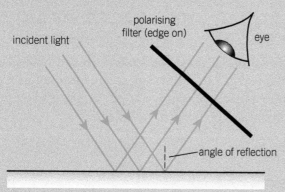

▲ Figure 5

Describe, referring to Figure 5, the experiment that you would perform using a polarising filter (a sheet of Polaroid) to determine whether the statement above is correct. Describe what you expect to observe. *(4 marks)*

June 2011 G482

7 Figure 6 shows the surface of water in a ripple tank at a given instant.

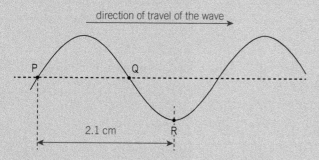

▲ Figure 6

a On a copy of Figure 6, show

   (i) the wavelength of the wave *(1 mark)*

   (ii) the amplitude of the wave. *(1 mark)*

b Compare and contrast the motion of particles at P and Q as the wave travels to the right. *(2 marks)*

c The distance between points P and R is 2.1 cm. The frequency of the wave is 20 Hz.

Calculate the speed of the wave. *(3 marks)*

d The speed of the waves in a ripple tank can be altered by changing the depth of the water.

State and explain the effect on the wavelength when the speed of the waves is doubled but the frequency is kept constant. *(2 marks)*

8 a A 1.2 mW laser emits light of wavelength 620 nm and a beam of diameter 0.82 mm.

Calculate

   (i) the frequency of the light *(2 marks)*

   (ii) the intensity of the beam of light. *(2 marks)*

b Figure 7 shows the path of a laser beam through a block of glass.

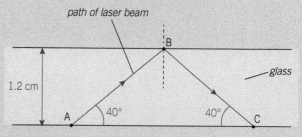

▲ Figure 7

The refractive index of glass is 1.50.

Calculate the time in ns it takes for the light to travel from A to B to C. *(4 marks)*

WAVES 2
12.1 Superposition of waves

Specification reference: 4.4.3

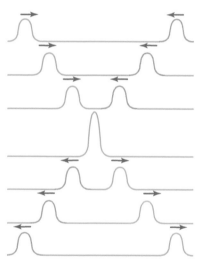

▲ **Figure 1** *Two pulses passing through each other superpose when they meet, in this case adding up to create a pulse with twice the amplitude of the individual pulses*

Noise-cancelling headphones

Some headphones offer noise cancellation. This feature relies on the **principle of superposition of waves** to remove unwanted sounds from the listener's surroundings, allowing them to focus on the music. A microphone on the outside of the headphones detects the background noise. The speakers inside the headphones then produce waves that aim to perfectly cancel out all external sounds. Some people use these headphones without music to allow them to sleep in noisy environments like aircraft cabins.

Superposition

When two waves of the same type meet, they pass through each other. Where the waves overlap, or **superpose**, they produce a single wave whose instantaneous displacement can be found using the principle of superposition of waves.

The principle of superposition states that when two waves meet at a point the resultant displacement at that point is equal to the sum of the displacements of the individual waves.

As displacement is a vector quantity, when the displacements of two waves are added together the resultant can be greater or smaller than the individual displacements of each wave. Some examples are shown in Table 1 and Figure 2.

▼ **Table 1** *Superposition of wave 1 and wave 2*

	A	B	C	D
Time /s	0.000	0.040	0.138	0.260
Displacement of wave 1 /m	0.20	0.00	0.08	−0.18
Displacement of wave 2 /m	−0.14	−0.14	0.08	0.18
Resultant displacement /m	0.06	−0.14	0.16	0.00

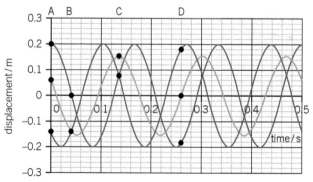

▲ **Figure 2** *More complex superposition of wave 1 (blue) and wave 2 (red) to produce the resultant wave (green)*

From superposition to interference

When two progressive waves continuously pass through each other they superpose and produce a resultant wave with a displacement equal to the sum of the individual displacements from the two waves. This effect is called **interference** and will be covered in more detail in Topic 12.2.

If the two waves are in phase then the maximum positive displacements (the peaks in a transverse wave) from each wave line up, creating a resultant displacement with increased amplitude. This is called **constructive interference** (Figure 3).

As intensity $\propto$ (amplitude)^2, the increase in amplitude resulting from constructive interference increases the intensity: sound waves are louder, and light is brighter.

If two progressive waves are in antiphase, then the maximum positive displacement (the peak in a transverse wave) from one wave lines up with the maximum negative displacement (the trough) from the other, and the resultant displacement is smaller than for each individual wave. This is called **destructive interference**.

If the waves have the same amplitude the resultant wave will have zero amplitude – it is cancelled out completely (Figure 3).

The reduction in the displacement results in a drop in intensity at that point. Sounds are quieter, and light is dimmer. If the resultant wave has zero amplitude, the intensity falls to zero.

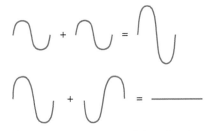

▲ **Figure 3** *Constructive (above) and destructive (below) interference*

Synoptic link

If you are unsure about the relationship between amplitude and intensity, look back at Topic 11.5, Intensity.

Summary questions

1 State and explain the type of interference used by noise-cancelling headphones. *(2 marks)*

2 Sketch two diagrams to illustrate the difference between constructive and destructive interference. *(2 marks)*

3 Determine the effect on the intensity when two waves of the same amplitude and perfectly in phase interfere. *(2 marks)*

4 Two sound waves are superposed. The first has a wavelength of 1.0 m and an amplitude of 5 mm. The second has a wavelength of 0.20 m and amplitude of 1 mm. Draw a displacement–distance graph for the resulting wave for distance in the range of 0 to 1.0 m. Assume the waves have a smooth sine-wave shape. *(6 marks)*

5 Two progressive waves travel towards and then pass through each other. Each wave travels at 1.0 m s^{-1}, and their starting positions are shown in Figure 4.

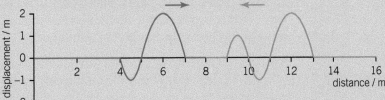

▲ **Figure 4**

Draw a series of six diagrams, each one second apart, to show how the resultant displacement changes as these two progressive waves superpose over 6 s. *(6 marks)*

12.2 Interference

Specification reference: 4.4.3

Learning outcomes

Demonstrate knowledge, understanding, and application of:

→ interference, coherence, path difference, and phase difference

→ constructive interference and destructive interference in terms of path difference and phase difference

→ two-source interference using sound and microwaves.

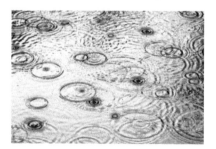

▲ **Figure 1** As the ripples overlap they interfere, sometimes constructively, other times destructively, creating patterns

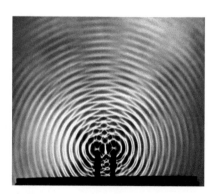

▲ **Figure 2** A stable interference pattern formed by two dippers in a ripple tank producing coherent water waves

Differences matter

If you watch the ripples on a pond when raindrops fall, you will see an **interference pattern** (Figure 1). As the wave caused by each raindrop travels outwards it overlaps with waves caused by other drops. At different points these superposed waves are in phase, interfering constructively, or out of phase, causing destructive interference. Random raindrops do not form a stable interference pattern but one that changes all the time. For a stable pattern, the waves must be **coherent** (Figure 2).

Forming stable interference patterns

Coherence refers to waves emitted from two sources having a *constant phase difference*. In order to be coherent the two waves must have the same frequency.

Filament lamps emit light of a range of different frequencies and ever-changing phase difference between different waves. In other words, they do not emit coherent light. Therefore, it is not possible to produce stable interference patterns using two filament lamps.

Path difference and phase difference

Interference patterns contain a series of **maxima** and **minima**. At a maximum the waves interfere constructively, at a minimum they interfere destructively. For example, a pair of loudspeakers emitting coherent sound waves produces an interference pattern with regions that are louder (maxima) and others that are quieter (minima) than the original waves. The same effect can be seen in Figure 2 with water waves. In places the water waves have increased amplitude (maxima) and in others the amplitude appears to be zero (minima).

In both cases these maxima and minima are a result of the two waves having travelled different distances from their sources. This difference in the distance travelled is called the **path difference**.

Figure 3 shows two sources emitting coherent waves. The wavelength of the progressive wave is λ. If the path difference to a point is zero or a whole number of wavelengths (0, λ, 2λ, …, $n\lambda$, where n is an integer), then the two waves will always arrive at that point in phase. This produces constructive interference at that point. The resultant wave at this point has maximum amplitude.

If the path difference to a point is an odd number of half wavelengths ($\frac{1}{2}\lambda$, $\frac{3}{2}\lambda$, …, $(n + \frac{1}{2})\lambda$, where n is an integer) the two waves will always arrive at that point in antiphase. This produces destructive interference. The resultant wave at this point has minimum amplitude.

At the central maxima, shown in Figure 4, the path difference is zero and so the phase difference is zero. At the first-order maxima the path difference is one whole wavelength, so the phase difference is 360° or

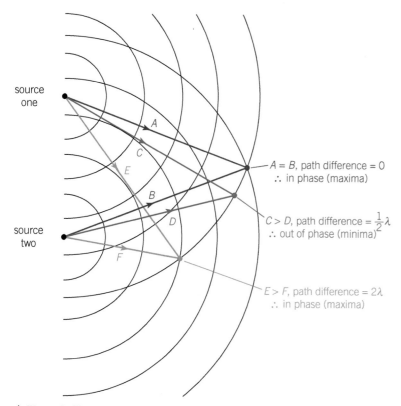

A = B, path difference = 0
∴ in phase (maxima)

C > D, path difference = $\frac{1}{2}\lambda$
∴ out of phase (minima)

E > F, path difference = 2λ
∴ in phase (maxima)

> ### Study tip
>
> Path difference is a distance, measured in m, cm, or mm. Do not confuse this with phase difference, which is measured in degrees or radians.

▲ **Figure 3** *The path difference determines the phase difference between the waves arriving at a given point*

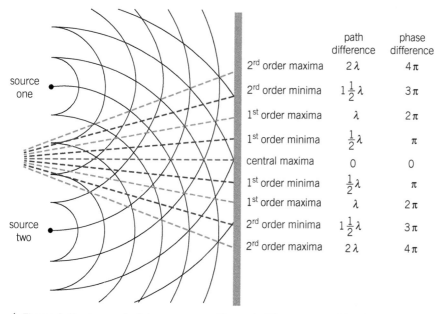

	path difference	phase difference
2rd order maxima	2λ	4π
2rd order minima	$1\frac{1}{2}\lambda$	3π
1st order maxima	λ	2π
1st order minima	$\frac{1}{2}\lambda$	π
central maxima	0	0
1st order minima	$\frac{1}{2}\lambda$	π
1st order maxima	λ	2π
2rd order minima	$1\frac{1}{2}\lambda$	3π
2rd order maxima	2λ	4π

▲ **Figure 4** *Maxima and minima are caused by path differences resulting in phase difference between the waves*

2π radians. The peaks from the first waves perfectly line up with the peaks of the second waves and so constructive interference occurs.

At the first-order minima the path difference is half a wavelength, a phase difference of 180° or π radians. The peaks from the first waves line up with the troughs of the second waves, resulting in destructive interference.

Interference in sound

Two loudspeakers connected to the same signal generator will emit coherent sound waves (Figure 5). The sound waves travel out from each loudspeaker and overlap, forming an interference pattern.

The interference pattern comprises a series of maxima (louder) and minima (softer). The positions of the maxima and minima can be detected with your ears or, more accurately, a microphone.

1   The speed of sound in air is approximately $340\,\text{m s}^{-1}$. The path difference at the first-order maxima is measured as 28 cm. Calculate the frequency of the sound.

2   Describe and explain what would happen to the interference pattern if the frequency of the sound from the loudspeakers were halved.

audio signal generator

speakers

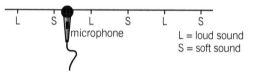

L   S   L   S   L   S

microphone

L = loud sound
S = soft sound

▲ **Figure 5** *The interference pattern from two loudspeakers forms alternating regions of louder and quieter sound*

Interference in microwaves

Producing coherent microwaves is more difficult than for sound. A single microwave source is used along with a pair of slits (a double slit), shown in Figure 6.

These diffracted microwaves overlap and form an interference pattern that can be detected with a microwave receiver connected to a voltmeter or an oscilloscope. Moving the receiver in an arc around the double slit detects the characteristic maxima and minima created as part of the interference pattern. The position of each can be carefully marked on paper.

1   Suggest a method to determine the wavelength of the microwaves used by measuring the path difference in an interference experiment. You should include details of any measurements taken and any subsequent calculations.

2   The frequency of the microwaves used is 24 GHz. With the help of a calculation, suggest a suitable width for each slit.

3   Describe how the equipment could be used to detect whether the microwaves are plane polarised.

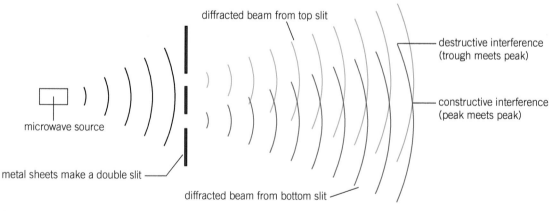

diffracted beam from top slit

destructive interference (trough meets peak)

constructive interference (peak meets peak)

microwave source

metal sheets make a double slit

diffracted beam from bottom slit

▲ **Figure 6** *The microwaves diffract at each slit, so each slit acts like a wave source and the diffracted waves from the two slits are coherent*

Thin film interference

The pattern of coloured light on thin oil films on water (Figure 7) is caused by interference. Light reflecting off the bottom surface of the oil interferes with the light reflected off the top surface. If the thickness of the oil results in a path difference that is a non-integer half number of wavelengths of light, the two sets of light waves are out of phase, destructive interference occurs, and the waves cancel out.

The colour results from the different wavelengths in white light and slight differences in the thickness of the oil layer. The distance the light travels through the oil before reflecting off the back surface differs. Different wavelengths of light are cancelled out by different thicknesses of oil. The wavelengths that are not cancelled out form the colours we observe.

▲ **Figure 7** *The thin film of oil on the top of a puddle can cause beautiful, colourful patterns to form by interference*

1 Draw a diagram to show the path taken by two rays of light, one reflecting off the top surface of the oil, the other entering the oil before reflecting off the bottom surface.
2 Use your diagram to show that there is a path difference between the two waves.
3 Explain why the oil needs to be thicker in order to provide destructive interference for red light compared with blue light.

Summary questions

1 State the type of interference formed at the following path differences:
   a 5λ; *(1 mark)*
   b 10λ; *(1 mark)*
   c 4.5λ. *(1 mark)*

2 Describe how the phase difference changes as you move away from the central maxima towards the third-order maxima. *(4 marks)*

3 Two coherent surface water waves are produced by a pair of dippers in a ripple tank. If the path difference to the second-order maxima is 9.0 cm, determine the wavelength of the water waves. *(2 marks)*

4 Two coherent sound waves of frequency 2.0 kHz form an interference pattern. The speed of sound in air is 340 m s⁻¹. Calculate:
   a the wavelength of the sound; *(1 mark)*
   b the path difference that causes a phase difference of 5π radians; *(2 marks)*
   c the path difference at the second-order minima. *(3 marks)*

12.3 The Young double-slit experiment

Specification reference: 4.4.3

Was Newton ever wrong?

Sir Isaac Newton is perhaps the greatest physicist who ever lived. His contributions to mathematics and physics included the co-invention of calculus, the formulation of laws on motion and gravity, and the design of the first reflecting telescope. Newton demonstrated that white light is composed of a spectrum of colours. He rejected the idea that light was a wave. Instead he developed his own corpuscular (or particle) theory, in which he described light as a stream of tiny particles.

Newton died in 1726, but his ideas remained the accepted scientific theory for 75 more years until Thomas Young's wonderfully simple experiment demonstrated that light can form an interference pattern, conclusive proof that light must be acting as a wave.

Young double-slit experiment

Two coherent waves are needed to form an interference pattern. Young devised the method to achieve this that now bears his name. He used a **monochromatic** source of light (which can be achieved using a colour filter that allows only a specific frequency of light to pass) and a narrow single slit to diffract the light.

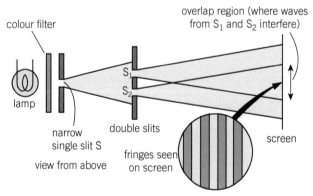

▲ **Figure 1** *The Young double-slit experiment made use of a single slit followed by a double slit to produce two sources of coherent light waves*

Light diffracting from the single slit arrives at the double slit in phase. It then diffracts again from the double slit. Each slit acts as a source of coherent waves, which spread from each slit, overlapping and forming an interference pattern that can be seen on a screen as alternating bright and dark regions called fringes.

The Young double-slit experiment successfully demonstrated the wave nature of light. Young used his experiment to determine the wavelength of various different colours of visible light.

A mathematical treatment

By considering the dimensions of different parts of a double-slit experiment we can derive a formula to calculate the wavelength λ of the light used to form the interference pattern.

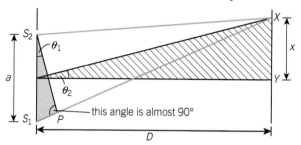

▲ **Figure 2** *The geometry of a typical double-slit experiment*

In Figure 2, the separation between the slits S_1 and S_2 is a. The interference pattern is observed on a distant screen a distance D from the slits, where $D \gg a$. A bright fringe is seen at Y, and the next adjacent bright fringe is observed at X, where the separation between the fringes is x. The path difference S_1P must be equal to one whole wavelength λ, since $D \gg a$. The two rays of light shown in blue are almost parallel to each other and the angles θ_1 and θ_2 are almost the same and very small. We can use the trigonometric approximation

$$\sin \theta_1 \approx \sin \theta_2 \approx \tan \theta_2$$

where $\sin \theta_1 = \lambda/a$ and $\tan \theta_2 = x/D$.

Therefore $\dfrac{\lambda}{a} \approx \dfrac{x}{D}$

This is often expressed as

$$\lambda = \frac{ax}{D}$$

This equation only applies if $a \ll D$ – the distance between the slits is much smaller than the distance from the screen to the double slits. Provided this is the case, the equation can be used to determine the wavelength of any wave producing an interference pattern from a double slit or two coherent sources, such as two loudspeakers connected to the same signal generator.

> **Study tip**
>
> The symbol $\approx$ means 'approximately equal to' and $\gg$ means 'very much greater than'.

Determining the wavelength of monochromatic light from a laser

A double slit can be used to determine the wavelength of light emitted from a laser [Figure 3]. There is no need for a filter or single slit as the light from the laser is already monochromatic and in phase.

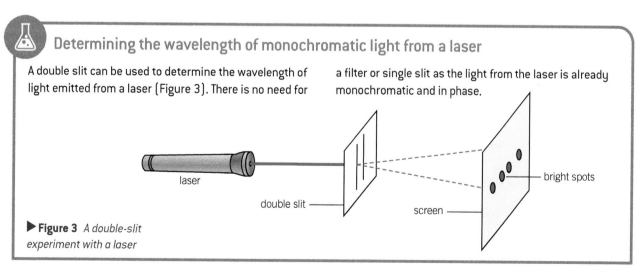

▶ **Figure 3** *A double-slit experiment with a laser*

By measuring the distance between several bright fringes, the separation x between adjacent fringes can be determined. If a is known (double slits are often labelled with this information) and D is measured directly, then λ can be calculated from:

$$\lambda = \frac{ax}{D}$$

1 Describe how the interference pattern would change if a laser emitting green light was used in place of the laser emitting red light.
2 Explain why it is better to measure the separation between multiple fringes to determine x, rather than between adjacent fringes.
3 A helium–neon laser is used to form an interference pattern on a screen 10 m from a double slit with a slit separation of 0.50 mm. Calculate the separation between adjacent fringes.

▲ **Figure 4** *The interference pattern formed from a helium–neon laser (λ = 632.8 nm)*

▲ Figure 5

Summary questions

1 Explain why it was necessary to use monochromatic light, along with a single slit and a double slit to produce a stable interference pattern for light. *(2 marks)*

2 The following measurements were taken during a double-slit experiment: separation between slits = 0.6 mm, separation between adjacent bright fringes = 1.4 mm, and distance from slits to screen = 1.6 m. Calculate the wavelength of light used. *(2 marks)*

3 The interference pattern in Figure 5 was obtained using a source of light of unknown wavelength. It is drawn to scale. The distance from the double slits with a slit separation of 1.0 mm to the screen was measured as 15 m. Determine the wavelength of the light. *(3 marks)*

4 The following measurements were taken during a double-slit experiment using light with a wavelength of 610 nm: separation between the slits = 0.40 mm and the separation between adjacent bright fringes = 1.8 mm. Calculate the distance from the slits to the screen. *(3 marks)*

5 Explain the effect on the interference pattern seen on the screen for a double-slit experiment when:
  a a light source with a longer wavelength is used; *(1 mark)*
  b the slit separation is doubled; *(2 marks)*
  c the distance from slits to screen is increased by a factor of three; *(2 marks)*
  d a light source with double the frequency is used. *(3 marks)*

12.4 Stationary waves

Specification reference: 4.4.4

Seeing UFOs

A lenticular cloud is a rare and beautiful natural phenomenon (Figure 1). They form over mountain ranges, but only if the conditions in the atmosphere are just right.

As the wind blows over the mountain peak a **stationary wave** (sometimes called **standing wave**) is formed. If the air contains enough moisture, the oscillations in the stationary wave make the water condense into the characteristic shape of this cloud.

Formation and properties of stationary waves

A stationary wave is not a single wave at all. It forms when two waves with the same frequency (and ideally the same amplitude) travelling in opposite directions are superposed (Figure 2). As they have the same frequency, at certain points they are in antiphase. At these points their displacements cancel out. This forms a **node**, a point where the displacement is always zero, and therefore the amplitude and the intensity are zero.

At other points when the two waves are always in phase, an **antinode** is formed – the point of greatest amplitude and therefore intensity.

▲ **Figure 2** *Stationary waves are formed by the superposition of two waves of the same frequency travelling in opposite directions, for example, on a stretched string or by microwaves*

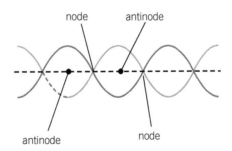

node   antinode

antinode   node

▲ **Figure 3** *A stationary wave is formed of a number of nodes and antinodes*

The separation between two adjacent nodes (or antinodes) is equal to half the wavelength of the original progressive wave (Figure 3), and the frequency is the same as that of the original waves. The wave profile for the stationary wave changes over time, creating the characteristic nodes and antinodes (Figure 4).

As the two progressive waves are travelling in opposite directions, there is no net energy transfer by a stationary wave, unlike a single progressive wave.

▲ **Figure 1** *Because of their lens-like shape, if they form at night lenticular clouds are sometimes reported as UFOs*

Study tip

Unlike a progressive wave, a stationary waves have nodes and antinodes.

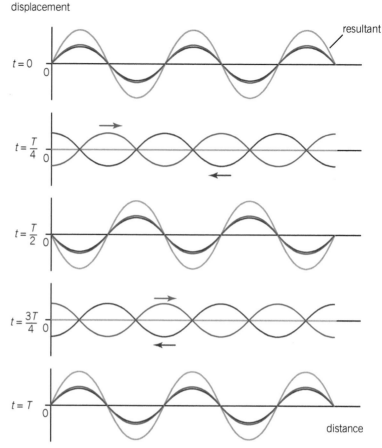

displacement

▲ **Figure 4** *A stationary wave (in green) is the resultant of two progressive waves travelling in opposite directions (red to the right, blue to the left)*

▲ **Figure 5** *Phase differences along a stationary wave are very different to phase differences along a progressive wave: particles A, B, and C are in phase with each other and in antiphase with particles D and E*

Phase differences along a stationary wave

In between adjacent nodes all the particles in a stationary wave are oscillating in phase with each other. They all reach their maximum positive displacement at the same time. However, their amplitudes differ, with the maximum amplitude at the antinode.

On different sides of a node the particles are in antiphase (they have a phase difference of π radians). The particles on one side of a node reach their maximum positive displacement at the same time as those on the other reach their maximum negative displacement.

Comparison of stationary and progressive waves

▼ **Table 1** *Summary of the properties of stationary and progressive waves*

	Progressive wave	Stationary wave
energy transfer	energy transferred in the direction of the wave	no net energy transfer
wavelength	minimum distance between two adjacent points oscillating in phase, for example, the distance between two peaks or two compressions	twice the distance between adjacent nodes (or antinodes) is equal to the wavelength of the progressive waves that created the stationary wave

▼ **Table 1** *(continued)*

	Progressive wave	Stationary wave
phase differences	the phase changes across one complete cycle of the wave	all parts of the wave between a pair of nodes are in phase, and on different sides of a node they are in antiphase
amplitude	all parts of the wave have the same amplitude (assuming no energy is lost to the surroundings)	maximum amplitude occurs at the antinode then drops to zero at the node

Forming a stationary wave using microwaves

A stationary wave can be formed by reflecting microwaves off a metal sheet so that two microwaves of the same frequency are travelling in opposite directions. A microwave receiver (Figure 6) will detect the changes in intensity between the nodes (lower/zero intensity) and antinodes (maximum intensity).

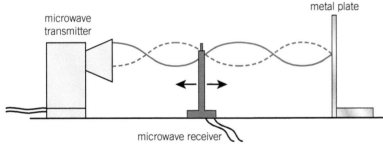

▲ **Figure 6** *Stationary microwaves*

The distance between the transmitter and metal sheet has to be adjusted until the receiver detects a series of nodes and antinodes. The distance between successive nodes or antinodes is equal to $\lambda/2$, where λ is the wavelength of the microwaves from the transmitter.

1   Describe how this experiment can be used to determine the wavelength of microwaves.
2   If the frequency of microwaves used in this experiment is 5.0 GHz, calculate the distance between adjacent nodes.
3   Explain why the node formed closest to the metal sheet provides the best cancellation whereas at other nodes the intensity does not fall to zero. (Hint: Think about the difference in distance travelled by the two waves forming the node, compared with the difference at the node closest to the transmitter.)

Summary questions

1   Describe two similarities between progressive and stationary waves. *(2 marks)*

2   Describe the phase difference between the two waves that form a stationary wave:
    a   at a node; *(1 mark)*
    b   at an antinode. *(1 mark)*

3   A standing wave has adjacent nodes 30 cm apart. Calculate the wavelength in metres of the progressive waves responsible for this standing wave. *(2 marks)*

4   Describe how the amplitude of a standing wave varies from one node to the next. *(3 marks)*

5   Figure 7 shows a standing wave formed on a stretched string.
    a   State the phase differences between:
        i   X and Y;   ii   X and Z. *(2 marks)*
    b   If the period of the wave is 6.0 s sketch the string after:
        i   1.5 s;   ii   3.0 s;
        iii   7.5 s. *(4 marks)*

▲ **Figure 7**

12.5 Harmonics

Specification reference: 4.4.4

Stringing out a note

Stringed instruments, like the violin, produce musical notes from stationary waves formed on their strings.

Each string has a **fundamental mode of vibration**. The frequency of this vibration is the **fundamental frequency**, and depends on the string's mass, tension, and length.

In addition, different **harmonics** are produced, with several different wavelengths. These wavelengths interfere with each other to produce the rich sound characteristic of stringed instruments. You can easily identify the type of stringed instrument – say guitar, violin, or piano – from its sound because of its unique combinations of harmonics.

Stationary waves on strings

If a string is stretched between two fixed points, these points act as nodes. When the string is plucked a progressive wave travels along the string and reflects off its ends. This creates two progressive waves travelling in opposite directions that then form a stationary wave.

▲ **Figure 1** *The strings of a violin are fixed at the ends so they are effectively nodes, resulting in the formation of a stationary wave when the bow rubs across a string*

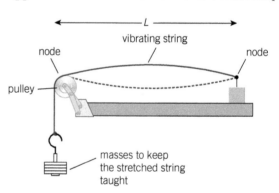

▲ **Figure 2** *A taut string vibrating in its fundamental mode of vibration, that is, at the fundamental frequency*

When the string is plucked, it vibrates in its fundamental mode of vibration (Figure 2), in which the wavelength of the progressive wave is double the length of the string ($2L$).

Harmonics and wavelength

The fundamental frequency f_0 is the minimum frequency of a stationary wave for a string. Along with this fundamental mode of vibration, the string can form other stationary waves called harmonics at higher frequencies (Table 1).

For a given string at a fixed tension, the speed of progressive waves along the string is constant. From the equation for progressive waves, $v = f\lambda$, you can see that as the frequency increases the wavelength must decrease in proportion. At a frequency of $2f_0$, the wavelength is half the wavelength at f_0.

▼ **Table 1** *The first five harmonics for a stationary wave between the fixed ends of a string — the frequency of each harmonic is an integer multiple of the fundamental frequency f_0 (in this case 20 Hz)*

Harmonic	Shape	Frequency / Hz	Frequency as a multiple of f_0	Wavelength of the progressive wave (where L is the length of the string)
1		20	f_0	$2L$
2		40	$2f_0$	L
3		60	$3f_0$	$\frac{2}{3}L$
4		80	$4f_0$	$\frac{1}{2}L$
5		100	$5f_0$	$\frac{2}{5}L$

 Investigating stationary waves on strings

Melde's experiment (Figure 3) is a simple way to investigate stationary waves on a string. A vibration generator can be used to change the frequency of the wave on the stretched string until a stable stationary wave is produced for a specific harmonic. The actual node on the right-hand side of the string is slightly to the right of the vibration generator (which cannot be a true node as it is vibrating).

In Figure 3, the string is vibrating at $2f_0$, approximately 214 Hz. If the generator is not set to an integer multiple of the fundamental frequency f_0 then no stationary wave will be formed. Instead the string will vibrate in non-distinct patterns.

With this apparatus the length of the string can be changed, or the fixed end can be replaced with a pulley and different masses attached to the string to vary its tension.

1   State the fundamental frequency (f_0) of the string in Figure 3.
2   The fundamental frequency of a string is 300 Hz. Sketch the shape of this string when the frequency of the vibration generator is set to:
    a   300 Hz;   b   600 Hz;   c   450 Hz.
3   The speed v of the progressive waves along a stretched string is given by $v = k\sqrt{T}$, where k is a constant and T is the tension in the string. Explain how increasing the tension affects the fundamental frequency.

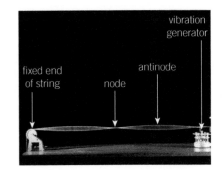

▲ **Figure 3** *Melde's string*

Summary questions

1 Describe how different harmonics on a string might be observed in a classroom. *(3 marks)*

2 Describe the effect on the fundamental frequency of a string if the length of the string doubles (assuming the tension remains constant). *(2 marks)*

3 The stationary wave in Figure 4 is produced at a frequency of 120 Hz and has a wavelength of 0.36 m.
   a State the fundamental frequency of the string. *(1 mark)*
   b Determine the wavelength of the original progressive wave. *(2 marks)*
   c Sketch the pattern produced if the frequency changes to:
   i 160 Hz;   ii 100 Hz. *(2 marks)*

▲ Figure 4

4 The graph in Figure 5 was recorded using data collected from a violin playing a single note. It shows how the intensity varies at different frequencies (the frequency spectrum).
   a State the fundamental frequency for the vibrating string. *(1 mark)*
   b Explain how the graph shows the existence of harmonics on the string. *(2 marks)*

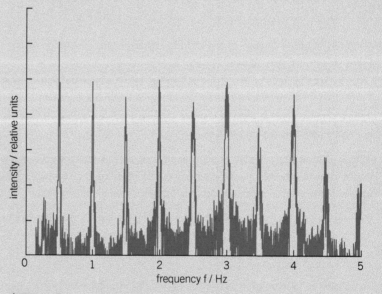

▲ Figure 5

5 A stationary wave is formed on a string of length 90 cm. At 3.6 kHz, six antinodes can be observed.
   a Determine the wavelength of the progressive waves on the string.
   b Calculate the speed of the progressive waves travelling along the string. *(4 marks)*

12.6 Stationary waves in air columns

Specification reference: 4.4.4

Playing the pipes

Stationary waves are formed not only by transverse waves like microwaves or waves on strings, but also by longitudinal waves, like sound. Most woodwind instruments, like these pan pipes (Figure 1), produce notes from stationary waves in air columns.

Gently blowing over the top of a tube creates a standing wave inside it, which produces a note at a particular frequency. The length of a tube determines the wavelength of the note it produces.

Stationary waves with sound

Sound waves reflected off a surface can form a stationary wave. The original wave and the reflected wave travel in opposite directions and superpose (Figure 2).

▲ **Figure 1** *Pan pipes are a traditional woodwind instrument played by Peruvians from the Andes*

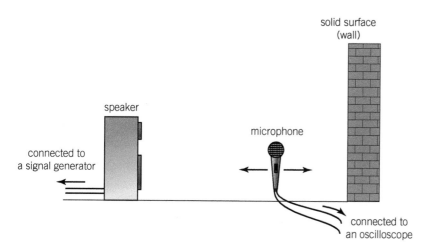

▲ **Figure 2** *Sound waves reflected off a solid surface can form a stationary wave in the same way as the stationary wave formed with microwaves described in Topic 12.4, Stationary waves*

Stationary sound waves can also be made in tubes by making the air column inside the tube vibrate at frequencies related to the length of the tube. The stationary wave formed depends on whether the ends of the tube are open or closed.

Stationary waves in a tube closed at one end

In order for a stationary wave to form in a tube closed at one end there must be an antinode at the open end and a node at the closed end. The air at the closed end cannot move, and so must form a node. At the open end, the oscillations of the air are at their greatest amplitude, so it must be an antinode.

The fundamental mode of vibration simply has a node at the base and an antinode at the open end. Harmonics are also possible (Figure 3).

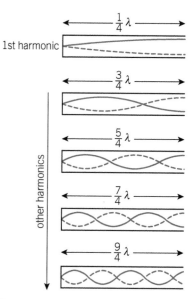

▲ **Figure 3** *Harmonics of a longitudinal wave in a tube closed at one end*

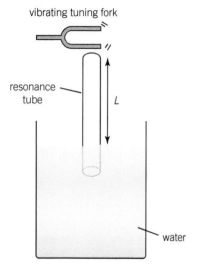

vibrating tuning fork

resonance
tube

L

water

▲ **Figure 4** *The top end of the tube is open. The level of water determines the position of the closed end of this tube.*

▼ **Table 1** *Measurements of f and L*

Frequency of tuning fork / Hz	Length of air column /cm
200	38.5
220	34.1
280	26.4
340	21.5
440	15.0
520	12.3
600	9.8

Unlike stationary waves on stretched strings, in a tube closed at one end it is not possible to form a harmonic at $2f_0$ – there is no second (or fourth, sixth, …) harmonic for a tube closed at one end. The frequencies of the harmonics in tubes closed at one end are always an odd multiple of the fundamental frequency ($3f_0$, $5f_0$, …).

Speed of sound in a resonance tube

Holding a tuning fork above a tube closed at one end can form a stationary wave inside the tube. The air vibrates at the same frequency as the tuning fork. If the frequency of the tuning fork is at the fundamental frequency for the air column, the sound becomes loud as the air inside the tube resonates.

In the apparatus in Figure 4 the length of the tube can be changed by raising and lowering it in the water. When the frequency of the tuning fork matches f_0, the length of the tube above water L must be equal to $\frac{1}{4}\lambda$ (see Figure 3).

The speed of sound in air can be calculated using $v = f\lambda = f \times 4L$, where f is the frequency of the tuning fork.

The length L corresponding to the fundamental mode of vibration is recorded for a number of different tuning forks. The results are shown in Table 1.

1 Copy the diagram of the experiment and draw in the stationary wave formed inside the tube. Label the position of any nodes and antinodes.

2 Plot a graph of L / m against $\frac{1}{f}$ / s.

a Show that the gradient of this graph is equal to $\frac{v}{4}$, where v is the speed of sound in air.

b Use your graph to find the speed of sound in air.

c Suggest a reason why the graph does not quite pass through the origin. (Hint: Look carefully at the diagram.)

Stationary waves in open tubes

A tube open at both ends must have an antinode at each end in order
to form a stationary wave, as explained above (Figure 5). Unlike
a tube closed at one end, harmonics at all integer multiples of the
fundamental frequency (f_0, $2f_0$, $3f_0$, ...) are possible in an open tube.

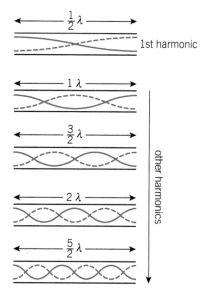

▲ **Figure 5** *Harmonics in an open tube*

Summary questions

1 Describe a simple method for the formation of a stationary sound
wave using a speaker, a microphone, and a solid surface, and how to use
it to determine the wavelength of sound produced from a
signal generator. *(4 marks)*

2 The air column inside a tube closed at one end of length 1.2 m vibrates
at its fundamental frequency. Calculate:
    a the wavelength of the sound; *(2 marks)*
    b the frequency of the sound produced (speed of sound
        through air = 340 m s^{-1}). *(1 mark)*

3 Repeat the calculations from question 2 for an open tube. *(3 marks)*

4 The air column inside an open tube is made to vibrate at $2f_0$,
where f_0 is the fundamental frequency. Identify the nodes and
antinodes in Figure 6 below. *(5 marks)*

▲ **Figure 6**

5 Explain why it is not possible to produce a harmonic at $2f_0$,
where f_0 is the fundamental frequency, in a tube closed at
one end and open at the other. *(2 marks)*

Practice questions

1 This question is about the Young double slit experiment. See Figure 1. The fringe pattern seen on the screen is shown to the right.

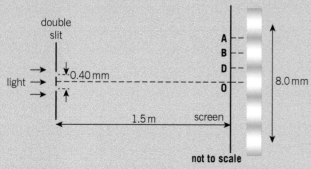

▲ Figure 1

Two parallel clear lines are scratched on a darkened glass slide 0.40 mm apart. When a beam of monochromatic visible light is shone through these slits, interference fringes are observed on a screen placed 1.5 m from the slide. The fringe at point **B** is bright and the fringe at point **D** is dark.

a Explain why this arrangement with two slits is used to produce visible fringes on the screen rather than two separate identical light sources. *(2 marks)*

b State the **phase difference** between the light waves from the two slits that meet on the screen in Figure 1 at points D and B. *(2 marks)*

c (i) Use Figure 1 to calculate the separation of adjacent bright fringes, the distance between **O** and **B**. *(1 mark)*

   (ii) Show that the wavelength λ of the monochromatic light is about 5×10^{-7} m. *(3 marks)*

d Calculate the **path difference**, in nanometers, between the light waves from the two slits that meet on the screen in Figure 1 at point **A**. *(2 marks)*

June 2013 G482

2 Figure 2 shows two loudspeakers connected to a signal generator, set to a frequency of 1.2 kHz. A person walks in the direction **P** to **Q** at a distance of 3.0 m from the loudspeakers.

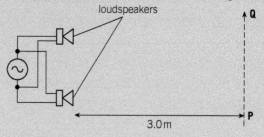

▲ Figure 2

(i) Calculate the wavelength λ of the sound waves omitted from the loudspeakers.

Speed of sound in air = 340 ms^{-1}

(2 marks)

(ii) Explain, either in terms of path difference or phase difference, why the intensity of the sound heard varies as the person moves along **PQ**.

(3 marks)

(iii) The distance x between adjacent positions of maximum sound is 0.50 m. Calculate the separation a between the loudspeakers. Assume that the equation used for the interference of light also applies to sound.

(2 marks)

(iv) The connections to one of the loudspeakers are reversed. Describe the similarities and differences in what the person hears. *(2 marks)*

Jan 2012 G482

3 Figure 3 shows a string stretched between two fixed points.

▲ Figure 3

The middle of the string is pulled up and then released. This creates a stationary wave on the string. The distance between the fixed supports is 32 cm.

a Explain how a stationary wave is produced on this stretched string. *(3 marks)*

b The string vibrates at its fundamental mode of vibration. Sketch the shape of the stationary wave produced. *(2 marks)*

c A stroboscope is used to determine the frequency of vibration of the string. The frequency is found to be 160 Hz. Calculate the speed of the transverse waves on the string. *(3 marks)*

4 a When used to describe stationary (standing) waves explain the terms

    (i) node, *(1 mark)*

    (ii) antinode. *(1 mark)*

b Figure 4 shows a string fixed at one end under tension. The frequency of the mechanical oscillator close to the fixed end is varied until a stationary wave is formed on the string.

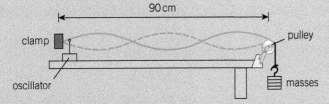

▲ **Figure 4**

    (i) Explain with reference to a progressive wave on the string how the stationary wave is formed. *(3 marks)*

    (ii) On a copy of Figure 4, label one node with the letter **N** and one antinode with the letter **A**. *(1 mark)*

    (iii) State the number of antinodes on the string in Figure 4. *(1 mark)*

    (iv) The frequency of the oscillator causing the stationary wave shown in Figure 4 is 120 Hz.

    The length of the string between the fixed end and the pulley is 90 cm.

    Calculate the speed of the progressive wave on the string. *(3 marks)*

c The speed v of a progressive wave on a stretched string is given by the formula

$$v = k\sqrt{W}$$

where k is a constant for that string. W is the tension in the string which is equal to the weight of the mass hanging from the end of the string.

In **(b)** the weight of the mass on the end of the string is 4.0 N. The oscillator continues to vibrate the string at 120 Hz. Explain whether or not you would expect to observe a stationary wave on the string when the weight of the suspended mass is changed to 9.0 N. *(3 marks)*

May 2009 G482

5 A vibrating tuning fork is held next to the open end of a horizontal tube. The tube is closed at the other end, see Figure 5. A stationary sound wave, of fundamental mode of vibration, is produced in the air column within the tube.

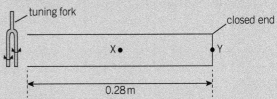

▲ **Figure 5**

The length of the air column in the tube is 0.60 m.

a On a copy of Figure 5, draw the stationary wave pattern produced in the air column within the tube.

Mark the positions of the node (**N**) and the antinode (**A**). *(2 marks)*

b State the oscillations of the air particles at points **X** and **Y**. *(2 marks)*

c State and explain the phase difference between air particles vibrating at the open end of the tube and at **X**. *(2 marks)*

d The stationary wave produced emits a note at its fundamental frequency f_0. The speed of sound in air is 340 m s^{-1}. Calculate the value of f_0. *(3 marks)*

Learning outcomes

Demonstrate knowledge, understanding, and application of:

→ the particulate nature (photon model) of electromagnetic radiation

→ photons as quanta of energy of electromagnetic radiation

→ energy of a photon $E = hf$ and $E = \dfrac{hc}{\lambda}$

→ the electronvolt

→ using LEDs and the equation $eV = \dfrac{hc}{\lambda}$ to estimate the value of the Planck constant h.

Imaging the brain

A single-photon emission computed tomography (SPECT) scan, like the one in Figure 1, relies on the idea of electromagnetic radiation as photons. In the previous chapter we saw evidence for the wave nature of electromagnetic radiation. **Quantum physics** explores how, at the very small scale, we use different models to describe electromagnetic radiation and subatomic particles like electrons. Quantum physics lies behind much modern technology, from medical scans like this to broadband connections.

Photons

One of the scientists instrumental in the development of quantum physics was the German Max Planck. In 1900 he discovered that electromagnetic energy could only exist in certain values – it appeared to come in little packets (quanta). His new model proposed that electromagnetic radiation had a particulate nature – it was tiny packets of energy, rather than a continuous wave. Einstein coined a new term for these packets, **photons**.

Developments in quantum physics have led to an understanding that we can use different models to describe electromagnetic radiation. For example, we use the photon model to explain how electromagnetic radiation interacts with matter, and the wave model to explain its propagation through space. These models are rigorously tested, and any model that does not match experimental data or makes incorrect predictions is modified or discarded, pushing our understanding forward.

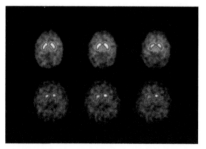

▲ **Figure 1** *This image was produced by recording single photons emitted as part of a brain scan – the top row shows normal brain activity, the bottom shows a patient with Parkinson's disease on one side*

Photon energy

The energy of each photon is directly proportional to its frequency. Specifically

$$E = hf$$

where E is the energy of the photon in J, f is the frequency of the electromagnetic radiation in Hz, and h is the **Planck constant**. It has an experimental value of $6.63 \times 10^{-34}\,\text{J s}$ and will be used throughout the quantum physics topic.

We can combine this equation with the wave equation $c = f\lambda$ to express the energy of a photon in terms of its wavelength and c (the speed of light through a vacuum, $3.00 \times 10^8\,\text{m s}^{-1}$).

$$E = \frac{hc}{\lambda}$$

Synoptic link

The wave equation $c = f\lambda$ was introduced in Topic 11.2, Wave properties.

This is an intriguing equation. It has both wave elements (because of the λ) and particulate elements (photon, because of the E). From this equation we can also see the energy of a photon is inversely proportional to its wavelength ($E \propto \frac{1}{\lambda}$). Short-wavelength photons, like X-rays, have much more energy than long-wavelength radio waves. This energy partly explains why X-rays can damage the cells of our bodies.

 Worked example: Photons of red light

A laser emits red light with a wavelength of 633 nm. Calculate the energy of each red photon emitted.

Step 1: Identify the correct equation to calculate the energy of a photon.

$$E = hf$$

We only know the wavelength of the photon, so we must use the wave equation to express frequency in terms of the wavelength and the speed of light.

$$c = f\lambda$$

$$f = \frac{c}{\lambda}$$

We can substitute this into our first equation, giving

$$E = \frac{hc}{\lambda}$$

Step 2: Substitute in the correct values, taking care with λ in nm.

$$E = \frac{6.63 \times 10^{-34} \times 3.00 \times 10^{8}}{633 \times 10^{-9}}$$

Calculate the energy of the photon in joules. $E = 3.14 \times 10^{-19}$ J

> **Study tip**
>
> A photon is a quantum of electromagnetic energy. The energy of a photon depends on frequency or wavelength of the electromagnetic radiation.

Questions of scale

At the subatomic scale of the quantum level, the SI unit of energy, the joule, is huge. Even the most energetic gamma photons only have energies of the order of tens of millijoules. A typical red photon has energy of around 3×10^{-19} J.

Just as we use the kWh as a unit of electrical energy when dealing with the billions of joules transferred to our homes, we often use another unit when measuring energies at the quantum scale, the **electronvolt** (eV).

The energy of 1 eV is defined as the energy transferred to or from an electron when it moves through a potential difference of 1 V.

We know that the work done on an electron moving through a p.d. is equal to the p.d. × the charge on the electron ($W = VQ = Ve$). Therefore, the work done on an electron as it moves through a p.d. of 1 V is given by

$$W = 1\,\text{V} \times 1.60 \times 10^{-19}\,\text{C} = 1.60 \times 10^{-19}\,\text{J}$$

> **Synoptic link**
>
> You can review the equation linking work and charge in Topic 9.2, Potential difference and electromotive force.

Therefore, 1 eV is equivalent to 1.60×10^{-19} J, or 1 J is equivalent to 6.25×10^{18} eV. 1 eV is a tiny amount of energy. It is very common to see energies expressed as keV, MeV, or even GeV for high-energy particles or photons.

A typical infrared photon has an energy of 1.5 eV. You can imagine this energy as the equivalent of the kinetic energy gained by an electron travelling through a p.d. of 1.5 V (between the terminals of a typical AA cell).

▼ **Table 1** *Some photon energies expressed in both J and eV*

To convert from J to eV, divide by 1.60×10^{-19}

Photon	Typical energy		Typical wavelength
	/ J	/ eV	/ m
radio	6.5×10^{-27}	4.0×10^{-8}	3.1×10^{1}
infrared	1.6×10^{-20}	0.1	1.2×10^{-5}
visible – red	3.0×10^{-19}	1.9	6.6×10^{-7}
visible – green	3.7×10^{-19}	2.3	5.4×10^{-7}
visible – blue	4.3×10^{-19}	2.7	4.6×10^{-7}
UV	9.6×10^{-19}	6.0	2.1×10^{-7}
X-ray	1.6×10^{-16}	1.0×10^{3}	1.2×10^{-9}
gamma	2.4×10^{-13}	1.5×10^{6}	8.3×10^{-13}

To convert from eV to J, multiply by 1.6×10^{-19}

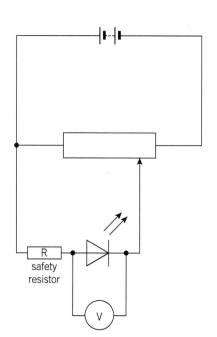

▲ **Figure 2** *A simple circuit containing an LED, a battery, a voltmeter, and a resistor (to protect the LED) can be used to give an approximate value for the Planck constant*

LEDs and the Planck constant

We can do a simple experiment with LEDs (Figure 2) to determine a value for the Planck constant by considering the energies of the photons they emit.

LEDs convert electrical energy into light energy. They emit visible photons when the p.d. across them is above a critical value (the threshold p.d.). Figure 3 shows the I–V characteristic for a typical LED.

When the p.d. reaches the threshold p.d. the LED lights up and starts emitting photons of a specific wavelength. At this p.d. the work done is given by $W = VQ$. This energy is about the same as the energy of the emitted photon. We can use the voltmeter to measure the minimum p.d. that is required to turn on the LED. A black tube placed over the LED helps to show exactly when the LED lights up. If we also know

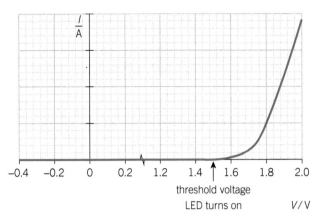

▲ **Figure 3** *The I–V characteristic and threshold p.d. for a typical LED*

the wavelength of the photons emitted by the LED we can determine the Planck constant.

At the threshold p.d., the energy transferred by an electron in the LED is approximately equal to the energy of the single photon it emits.

threshold p.d. × charge on electron ≈ energy of emitted photon
$$Ve = hf$$

Expressing this in terms of the wavelength of the emitted photon λ gives

$$eV = \frac{hc}{\lambda}$$

We can use this equation for a single LED and calculate h, but in order to obtain a more accurate value we should gather data using a variety of different-wavelength LEDs (the threshold p.d. dictates the colour of the LED). We can then plot a graph of V against $\frac{1}{\lambda}$ (Figure 4). The equation for a straight-line graph, $y = mx + c$, is equivalent to $V = \frac{hc}{\lambda e}$ here, so the Planck constant can be determined from the gradient of the graph, $\frac{hc}{e}$.

Synoptic link

For more information about the gradient of straight-line graphs, see Appendix A2, Recording results.

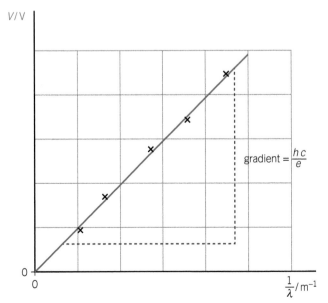

▲ **Figure 4** *A plot of V against $\frac{1}{\lambda}$*

Summary questions

1 Calculate the energy of each photon from the following frequencies:
   a infrared at 1.02×10^{14} Hz;   (1 mark)
   b a radio wave at 97.0 MHz;   (1 mark)
   c visible light at 6.00×10^{14} Hz.   (1 mark)

2 State and explain which part of the visible spectrum has
   photons with the highest energy.   (3 marks)

3 A photon has energy of 3.32×10^{-18} J. Calculate its
   wavelength.   (3 marks)

4 Convert the following energies into eV:
   a 1.0 J;   (1 mark)
   b 3.30×10^{-14} J;   (1 mark)
   c 600 nJ.   (1 mark)

5 Calculate the frequencies of the photons in Table 1.   (8 marks)

6 Calculate the photon energies in eV of each of the following
   wavelengths:
   a X-rays at 4.50×10^{-10} m;   (2 marks)
   b orange light at 600 nm.   (2 marks)

7 a An LED emits orange photons with a wavelength of 620 nm.
      Calculate the threshold p.d.   (2 marks)
   b Sketch the $I-V$ characteristics for an LED that emits red
      photons and one that emits blue photons. Use the graphs
      to explain the differences in the threshold p.d.   (6 marks)

8 A laser emits blue light of wavelength 405 nm and
   radiant power 10 mW. Calculate the number of photons
   emitted per second.   (3 marks)

9 A student is given an LED that emits monochromatic light of
   wavelength λ_R. Plan an experiment that uses this LED to determine
   a value for the Planck constant h. Suggest how the precision of the
   experiment can be improved.   (6 marks)

13.2 The photoelectric effect

Specification reference: 4.5.2

Surface charging

Surface charging is a phenomenon experienced by all spacecraft. Outside the Earth's protective atmosphere, high-energy electromagnetic radiation causes electrons to be emitted from the metal parts of the spacecraft facing the Sun. This is called the **photoelectric effect**.

Surface charging results in some parts of the spacecraft carrying a positive charge, potentially leading to a damaging flow of charge through key electronic components inside the spacecraft. Engineers have to design solutions to this problem to ensure that charges cannot build up to potentially damaging levels.

The photoelectric effect

In 1887 Heinrich Hertz reported that when he shone UV radiation onto zinc, electrons were emitted from the surface of the metal.

This is the photoelectric effect. The emitted electrons are sometimes called **photoelectrons**. They are normal electrons, but their name describes their origin – emitted through the photoelectric effect (Figure 2).

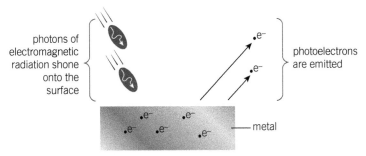

▲ **Figure 2** *The photoelectric effect occurs when electromagnetic radiation incident on the surface of a metal causes electrons to be emitted*

▲ **Figure 1** *The European Space Agency and its Japanese equivalent plan to launch the BepiColumbo probe, designed to minimise the hazards caused by surface charging, to Mercury in 2016*

The gold-leaf electroscope

A simple demonstration of the effect can be seen with a **gold-leaf electroscope**. These were originally designed to measure p.d. (an early voltmeter). However, we can use them to demonstrate how like electrical charges repel each other.

Briefly touching the top plate with the negative electrode from a high-voltage power supply will charge the electroscope. Excess electrons are deposited onto the plate and stem of the electroscope. Any charge developed on the plate at the top of the electroscope spreads to the stem and the gold leaf. As both the stem and gold leaf have the same charge, they repel each other, and the leaf lifts away from the stem (Figure 3). If a clean piece of zinc is placed on top of a negatively charged

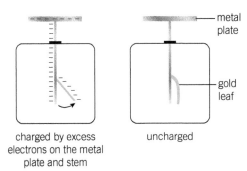

charged by excess electrons on the metal plate and stem

uncharged

▲ **Figure 3** *Charged and uncharged electroscopes*

243

electroscope and UV radiation shines onto the zinc surface, then the gold leaf slowly falls back towards the stem. This shows that the electroscope has gradually lost its negative charge, because the incident radiation (in this case UV) has caused the free electrons to be emitted from the zinc. These electrons are known as photoelectrons.

Three key observations from the photoelectric effect

The electroscope experiment is simple, but it was revolutionary. When different frequencies of incident radiation were investigated in more detail, scientists at the time made three key observations.

1 Photoelectrons were emitted only if the incident radiation was above a certain frequency (called the **threshold frequency** f_0) for each metal. No matter how intense the incident radiation (how bright the light), not a single electron would be emitted if the frequency was less than the threshold frequency.

2 If the incident radiation was above the threshold frequency, emission of photoelectrons was instantaneous.

3 If the incident radiation was above the threshold frequency, increasing the intensity of the radiation did not increase the maximum kinetic energy of the photoelectrons. Instead more electrons were emitted. The only way to increase the maximum kinetic energy was to increase the frequency of the incident radiation.

These observations could not be explained using wave model of electromagnetic radiation. For example, if the threshold frequency for a particular metal is in the green part of the visible spectrum, bright red light does not cause emission, yet very dim blue light would. This does not fit with the wave model, in which the rate of energy transferred by the radiation is dependent on its intensity (brightness). The more intense the radiation, the more energy is transferred to the metal per second, and bright red light transfers more energy per second than dim blue light. Clearly a new model for electromagnetic radiation was needed to explain the observations.

Using photons to illuminate the photoelectric effect

The observations in the photoelectric effect can be explained if the wave model of light is replaced with the photon model. In 1905 Einstein published an explanation of the effect. Building on Planck's work, he proposed the idea of electromagnetic radiation as a stream of photons, rather than continuous waves.

He suggested that each electron in the surface of the metal must require a certain amount of energy in order to escape from the metal, and that each photon could transfer its exact energy to one surface electron in a one-to-one interaction.

As the energy of the photon is dependent on its frequency ($E = hf$), if the frequency of the photon is too low, the intensity of the light – that is, the number of photons per second – does not matter, as a single photon delivers its energy to a single surface electron in a one-to-one interaction. If a photon does not carry enough energy on its own to free an electron, the number of photons makes no difference. However, when the frequency of the light is above the threshold frequency f_0 for the metal, then each individual photon has enough energy to free a single surface electron and so photoelectrons are emitted (Figure 4).

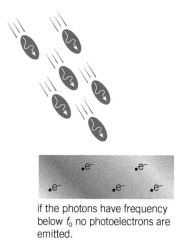

if the photons have frequency below f_0 no photoelectrons are emitted.

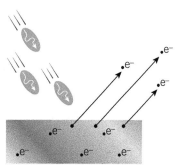

if the photons have frequency above f_0 photoelectrons are emitted (even when the light is dimmer).

▲ **Figure 4** *Replacing the wave model with the photon model allowed Einstein to explain the photoelectric effect*

Study tip

The threshold frequency and work function are properties of the metal surface. They are not properties of electrons or photons.

This also explained why there was no time delay. As long as the incident radiation has frequency greater than, or equal to, the threshold frequency, as soon as photons hit the surface of the metal, photoelectrons are emitted. Electrons cannot accumulate energy from multiple photons. Only one-to-one interactions are possible between photons and electrons.

Einstein was also able to explain the third observation. Depending on their position relative to the positive ions in the metal, electrons would require different amounts of energy to free them. Einstein defined a constant for each metal which he called the **work function** ϕ. This is the *minimum* energy required to free an electron from the surface of the metal.

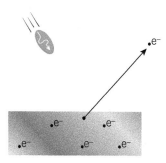

Increasing the intensity of the radiation means more photons per second hit the metal surface. As each photon interacts one-to-one with a single surface electron, as long as the radiation has frequency above the threshold frequency for the metal, more photons per second means a greater rate of photoelectrons emitted from the metal. The rate of emission of photoelectrons is directly proportional to the intensity of the incident radiation. Double the intensity and you double the number of photons per second, leading to a doubling in the number of electrons emitted from the metal per second.

Using the principle of conservation of energy, Einstein deduced that the kinetic energy of each photoelectron depends on how much energy was *left over* after the electron was freed from the metal (more

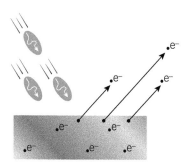

▲ **Figure 5** *Intense radiation means a higher rate of photons landing and so a higher rate of electrons escaping*

on this in Topic 13.3, Einstein's photoelectric effect equation). At a given frequency all photons have the same amount of energy, and the metal a specific work function, so there is a maximum value of kinetic energy that any emitted photoelectrons can have. Increasing the intensity results in a greater rate of emission, but none of the emitted photoelectrons will move faster.

The only way to increase the maximum kinetic energy of the emitted photoelectrons is to increase the frequency of the radiation. In this case each photon has more energy and so each electron has more kinetic energy after it has been freed from the metal.

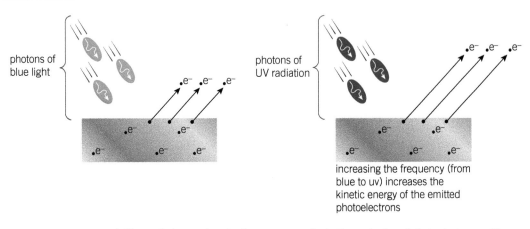

increasing the frequency (from blue to uv) increases the kinetic energy of the emitted photoelectrons

▲ **Figure 6** *Increasing the frequency results in the emission of photoelectrons with a higher kinetic energy (as there is more energy to spare after the electrons have been freed from the metal)*

Summary questions

1 Describe what would happen to an uncharged gold-leaf electroscope if its top surface were to come into contact with a positive electrode. *(2 marks)*

2 If a particular metal had a threshold frequency in the red part of the visible spectrum, explain what would happen to the metal if radiation was incident on its surface from:
   a the infrared part of the spectrum; *(2 marks)*
   b the blue part of the visible spectrum. *(1 mark)*

3 Explain why the maximum kinetic energy of photoelectrons emitted during the photoelectric effect depends on the frequency of the incident radiation. *(3 marks)*

4 The threshold wavelength λ_0 is the *longest* wavelength that will give rise to the photoelectric effect. Derive an expression for λ_0 in terms of the threshold frequency f_0. *(2 marks)*

5 State and explain the effect of quadrupling the intensity of incident radiation (keeping the frequency constant) on a metal surface emitting photoelectrons. *(3 marks)*

Seeing in the dark

In some kinds of night vision goggles the photoelectric effect is used to amplify light. In the construction of the goggles it is important to select materials with an appropriate work function. Often a semiconductor material such as gallium arsenide is used.

An image intensifier produces an image even in conditions that appear pitch black to the human eye. The emission of individual photoelectrons due to the photoelectric effect is used to build up a clear picture of the surroundings.

Conservation of energy and the photoelectric effect

In his model of the photoelectric effect Einstein applied the idea of conservation of energy to the photons and photoelectrons. By thinking about the energies involved he derived what is now simply known as **Einstein's photoelectric effect equation**. He realised that the energy of each individual photon must be conserved. This energy does two things:

- it frees a single electron from the surface of the metal in a one-to-one interaction
- any remainder is transferred into the kinetic energy of the photoelectron.

By using his idea of work function ϕ as the *minimum* energy to free the electron from a particular metal, he produced a general equation relating the energy of each photon, the work function of the metal, and the maximum kinetic energy of the emitted photoelectron.

According to the principle of conservation of energy, we have

energy of a single photon	=	minimum energy required to free a single electron from the metal surface	+	maximum kinetic energy of the emitted electron

$$hf = \phi + KE_{max}$$

All the terms in the equation, hf, ϕ, and KE_{max}, are energies, so all should be measured in joules, or consistently in electronvolts.

Learning outcomes

Demonstrate knowledge, understanding, and application of:

→ Einstein's photoelectric equation $hf = \phi + KE_{max}$

▲ **Figure 1** *As well as military applications, night vision technologies can be used in conservation to monitor animals in complete darkness, like these rhinos at a waterhole*

 Worked example: Emitting photoelectrons from a metal surface

Photoelectrons with a maximum kinetic energy of 9.34×10^{-19} J are emitted from a metal with a work function of 2.40 eV. Calculate the frequency of the incident radiation. ⊙➔

Step 1: Identify the correct equation to calculate the frequency of the radiation.

$$hf = \phi + KE_{max}$$

Convert the work function into joules.

$$2.40\,eV \times 1.60 \times 10^{-19} = 3.84 \times 10^{-19}\,J$$

Step 2: Substitute the values into the equation to calculate the energy of a single photon.

$$hf = 3.84 \times 10^{-19} + 9.34 \times 10^{-19} = 1.32 \times 10^{-18}\,J$$

Using this value we can determine the frequency of the radiation.

$$f = \frac{1.32 \times 10^{-18}}{h} = \frac{1.32 \times 10^{-18}}{6.63 \times 10^{-34}} = 1.99 \times 10^{15}\,Hz$$

Why *maximum* kinetic energy?

Some electrons in the surface of the metal are closer to the positive metal ions than others. Their relative positions affect how much energy is required to free them. The work function is the minimum energy required to free an electron from the metal – most electrons need a little more energy than the work function to free them.

An electron that requires the minimum amount of energy to free it (the work function of the metal) would have the most energy left over from the incident photon. Only a few of the emitted photoelectrons have this *maximum* kinetic energy – most have a little bit less, and so travel a little slower.

If a photon strikes the surface of the metal at the threshold frequency f_o for the metal then it will only have enough energy to free a surface electron, with none left over to be transferred into kinetic energy of the electron.

In this case Einstein's photoelectric effect equation becomes

$$hf_0 = \phi + 0 \text{ or simply } hf_0 = \phi$$

A graph of KE_{max} against incident frequency

In Topic 13.2, The photoelectric effect, we learnt that the only way to increase the maximum kinetic energy of photoelectrons is to increase the frequency of the incident radiation. A graph of the maximum kinetic energy of the photoelectrons plotted against the frequency of the radiation on the surface can be seen in Figure 2.

Considering the general equation for a straight-line graph, $y = mx + c$, and rearranging $hf = \phi + KE_{max}$ to match, we get $KE_{max} = hf - \phi$. We can see that the gradient of this graph must equal the Planck constant h and the y-axis intercept is equal to $-\phi$, where ϕ is the work function of the metal. See Figure 2.

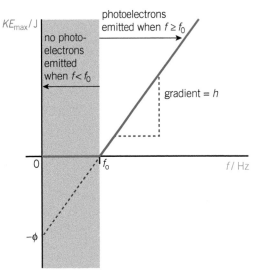

▲ **Figure 2** *A graph to show how the frequency f of the incident radiation on a metal surface affects the maximum kinetic energy KE of the emitted photoelectrons*

▼ **Table 1**

Metal	$\phi\,/\,eV$
caesium	2.14
sodium	2.36
aluminium	4.08
zinc	4.30

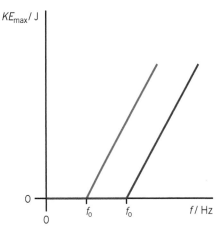

▲ **Figure 3** *Different metals have different work functions leading to different threshold frequencies*

Different metals

Every metal has a different work function, so the threshold frequency for each metal is different. Table 1 shows the work function for some metals.

In each case a graph of maximum kinetic energy against frequency will have the same gradient. This gradient is equal to h (Figure 3).

Summary questions

1 Photoelectrons with a maximum kinetic energy of 2.68×10^{-19} J are emitted from a metal with a work function of 3.77×10^{-19} J. Calculate the energy of the incident photons. *(2 marks)*

2 Aluminium has a work function of 4.08 eV. Calculate the maximum kinetic energy in eV of photoelectrons emitted from aluminium when illuminated with photons with energy of:
   a 5.20 eV;   b 3.10 eV.   *(3 marks)*

3 Using Table 1, calculate the threshold frequency for zinc and for sodium. *(4 marks)*

4 Explain why under monochromatic radiation, emitted photoelectrons have a range of different kinetic energies. *(3 marks)*

5 Calculate the maximum velocity of photoelectrons emitted from sodium when the incident radiation has a frequency of 1.48×10^{15} Hz. *(4 marks)*

6 The data in Table 2 was collected during an investigation into the photoelectric effect. Use this data to determine the threshold frequency and work function for the metal. *(4 marks)*

▼ Table 2

Frequency of incident radiation / Hz	Maximum kinetic energy of photoelectrons / eV
8.0×10^{14}	1.36
1.1×10^{15}	2.61

13.4 Wave–particle duality

Learning outcomes

Demonstrate knowledge, understanding, and application of:

→ electron diffraction

→ diffraction of electrons by the atoms of graphite

→ the de Broglie equation $\lambda = \dfrac{h}{p}$.

Diffracting particles

In 1924 the French physicist Louis de Broglie proposed that all matter can have both wave and particle properties. He suggested that tiny subatomic particles, like electrons, have wavelengths and can be made to exhibit wave properties like diffraction. We have already seen in Topic 13.1, The photon model, that the particulate model of electromagnetic radiation can explain the photoelectric effect. So photons can be thought of as particles, yet electromagnetic radiation is also able to diffract, an example of wave behaviour.

This **wave–particle duality** is a model used to describe how all matter has both wave and particle properties. De Broglie, who was awarded the 1929 Nobel Prize in Physics for his insight, realised that all particles travel through space as waves. Anything with mass that is moving has wave-like properties. These waves are referred to as matter waves or de Broglie waves.

One of the largest objects to show its wave nature is the carbon-60 molecule (buckminsterfullerene, Figure 1). In 1999 it was used to form a diffraction pattern. The C_{60} had a wavelength of just 2.5×10^{-12} m, a hundred times smaller than the diameter of an atom.

▲ **Figure 1** *Carbon-60 is one of the largest particles that has been successfully diffracted to date*

Electron diffraction

We would normally describe electrons as particles, as they have mass and charge. As a result they can be accelerated and deflected by electric and magnetic fields. This behaviour is only associated with particles. However, under certain conditions we can make electrons diffract. They spread out like waves as they pass through a tiny gap, and can even form diffraction patterns in the same way as light.

If an electron gun fires electrons at a thin piece of **polycrystalline graphite**, which has carbon atoms arranged in many different layers, the electrons pass between the individual carbon atoms in the graphite. The gap between the atoms is so small ($\sim 10^{-10}$ m) that it is similar to the wavelength of the electrons and so the electrons diffract, as waves, and form a diffraction pattern seen on the end of the tube (Figures 2 and 3).

We do not normally notice the wave nature of electrons because we need a tiny gap in order to observe electrons diffracting. For diffraction to occur the size of the gap through which the electrons pass must be similar to their wavelength.

This experiment beautifully demonstrates both the particle and wave nature of electrons. They are behaving as particles when they are accelerated by the high potential difference, they behave as waves when they diffract, and then they behave as particles again as they hit the screen with discrete impacts.

Synoptic link

Topic 9.3, The electron gun, explains how electron guns work.

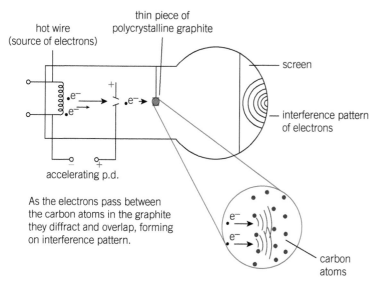

As the electrons pass between the carbon atoms in the graphite they diffract and overlap, forming on interference pattern.

▲ **Figure 2** *When electrons pass through a piece of polycrystalline graphite they form a diffraction pattern*

▲ **Figure 3** *These rings are the result of the diffraction of electrons*

The de Broglie equation

In developing wave–particle duality, de Broglie realised that the wavelength λ of a particle was inversely proportional to its momentum p. As the momentum of a particle increases by a certain factor its wavelength reduces by the same factor.

$$\lambda \propto \frac{1}{p}$$

Further investigation led to the development of what is now called the **de Broglie equation**.

$$\lambda = \frac{h}{p}$$

where λ is the wavelength in m, h is the Planck constant, and p is the momentum of the particle in $kg\,m\,s^{-1}$.

 Worked example: The wavelength of a fast-moving electron

Calculate the wavelength of an electron travelling at $2.00 \times 10^8\,m\,s^{-1}$.

Step 1: Identify the correct equation to calculate the wavelength.

$$\lambda = \frac{h}{p}$$

As momentum is $p = mv$ we can substitute this into the previous equation giving

$$\lambda = \frac{h}{mv}$$

> **Synoptic link**
>
> You have already learnt about the diffraction of waves in detail as part of Topic 11.4, Diffraction and polarisation. The same principles apply to the diffraction of particles like electrons.

> **Study tip**
>
> It's important to be able to give examples of experiments which show the wave and particle natures of both light and electrons.

> **Synoptic link**
>
> Momentum is explored in Topic 7.2, Linear momentum.

Step 2: Substitute in known values in SI units and calculate the wavelength in metres.

$$\lambda = \frac{6.63 \times 10^{-34}}{9.11 \times 10^{-31} \times 2.00 \times 10^{8}} = 3.64 \times 10^{-12}\,\text{m}$$

Notice how small this wavelength is (less than the diameter of an atom). In order to diffract electrons tiny gaps are needed.

Accelerated particles

We can relate an object's kinetic energy E_k to its wavelength by considering both the de Broglie equation and our general equation for kinetic energy. From the worked example above

$$\lambda = \frac{h}{mv}$$

If we can find an expression for mv that includes E_k we can determine the relationship between a particle's kinetic energy and its wavelength.

$$E_k = \frac{1}{2}mv^2$$

$$2E_k = mv^2$$

$$2E_k m = m^2 v^2$$

$$\sqrt{2E_k m} = mv$$

We can then substitute this expression into the de Broglie equation.

$$\lambda = \frac{h}{mv} = \frac{h}{\sqrt{2E_k m}}$$

As h and m are both constants, we can see that the wavelength of a particle is inversely proportional to the square root of its kinetic energy.

$$\lambda \propto \frac{1}{\sqrt{E_k}}$$

If the kinetic energy of a particle decreases by a factor of two, its wavelength increases by $\sqrt{2}$.

1 State the effect on the wavelength if the kinetic energy of a particle: **a** increases; **b** increases by a factor of three; **c** decreases by a factor of 10.
2 Calculate the wavelength of: **a** a proton that has a kinetic energy of 3.20×10^{-17} J; **b** an electron that has a kinetic energy of 600 eV.
3 An electron is accelerated through a p.d. of 100 V. Calculate its wavelength.

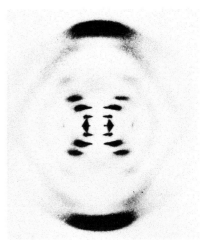

Crystallography

Understanding of the wave properties of electrons has led to many applications, including **crystallography**.

Crystallography is a method for determining the arrangement of atoms within a compound. The process often uses X-rays, which are fired at a crystal of the compound. The X-rays are diffracted as they pass through the gaps between the individual atoms. By studying the patterns formed by these diffracted beams, a crystallographer is able to produce a precise 3D model of the arrangement of the atoms within the substance.

Electrons are often used in a very similar way to X-rays. Using electrons instead of X-rays for crystallography offers advantages.

● Their wavelength can be tuned to be very similar in size to the gap between atoms, resulting in strong diffraction and a clear pattern for analysis.

● They can be used with very thin sheets or layers of material.

1 Describe how the wavelength of an electron can be altered using an electron gun.
2 Suggest why X-rays cannot be used to study thin sheets of material.
3 Calculate the wavelength of an electron with a kinetic energy of 40 eV and explain why such an electron would be useful for studying crystal structures.

▲ **Figure 4** *Perhaps the most famous example of X-ray crystallography was Watson and Crick's use of Rosalind Franklin's image to determine the structure of DNA in 1953*

Beyond electrons

The de Broglie equation can be applied to all particles. Like electrons, protons and neutrons have been shown to have wave properties – they form diffraction patterns. However, as particles become larger their wave properties become harder to observe. The mass of individual protons is much greater than electrons, so at the same speed their momentum is significantly greater and therefore their wavelength is much smaller, and much harder to observe.

Summary questions

1 Calculate the wavelength of a proton that has a momentum of $1.67 \times 10^{-19}\,kg\,m\,s^{-1}$. *(2 marks)*

2 Explain why, when travelling at the same speed, a proton has a much smaller wavelength than an electron. *(3 marks)*

3 Explain why the wave properties of electrons are not evident in most experiments. *(3 marks)*

4 Calculate the speeds of electrons with the following wavelengths:
   a $3.63 \times 10^{-10}\,m$;   b $4.85\,pm$. *(4 marks)*

5 An electron is accelerated in an electron gun to a speed of $4.20 \times 10^{7}\,m\,s^{-1}$. Calculate:
   a the wavelength of the electron; *(2 marks)*
   b the speed of a neutron with the same wavelength as **a** (mass of neutron = $1.67 \times 10^{-27}\,kg$). *(2 marks)*

6 Calculate the wavelength of an electron travelling at a speed of $0.25\,c$ ($c = 3.00 \times 10^{8}\,m\,s^{-1}$). *(2 marks)*

Practice questions

1   a   State what is meant by a photon. (*1 mark*)

   b   Describe and explain the photoelectric effect in terms of photons and surface electrons of a metal. (*4 marks*)

   c   A laser emits light of wavelength 6.3×10^{-7} m and of radiant power 2.0 mW. Calculate

     (i)   the energy of each photon in electronvolts (eV), (*3 marks*)

     (ii)   the number of photons emitted per second from the laser. (*2 marks*)

2   a   Electromagnetic waves of wavelength 220 nm are incident on the surface of a metal. Electrons emitted from the surface of the metal have maximum speed 1.2×10^{6} m s^{-1}. Calculate

     (i)   the work function of the metal, (*4 marks*)

     (ii)   the threshold frequency of the metal. (*2 marks*)

   b   The intensity of the electromagnetic waves incident on the metal in (**a**) is increased. State and explain the effect this has on the maximum kinetic energy of the electrons emitted from the metal surface. (*2 marks*)

3   A negatively charged metal plate is exposed to electromagnetic waves of a range of frequencies. Figure 1 shows the variation of the maximum kinetic energy E_{max} of the emitted photoelectrons with frequency f of the incident electromagnetic waves.

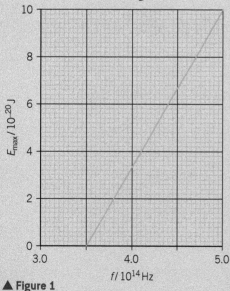

▲ **Figure 1**

a   Explain why the straight line graph shown in Figure 1 does not pass through the origin. (*2 marks*)

b   Determine the work function of the metal. (*2 marks*)

c   The metal plate is replaced with one having a work function greater than the value calculated in (b). State and explain the change to the shape of the graph of E_{max} against f. (*4 marks*)

4   a   Explain what is meant by the de Broglie wavelength of a particle. (*2 marks*)

   b   Electrons are accelerated through a potential difference of 200 V. Calculate the final de Broglie wavelength of these electrons. (*5 marks*)

   c   The electrons from (**b**) can be diffracted by the atoms in a solid. Suggest what you can deduce about the arrangement of the atoms in solids. (*1 mark*)

5   In a demonstration experiment of the photoelectric effect, light of wavelength 440 nm incident on a clean metal surface causes electrons to be emitted. No electrons are emitted from the surface when the wavelength of the incident light is greater than 550 nm.

   a   (i)   Define the term *work function*. (*2 marks*)

     (ii)   Explain how the work function is related to the threshold frequency. (*2 marks*)

     (iii)   Calculate the value of the work function for this metal. (*2 marks*)

   b   (i)   Show that the maximum speed of the emitted electrons in the experiment is about 4.5×10^{5} m s^{-1}. (*3 marks*)

     (ii)   Calculate the minimum de Broglie wavelength of an emitted electron. (*2 marks*)

6   a   State two properties of a photon. (*2 marks*)

   b   Figure 2 shows a graph of E against $\frac{1}{\lambda}$, where E is the energy of a photon and λ is its wavelength.

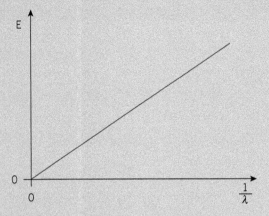

▲ **Figure 2**

(i) Explain why the graph of Figure 2 is a straight line. *(2 marks)*

(ii) State and explain what the gradient of the graph represents. *(2 marks)*

c The Sun emits electromagnetic radiation of average wavelength 5.5×10^{-7} m. The radiant power emitted from its surface is 6.3×10^{7} W m^{-2}. The radius of the Sun is 7.0×10^{8} m.

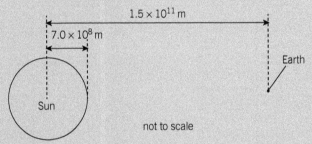

▲ **Figure 3**

(i) Estimate the total number of photons emitted per second from the surface of the Sun. *(4 marks)*

(ii) The Earth is about 1.5×10^{11} m from the Sun. Determine the average intensity of the radiation at the Earth from the Sun. *(3 marks)*

7 a State how the Planck constant h is used to model photons and the wave-like properties of particles. *(2 marks)*

b Explain what is meant by the photoelectric effect. *(1 mark)*

c Define threshold frequency of a metal. *(1 mark)*

d Electromagnetic radiation of wavelength 4.0×10^{-7} m is incident on the surface of a metal plate. The maximum energy of the emitted photoelectron is 1.6 eV.

Calculate the work function of the metal in eV. *(4 marks)*

8 A negatively charged metal plate is exposed to electromagnetic radiation of frequency f.

Figure 8 shows the variation of the maximum kinetic energy E_k of the photoelectrons emitted from the metal with frequency f.

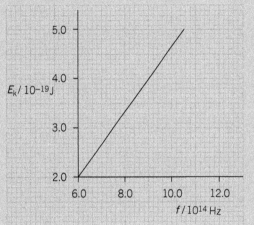

▲ **Figure 4**

a Use Figure 4 to determine the Planck constant. Explain your answer. *(3 marks)*

b State and explain the effect on the shape of the graph when the intensity of the incident radiation on the surface of the metal is increased. *(2 marks)*

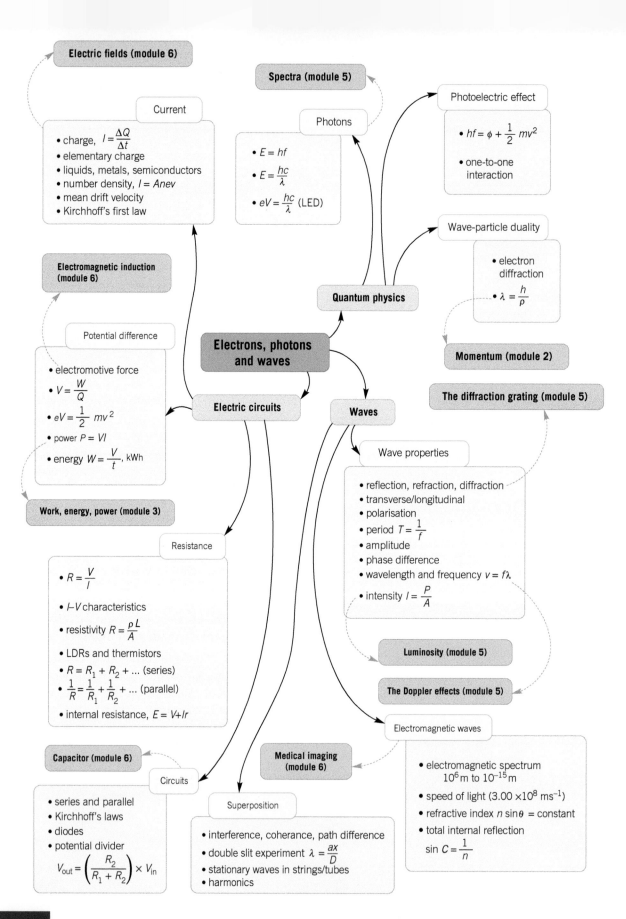

Electric fields (module 6)

Current
- charge, $I = \dfrac{\Delta Q}{\Delta t}$
- elementary charge
- liquids, metals, semiconductors
- number density, $I = Anev$
- mean drift velocity
- Kirchhoff's first law

Spectra (module 5)

Photons
- $E = hf$
- $E = \dfrac{hc}{\lambda}$
- $eV = \dfrac{hc}{\lambda}$ (LED)

Photoelectric effect
- $hf = \phi + \dfrac{1}{2}mv^2$
- one-to-one interaction

Quantum physics

Wave-particle duality
- electron diffraction
- $\lambda = \dfrac{h}{\rho}$

Momentum (module 2)

Electromagnetic induction (module 6)

Potential difference
- electromotive force
- $V = \dfrac{W}{Q}$
- $eV = \dfrac{1}{2}mv^2$
- power $P = VI$
- energy $W = \dfrac{V}{t}$, kWh

Electrons, photons and waves

Electric circuits

Waves

Work, energy, power (module 3)

The diffraction grating (module 5)

Wave properties
- reflection, refraction, diffraction
- transverse/longitudinal
- polarisation
- period $T = \dfrac{1}{f}$
- amplitude
- phase difference
- wavelength and frequency $v = f\lambda$
- intensity $I = \dfrac{P}{A}$

Resistance
- $R = \dfrac{V}{I}$
- I–V characteristics
- resistivity $R = \dfrac{\rho L}{A}$
- LDRs and thermistors
- $R = R_1 + R_2 + \ldots$ (series)
- $\dfrac{1}{R} = \dfrac{1}{R_1} + \dfrac{1}{R_2} + \ldots$ (parallel)
- internal resistance, $E = V + Ir$

Luminosity (module 5)

The Doppler effects (module 5)

Capacitor (module 6)

Medical imaging (module 6)

Electromagnetic waves
- electromagnetic spectrum 10^6 m to 10^{-15} m
- speed of light (3.00×10^8 ms^{-1})
- refractive index $n\sin\theta$ = constant
- total internal reflection $\sin C = \dfrac{1}{n}$

Circuits
- series and parallel
- Kirchhoff's laws
- diodes
- potential divider
$V_{\text{out}} = \left(\dfrac{R_2}{R_1 + R_2}\right) \times V_{\text{in}}$

Superposition
- interference, coherence, path difference
- double slit experiment $\lambda = \dfrac{ax}{D}$
- stationary waves in strings/tubes
- harmonics

Feeling the strain

How can we monitor extremely small changes in distance, such as small changes in the length of a load-bearing section of a bridge or tiny ground movement caused by seismic activity? A strain gauge can be used. It could be attached to the bridge or secured to the ground. The strain gauge consists of thin wire mounted on a non-conductive material (see Figure 1). Stretching the gauge decreases the cross-sectional area of the wire and increases its length. The resistance of the wire increases by a tiny amount. It would not be possible to measure these tiny changes in resistance using an ohmmeter. However, a circuit known as a Wheatstone bridge circuit can be used to monitor these tiny changes in resistance (see Figure 2). The variation in the voltage V can be monitored remotely using dataloggers and even using mobile phones.

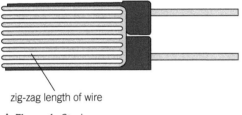

zig-zag length of wire

▲ **Figure 1** *Strain gauge*

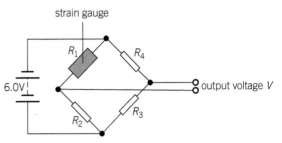

▲ **Figure 2** *Wheatstone circuit with a strain gauge*

The output voltage V is related to the values of the resistances R_1, R_2, R_3, and R_4 by the equation

$$V = \left(\frac{R_1}{R_1 + R_2} - \frac{R_4}{R_3 + R_4} \right) \cdot 6.0$$

1 List situations, other than the two mentioned here, where a strain gauge might be useful.
2 Show that the resistance of a strain gauge is directly proportional to length^2 of the wire.
3 Consider a Wheatstone bridge circuit where initially all the values of the resistances are the same and equal to $100 \, \Omega$. What is V equal to? Now imagine that there is 0.001% change in the resistance R_1 of the strain gauge. What is the output voltage now?

Energy and electrons

In Chapters 11, 12 and 13 you learnt about progressive waves, stationary waves and quantum physics. These are important topics in physics. Physicists have used these simple ideas to explain why the energy of electrons within an atom is quantised. You now have the knowledge to recreate the modelling process used in the development of quantum physics. This is an important aspect of How Science Works – theories and models are amalgamated to reveal something new about nature.

Here are some extension questions you can tackle to further improve your knowledge of atoms. These ideas will give you a head start when you study *Energy levels in atoms* in 6.4.

Imagine the hydrogen atom is one-dimensional, just like a string fixed at both ends. The size of the atom is about 10^{-10} m. You know that a moving electron has a de Broglie wavelength. The electron bound to the atom will produce a stationary wave.

1 What is the longest de Broglie wavelength of an electron that is bound to the atom?
2 Calculate the kinetic energy of this electron in joules and in eV.
3 Determine the other possible kinetic energies this electron can have within the atom.
4 Explain what is meant by the statement: *The energy of the electron within the atom is quantised*.

Paper 1 style questions

SECTION A

Answer **all** the questions.

1   A student investigating an electrical experiment records the following measurements in the lab book.

   - current in the LED = $120 \pm 8\,mA$

   - potential difference across the LED
     $= 1.8 \pm 0.2\,V$

   What is the percentage uncertainty in the resistance of the LED?

   A   4.4 %

   B   6.7%

   C   11%

   D   18%

   (1 mark)

2   A spring of original length 3.0 cm and force constant $100\,N\,m^{-1}$ is placed on a smooth horizontal surface. Its length is changed from 6.0 cm to 8.0 cm.

   What is the change in the energy stored by the spring?

   A   0.020 J

   B   0.080 J

   C   0.140 J

   D   1.00 J

   (1 mark)

3   A wooden block is held under water and then released, as shown in Figure 1.

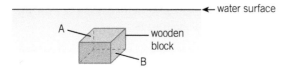

▲ Figure 1

   The wooden block moves towards the surface of the water.

   Which of the following statements is/are true about the block as soon as it is released?

   1   The force experienced by the face B due to water is greater than the force experienced by the face A.

   2   The upthrust on the block is equal to its weight.

3   The mass of the water displaced is equal to the weight of the block.

   A   1, 2 and 3 are correct

   B   Only 1 and 2 are correct

   C   Only 2 and 3 are correct

   D   Only 1 is correct

   (1 mark)

4   Figure 2 shows a stationary wave pattern formed in an air column.

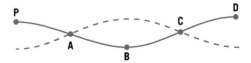

▲ Figure 2

   Which point **A**, **B**, **C**, or **D** has a phase difference of 180° with reference to **P**?

   (1 mark)

5   A ray of monochromatic light is incident at a boundary between two transparent materials. The refractive index of the materials is 1.30 and 1.50. The angle of refraction for the emergent ray is 60°.

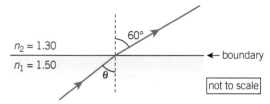

▲ Figure 3

   What is the angle θ of incidence?

   A   42°

   B   49°

   C   60°

   D   88°

   (1 mark)

6   Figure 4 shows the cross-section of a metal wire connected to a power supply. The charge carriers within the metal wire move from right to left.

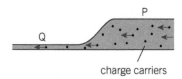

▲ Figure 4

The section Q of the wire is thinner than section P.

Which statement is correct?

A The direction of the conventional current is from right to left.

B The section Q of the wire has fewer charge carriers per unit volume.

C The current in both sections is the same.

D The charge carriers are negative ions.

(1 mark)

7 A resistor **R** is connected in parallel with a resistor of resistance $10\,\Omega$. The total resistance of the combination is $6.0\,\Omega$. What is the resistance of resistor **R**?

A $0.067\,\Omega$

B $3.8\,\Omega$

C $4.0\,\Omega$

D $15\,\Omega$

(1 mark)

8 What is a reasonable estimate for the energy of a photon of visible light?

A $4 \times 10^{-19}\,J$

B $4 \times 10^{-18}\,J$

C $4 \times 10^{-16}\,J$

D $4 \times 10^{-11}\,J$

(1 mark)

9 Students A and B use micrometer screw gauges to measure the diameter of a copper wire in three different places along its length. The diameter of the wire according to the manufacturer is $0.278\,mm$. The results recorded by students A and B are shown in Figure 5.

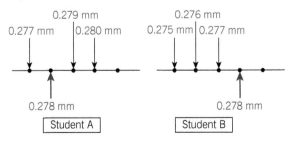

▲ Figure 5

Which statement is correct about the measurements made by the student B compared with those of student A?

A The measurements are more accurate.

B The measurements are not as precise.

C The measurements are both more accurate and more precise.

D The measurements are not accurate but are more precise.

(1 mark)

10 The circuit in Figure 6 is constructed by a student in the laboratory.

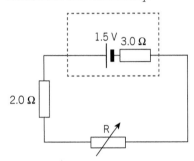

▲ Figure 6

The e.m.f. of the cell is $1.5\,V$ and it has an internal resistance of $3.0\,\Omega$. A resistor of resistance $2.0\,\Omega$ and a variable resistor R are connected in series to the terminals of the cell. The variable resistor is set to a resistance value of $7.0\,\Omega$.

What is the value of the ratio

$$\frac{\text{power dissipated in R}}{\text{power supplied by the cell}}?$$

A 0.17

B 0.25

C 0.58

D 0.75

(1 mark)

SECTION B

Answer **all** the questions

11 a Define *velocity*. *(1 mark)*

b The mass of an ostrich is $130\,kg$. It can run at a maximum speed of 70 kilometers per hour.

(i) Calculate the maximum kinetic energy of the ostrich when it is running.

(3 marks)

(ii) Scientists have recently found fossils of a prehistoric bird known as Mononykus. Figure 7 shows what the Mononykus would have looked like.

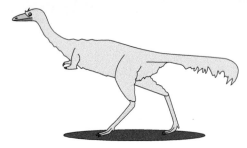

▲ Figure 7

According to a student, the Mononykus looks similar to our modern day ostrich. The length, height and width of the Mononykus were all **half** that of an ostrich. Estimate the mass of the Mononykus. Explain your reasoning. *(2 marks)*

G481 June 2014

12 Figure 8 shows a block of wood held at rest at the top of a smooth ramp.

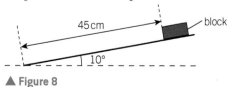

▲ Figure 8

The ramp makes an angle of 10° to the horizontal. The block is released and it slides down the ramp.

a Calculate the acceleration of the block along the length of the ramp. *(2 marks)*

b The block travels a total distance of 45 cm down the ramp. Calculate the time it takes to reach the bottom of the ramp. *(3 marks)*

c The speed of the block at the bottom of the ramp is *v*. Describe a simple experiment a student can carry out to determine an approximate value of the speed *v*. The student only has a metre rule and a stopwatch. *(3 marks)*

13 Figure 9 shows a metal wire stretched horizontally between two supports on a laboratory bench.

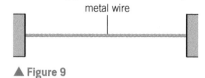

▲ Figure 9

The tension in the wire is 15 N and it has a cross-sectional area $3.1 \times 10^{-7}\,\text{m}^2$. The Young modulus of the wire is 4.2×10^{10} Pa.

a Calculate the strain of the wire. *(3 marks)*

b Design a simple electrical circuit to determine the resistivity of the metal used to make the wire. *(4 marks)*

c Figure 10 shows an electrical circuit. Calculate the total resistance between **A** and **B**. *(3 marks)*

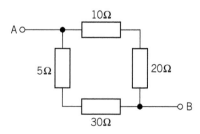

▲ Figure 10

14 a Explain what is meant by coherent waves. *(1 mark)*

b State two ways in which a stationary waves differs from a progressive wave. *(2 marks)*

c Figure 11 shows a stationary pattern on a length of stretched string.

drawn to scale

▲ Figure 11

The distances shown in Figure 11 are **drawn to scale**. The frequency of vibration of the string is 110 Hz.

(i) By taking measurements from Figure 11, determine the wavelength of the progressive waves on the string. *(2 marks)*

(ii) Calculate the speed of the progressive waves on the string. *(2 marks)*

15 a Sketch a graph of energy *E* of a photon against frequency *f* of the electromagnetic radiation. *(1 mark)*

b Electromagnetic waves of frequency 8.93×10^{14} Hz are incident on the surface of metal. The work function of the metal is 3.20×10^{-19} J.

(i) Calculate the energy of the photons. *(2 marks)*

(ii) Calculate the maximum speed v_{max} of the photoelectrons emitted from the metal surface. *(3 marks)*

16 a Define *refractive index* of a material. (*1 mark*)

b Figure 12 shows the path of a ray of light as it crosses the boundary between two materials A and B.

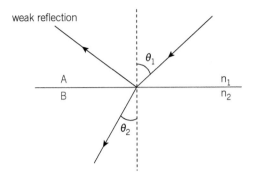

▲ **Figure 12**

The refractive index of material A is n_1 and the angle of incidence of the ray of light is θ_1. The angle of refraction in material B is θ_2 and the refractive index of material B is n_2. Write an equation that relates n_1, n_2, θ_1 and θ_2. (*1 mark*)

c A student is investigating the refraction of light by a transparent material by measuring the angles of incidence *i* and refraction *r*. Figure 13 show the results from the experiment.

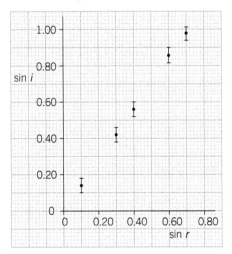

▲ **Figure 13**

Use Figure 13 to determine

(i) the refractive index of the material (*2 marks*)

(ii) the critical angle for this material. (*2 marks*)

d You are provided with a semi-circular glass block and a ray-box with a suitable supply.

Design a laboratory experiment to determine the critical angle of the glass of the semi-circular block and hence the refractive index of the glass. You may use other equipment available in the laboratory. In your description pay particular attention to

- how the apparatus is used
- what measurements are taken
- how the data is analysed. (*4 marks*)

17 a State *Ohm's law*. (*1 mark*)

b The *I-V* characteristic of a particular component is shown in Figure 14.

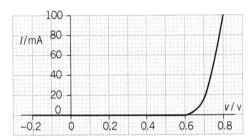

▲ **Figure 14**

(i) Use Figure 14 to describe how the resistance of this component depends on the potential difference (p.d.) across it. You may do calculations to support your answer. (*3 marks*)

(ii) Draw a circuit diagram for an arrangement that could be used to collect results to plot the graph shown in Figure 14. (*3 marks*)

c Figure 15 shows an electrical circuit.

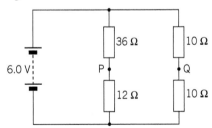

▲ **Figure 15**

The e.m.f. of the battery is 6.0 V and it has negligible internal resistance.

Calculate

(i) the current in the 36 Ω resistor (*2 marks*)

(ii) the potential difference across the 12 Ω resistor (*1 mark*)

(iii) the potential difference between points **P** and **Q**. (*2 marks*)

Paper 2 style questions

Answer **all** the questions

1 a Figure 1a shows a 500 g mass suspended from two strings. The mass hangs vertically and is in equilibrium.

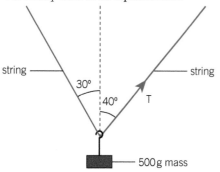

▲ **Figure 1a**

(i) Determine the tension T in one of the strings. (*4 marks*)

(ii) Describe how a student could determine the value of T experimentally in the laboratory. State one possible limitation of the experiment. (*2 marks*)

b Figure 1b shows an experiment designed by a student.

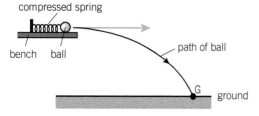

▲ **Figure 1b**

A metal ball is pushed against a compressible spring and then released. The ball has a horizontal velocity of 1.5 m s^{-1}. The ball leaves the horizontal bench and lands on the ground below at point **G**.

Assume friction has negligible effect on the motion of the ball.

(i) Describe the energy changes of the **ball** from the instant it is held against the compressed spring to the instant just before it lands at **G**. (*4 marks*)

(ii) The ball takes 0.42 s to travel from top of the bench to **G**.

Calculate the height of the bench from the ground. (*3 marks*)

2 Figure 2 shows a simple pendulum. It consists of a metal ball of diameter 2.00 cm and a thin string.

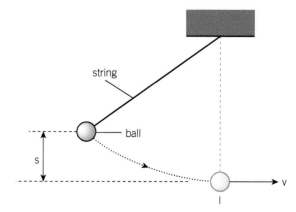

▲ **Figure 2**

The ball is raised to a vertical height s and then released.

a Show that its speed v at the bottom of its swing is given by the equation $v^2 = 2gs$, where g is the acceleration of free fall. (*3 marks*)

b Describe how a student could determine the speed v at the bottom of the pendulum's swing in the laboratory. State one possible limitation of your method. (*4 marks*)

c The table below shows **some** of the results obtained by a student.

s/m	v/m s^{-1}	v^2/m^2 s^{-2}
0.210	2.03 ± 0.15	4.12 ± 0.61
0.330	2.54 ± 0.18	
0.410	2.83 ± 0.25	8.01 ± 1.42
0.490	3.10 ± 0.30	9.61 ± 1.86

(i) Copy and complete the table by determining the missing value for v^2 and the absolute uncertainty in this value. *(3 marks)*

(ii) The student plots a graph of v^2 against s. Explain how the graph may be used to determine the acceleration of free fall g. *(2 marks)*

3 Figure 2 shows the *I-V* characteristic of a blue light-emitting diode (LED).

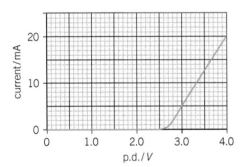

▲ **Figure 3a**

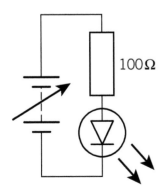

▲ **Figure 3b**

a (i) The data for plotting the *I-V* characteristic is collected using the components shown in Figure 3b. By drawing on a copy of Figure 3b complete the circuit showing how you would connect the two meters needed to collect these data. *(1 mark)*

(ii) When the current in the circuit of Figure 3b is 20 mA calculate the terminal potential difference across the supply.

(3 marks)

b The energy of each photon emitted by the LED comes from an electron passing through the LED. The energy of each blue photon emitted by the LED is 4.1×10^{-19} J.

(i) Calculate the energy of a blue photon in electron volts.

(1 mark)

(ii) Explain how your answer to (i) is related to the shape of the curve in Figure 3a. *(2 marks)*

c Calculate for a current of 20 mA

(i) the number n of electrons passing through the LED per second,

(2 marks)

(ii) the total energy of the light emitted per second,

(2 marks)

(iii) the efficiency of the LED in transforming electrical energy into light energy.

(2 marks)

d The energy of a photon emitted by a red LED is 2.0 eV. The current in this LED is 20 mA when the p.d. across it is 3.4 V. Draw the *I-V* characteristic of this LED on a copy of Figure 3a. *(2 marks)*

Q4 G482 June 2014 paper

4   a   Interference of waves from two sources can only be observed when the waves are coherent.

Explain the meaning of

(i)   *interference*,   (*2 marks*)

(ii) *coherence*.   (*1 mark*)

   b   Figure 4 shows two microwave transmitters **A** and **B** 0.20 m apart. The transmitters emit microwaves of equal amplitude in phase and of wavelength 30 mm. A detector, moved along the line **PQ** at a distance of 5.0 m from **AB**, detects regions of high and low intensity forming an interference pattern.

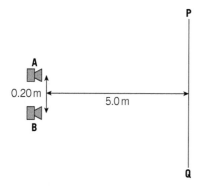

▲ Figure 4

(i)   Use the ideas of path difference or phase difference to explain how the interference pattern is formed.   (*3 marks*)

(ii)   Calculate the separation between one region of high intensity and the next along the line **PQ**.
                                         (*2 marks*)

(iii) State the effect, if any, on the position and intensity of the maxima when each of the following changes is made, separately, to the experiment.

   1   The amplitude of the transmitted waves is doubled.
                                         (*2 marks*)

   2   The separation between the transmitters is halved.
                                         (*2 marks*)

   3   The phase of transmitter **A** is reversed so that there is now a phase difference of 180° between the waves from **A** and **B**.
                                         (*2 marks*)

Q6 G482 Jan 2012 paper

5   a   Figure 5 shows a ball of mass 0.050 kg resting on the strings of a tennis racket held horizontally.

▲ Figure 5

(i)   On a copy of Figure 5, draw and label arrows to represent the **two** forces acting on the ball.   (*2 marks*)

(ii)   Calculate the difference in magnitude between the two forces on the ball when the racket is accelerated upwards at 2.0 m s⁻².
                                         (*2 marks*)

   b   The ball is dropped from rest at a point 0.80 m above the racket head. The racket is fixed rigidly. Assume that the ball makes an elastic collision with the strings and that any effects of air resistance are negligible.

Calculate

(i)   the speed of the ball just before impact,
                                         (*2 marks*)

(ii)   the momentum of the ball just before impact,
                                         (*1 mark*)

(iii) the change in momentum of the ball during the impact,

(*1 mark*)

(iv) the average force during the impact for a contact time of 0.050 s.

(*1 mark*)

c The two forces you have drawn in **(a)(i)** are not a pair of forces as required by Newton's third law of motion. However each of these forces does have a corresponding equal and opposite force to satisfy Newton's third law. Describe these equal and opposite forces and state the objects on which they act. (*4 marks*)

Q1 2824 Jan 2010 paper

6 **a** Show that the momentum p of a particle is given by the equation $p = \sqrt{2Em}$, where m is the mass of the particle and E is its kinetic energy. (*3 marks*)

b Slow-moving neutrons from a nuclear reactor are used to investigate the structure of complex molecules such as DNA. Neutrons can be diffracted by DNA. The mass of a neutron is 1.7×10^{-27} kg.

(i) Calculate the de Broglie wavelength of a neutron of kinetic energy 6.2×10^{-21} J. (*4 marks*)

(ii) Suggest why these slow-moving neutrons can be diffracted by DNA. (*1 mark*)

c Charged particles are accelerated in a laboratory by a group of scientists. The de Broglie wavelength of the particles is λ and their kinetic energy is E. Figure 6 shows a graph of λ^2 against $\frac{1}{E}$ for these accelerated particles.

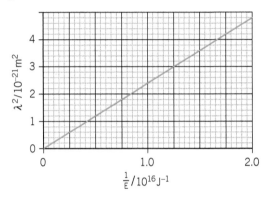

▲ Figure 6

Use Figure 6 to determine the mass m of the particles. (*4 marks*)

A1 - Physical quantities and units

Being sensible

You already know that all physical quantities have a numerical value and a unit. For example, speed is a physical quantity and its SI base unit is m s^{-1}.

When carrying out calculations or experiments, it is always sensible to have a close look at the answer to see if it is reasonable. Experience of doing many calculations in physics will help you gauge whether or not an answer is reasonable. The list below provides helpful benchmarks in mechanics for you to make well-reasoned estimates of other physical quantities. Try to add to this list as you go through your physics course.

Synoptic link

You first met physical quantities in Topic 2.1, Quantities and units.

▼ **Table 1**

Length of a ruler = 30 cm	walking speed = 1 m s^{-1}
height of a person = 1.5 m	speed of a car on a motorway = 30 m s^{-1}
mass of an apple = 0.1 kg	acceleration of free fall = 10 m s^{-2}
mass of a person = 70 kg	density of air = 1 kg m^{-3}
mass of a car = 1000 kg	density of water = 1000 kg m^{-3}

Making estimates

What is the kinetic energy of the car shown in Figure 1?

You can make an estimate of the energy using the equation for kinetic energy and some of the values listed in Table 1. Estimates use known facts and sensible assumptions to arrive at an answer, so they are not just wild guesses. You should not be plucking random values from your head when making estimates.

The kinetic energy E_k of a car is given by the equation

$$E_k = \frac{1}{2} \text{ mass} \times \text{speed}^2$$

$$E_k = \frac{1}{2} \times 1000 \times 30^2 = 4.5 \times 10^5 \text{ J} = 5 \times 10^5 \text{ J (1 s.f.)}$$

Assumptions are needed for estimates. In the estimation above, we assumed that the car is travelling at a constant speed.

▲ **Figure 1** *Estimating the kinetic energy of a car*

Working out units

Most physical quantities can be expressed by combinations of six base units – kg, m, s, A, K, and mol.

The area of a rectangular room can be found by multiplying its length and width together. This means that the derived unit for area is m × m = m^2.

All derived units can be worked out using an appropriate equation and then multiplying and/or dividing the base units. For example, speed is distance divided by time. This means that the derived unit for speed is m ÷ s = m s^{-1}.

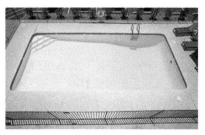

▲ **Figure 2** *Estimating the mass of water in a swimming pool*

Checking if an equation is correct in terms of units

An equation describing a relationship between physical quantities can only be correct if both sides have the same SI units. If the left-hand side of an equation has the units $kg\,m^{-2}\,s^{-1}$, then the right-hand side of the equation must also have the same units, $kg\,m^{-2}\,s^{-1}$. The equation is **homogeneous** in terms of units, where homogeneous means identical. Checking the homogeneity of physical equations using SI units is a powerful method of assessing whether or not an equation is correct.

 All balanced out

For an object that started from rest, the final velocity is given by the equation $v^2 = 2as$, where a is the acceleration and s is the displacement of the object. You can show that the equation is homogeneous by determining the units on the left- and right-hand sides of the equal sign and showing them to be identical.

Since numbers have no units, you will not need to worry about the 2 in this equation.

▼ Table 2

quantity	v	a	s	'left hand side' v^2	'right hand side' $2as$
base unit	$m\,s^{-1}$	$m\,s^{-2}$	m	$(m\,s^{-1})^2 = m^2\,s^{-2}$	$(m\,s^{-2}) \times (m) = m^2\,s^{-2}$

 Using an equation to find SI units

The drag force F acting on an object falling through the air at a speed v is given by the equation $F = kAv^2$, where A is the area of the object. To determine the units for k you must rearrange the equation with k as the subject, then substitute all the known SI units for the quantities.

$$k = \frac{F}{Av^2}$$

The quantity k has the SI units $kg\,m^{-3}$.

▲ **Figure 3** *Are these equations homogeneous?*

▼ Table 3

quantity	F	A	v	$k = \dfrac{F}{Av^2}$
base unit	$kg\,m\,s^{-2}$	m^2	$m\,s^{-1}$	$\dfrac{kg\,m\,s^{-2}}{m^2 \times (m\,s^{-1})^2} = \dfrac{kg\,m\,s^{-2}}{m^2 \times m^2\,s^{-2}} = kg\,m^{-3}$

Table of results

Tables are useful for displaying a lot of information at once. This is usually numerical data, for example, readings taken from measuring devices during experiments. You can clearly see the values of each quantity in a table, although it may be difficult to spot patterns or trends in the data.

▼ **Table 1**

V/V	I/A	R/Ω
0.15	1.01	0.15
0.32	2.12	0.15
0.38	2.42	0.16

Table 1 shows a section of a typical results table from an investigation of an electrical component. The measured quantities are potential difference V (or p.d.) and the current I. The units are also included. The first two columns have headings V/V and I/A. The slash is used to separate the quantity from its unit. The final column is for resistance R of the component, which is calculated by dividing V by I. The unit of electrical resistance, Ω, for this processed data is also included.

You have to think about the exactness to which the quantities are measured. All the values for V and I are measured to 2 decimal places. Since the values in the final column are calculated from V and I, values of resistance must be written to the correct number of significant figure. The V values are written to 2 significant figures and the I values are all to 3 significant figures. Therefore the values of R can only be quoted to 2 significant figures – equal to the *lowest* significant figures used in the calculation.

Plotting graphs

In physics, you often plot a graph of your results in order to identify any trends or patterns. What do you plot on the vertical axis (y-axis) and on the horizontal axis (x-axis)? As a general rule, you plot the **independent variable**, the one you deliberately change in an experiment, on the x-axis, and the **dependent variable**, the variable which changes as a result, on the y–axis.

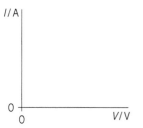

▲ **Figure 1** *The axes of the graph must have the correct labels*

You must label both axes correctly with the quantities and their units. The labels are simply those you would have used in your column headings (see Table 1).

Here are a few useful tips when plotting graphs:

- Never choose scales which are multiples of 3, 7, 11, or 13.
- Ensure that you are using most of the graph paper to plot your results.
- Plot your points using small crosses.
- For a straight line graph, draw the straight **best fit line** using a long ruler.
- When calculating the gradient (see below), clearly show the triangle used – the triangle should be large, using at least half of the line drawn.

Gradients

You can determine the **gradient** of a straight line graph using a large triangle. For a curve, you have to draw a tangent first and then determine its gradient using a large triangle. The gradient is worked out using the following equation:

$$\text{gradient} = \frac{\text{change in } y}{\text{change in } x} = \frac{\Delta y}{\Delta x}$$

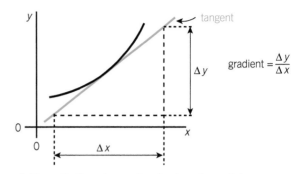

▲ **Figure 2** Use a large triangle when determining the gradient

Straight lines

The general equation for a straight line is written in the form

$$y = mx + c$$

where m is the gradient and c is the y-intercept (Figure 3).

In your physics work, you will come across equations that can easily be presented in the form of a straight line equation. For example, $v = u + at$. This is the equation for an object accelerating in a straight line. The initial velocity of the object is u, the acceleration is a, and v is the final velocity after a time t. If you plot t on the x-axis and v on the y-axis, then the gradient of the line must be acceleration a, and the y-intercept must be u (Figure 4).

$$v = at + u$$
$$y = mx + c$$

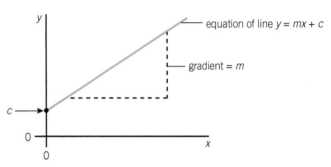

▲ **Figure 3** $y = mx + c$

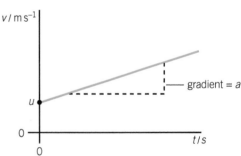

▲ **Figure 4** The y-intercept is u and the gradient is a

Curves to straight lines

Not all the graphs you plot in physics produce straight lines. Consider the equation $s = \frac{1}{2}at^2$ for an object accelerating from rest. The constant acceleration of the object is a and its displacement after a time t is s. If you plot s against t you will get a curve – a parabola. It is difficult to extract useful information from such a curve. However, you can turn this equation into a straight line equation as shown below.

$$s = \frac{1}{2}at^2 + 0$$
$$y = mx + c$$

You will get a straight line graph by plotting t^2 on the x-axis and s on the y-axis. The gradient of the line is $\frac{1}{2}a$ (Figure 5). By plotting the data in this way, you can determine the acceleration a of the object by multiplying the gradient by 2.

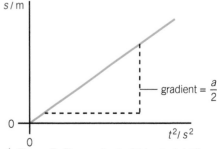

▲ **Figure 5** The gradient of this straight line is $\frac{a}{2}$ and the line passes through the origin

Errors are not mistakes

Making measurements and using instruments is a key part of scientific activity. No measurement can ever be perfect. When you think about errors you may think about mistakes. If you make a mistake in an experiment, you may be may be able to do the experiment again without the mistake. However, an experiment will still contain errors. An error in a scientific sense is the difference between the result you get and the correct result. Errors are usually caused by measuring devices, even if they are used correctly, or by the design of the experiment itself.

▲ **Figure 1** *An error made when you use a calculator is a mistake that can be corrected but measurement errors are more difficult to deal with*

Measurement errors

A **true value** is the value that would be obtained in an ideal measurement. True values may be values found in a data book, or the values you expect to get in your experiments. A **measurement error** is the difference between a measured value and the true value for the quantity being measured. Remember that mistakes are not counted as errors here.

Random errors

Random errors can happen when any measurement is being made. They are measurement errors in which measurements vary unpredictably. There can be many reasons for this, including

* factors that are not controlled in the experiment
* difficulty in deciding on the reading given by a measuring device.

Random errors cannot be corrected. All you can hope to do is reduce their effect by making more measurements and reporting the mean value.

Systematic errors

Systematic errors are measurement errors in which the measurements differ from the true values by a consistent amount each time a measurement is made. Reasons for this include

* the way in which measurements are taken
* faulty measuring devices.

For example, poor contact between a thermometer and the object whose temperature is being measured will cause systematic errors. A faulty measuring device may give readings that are consistently too high or too low. This may be because it has not been calibrated correctly. A faulty device may give a **zero error**, in which the reading is not zero when the quantity being measured is zero (Figure 2).

Unlike random errors, it may be possible to correct for systematic errors.

▲ **Figure 2** *Zero error on an ammeter*

Precision and accuracy

In everyday life, people often use the words *precise* and *accurate* to mean the same thing – this is not the case in physics.

- **Accuracy** is to do with how close a measurement result is to the true value – the closer it is, the more **accurate** it is.
- **Precision** is to do with how close repeated measurements are to each other – the closer they are to each other, the more **precise** the measurement is.

Figure 3 is a visual way of appreciating the terms accuracy and precision using a dartboard as an example. You are aiming for bullseye – it represents the true value.

not accurate
not precise

accurate
not precise

not accurate
precise

accurate
precise

▲ **Figure 3** *Accuracy and precision*

Uncertainty

Random and systematic errors mean that you rarely obtain the same value for a particular measurement. Consider using a micrometer to measure the diameter of a supposedly uniform copper wire. The readings below show the readings along the length of the copper wire:

$$0.53\,\text{mm}, 0.49\,\text{mm}, 0.52\,\text{mm}, 0.51\,\text{mm}$$

The smallest scale division of this instrument is 0.01 mm but the readings above show a spread much greater than this.

The **mean** value for the diameter can be calculated by adding together the values for each repeat reading, then dividing by the number of readings. The mean diameter of the copper wire is

$$\frac{0.53\,\text{mm} + 0.49\,\text{mm} + 0.52\,\text{mm} + 0.51\,\text{mm}}{4} = 0.51\,\text{mm}.$$

The **range** of the measurements is 0.04 mm. This is the difference between the smallest and largest readings (0.53 mm – 0.49 mm).

The **uncertainty** in the measurement is an interval within which the true value can be expected to lie. The **absolute uncertainty** in the mean value of a measurement can be approximated as half the range. This is often expressed as ± value. In this example, you can write the diameter as its mean value ± absolute uncertainty

$$\text{diameter} = 0.51 \pm 0.02\,\text{mm}$$

The **percentage uncertainty** in the diameter can be calculated from its absolute uncertainty and mean value as follows:

$$\% \text{ uncertainty in diameter} = \frac{\text{absolute uncertainty}}{\text{mean value}} \times 100$$
$$= \frac{0.02}{0.51} \times 100 = 3.9\%$$

Finally, what do you do when repeat measurements give identical values or you have just taken a single measurement? In this situation you can approximate the absolute uncertainty to be equal to the **resolution** of the measuring instrument. This is the smallest change in the measured quantity that the instrument can show. The micrometer readings for the copper wire are written to 2 decimal places, with ±0.01 mm being the smallest change it could show.

Therefore, if all the readings were 0.53 mm, or you just had a single reading of 0.53 mm, the diameter of the copper wire may be written as

$$\text{diameter} = 0.53 \pm 0.01 \text{ mm}$$

Analysing uncertainties

Uncertainties can help you identify where the greatest errors in an experiment are, giving you the chance to improve it using a different method or measuring instrument.

The final uncertainty in an answer depends on how quantities are combined. Here are three important rules about the way uncertainties propagate.

1 **Adding or subtracting quantities**

    When you add or subtract quantities in an equation, you add the absolute uncertainties for each value.

 What is the extension?

The original length of a spring is 2.5 ± 0.1 cm and the final length is 15.0 ± 0.2 cm. Calculate the extension of the spring and the absolute uncertainty.

Step 1: Calculate the extension by subtracting the lengths.

$$\text{extension} = 15.0 - 2.5 = 12.5 \text{ cm}$$

Step 2: Add the absolute uncertainties.

$$\text{absolute uncertainty} = 0.1 + 0.2 = 0.3$$

Step 3: Write the answer in the normal convention.

$$\text{extension} = 12.5 \pm 0.3 \text{ mm}$$

2 **Multiplying or dividing quantities**

    When you multiply or divide quantities, you add the percentage uncertainties for each value.

 What is the resistance?

The current I in a resistor is 1.60 ± 0.02 A and the potential difference V across the resistor is 6.00 ± 0.20 V. Calculate the resistance and the absolute uncertainty.

Step 1: Calculate the resistance R of the resistor.

$$R = \frac{V}{I} = \frac{6.00}{1.60} = 3.75 \, \Omega$$

Step 2: Calculate the percentage uncertainty in each measurement.

$$\% \text{ uncertainty in } I = \frac{0.02}{1.60} \times 100 = 1.25\%$$

$$\% \text{ uncertainty in } V = \frac{0.20}{6.00} \times 100 = 3.33\%$$

Step 3: Add the percentage uncertainties.

% uncertainty in $R = 1.25 + 3.33 = 4.58\%$

Step 4: Calculate the absolute uncertainty in R.

absolute uncertainty in $R = 0.0458 \times 3.75 = 0.17\,\Omega$

Step 5: The values of V and I are quoted to 3 significant figures, therefore the final answer for the resistance must also be written to 3 significant figures.

$R = 3.75 \pm 0.17\,\Omega$ (3 s.f.)

3   Raising a quantity to a power

When a measurement in a calculation is raised to a power n, your percentage uncertainty is increased n times. The power n can be an integer or a fraction.

 ## Cross-sectional area of a wire

The diameter of a wire is recorded as $0.51 \pm 0.02\,$mm. Calculate the cross-sectional area of the wire and the absolute uncertainty.

Step 1: Calculate the cross-sectional area A of the wire.

$$A = \frac{\pi d^2}{4} = \frac{\pi \times (0.51 \times 10^{-3})^2}{4} = 2.04 \times 10^{-7}\,\text{m}^2$$

Step 2: The percentage uncertainty in A is equal to 2 times the percentage uncertainty in d.

(The π and the 4 are numbers and therefore have no uncertainty associated with them.)

% uncertainty in $A = 2 \times \left(\dfrac{0.02}{0.51} \times 100\right) = 7.84\%$

Step 3: Calculate the absolute uncertainty in A.

absolute uncertainty in $A = 0.0784 \times 2.04 \times 10^{-7} = 0.16 \times 10^{-7}\,\text{m}^2$

Step 4: The diameter of the wire is quoted to 2 significant figures, therefore the final answer for the cross-sectional must also be written to 2 significant figures.

$A = (2.0 \pm 0.2) \times 10^{-7}\,\text{m}^2$ (2 s.f.)

Graphs

Straight line graphs are important in physics because you can use them to formulate relationships between physical quantities. As indicated in Appendix A2, Recording results, you plot points using small crosses. If the points appear to lie on a straight line, then you can draw your straight line of best fit using a long ruler. You must ignore a point that is much further than any other point from the best fit line. This point is referred to as being **anomalous**.

The uncertainty in a measurement can be used to give a small range or **error bar** for each measurement. Instead of plotting just the points on a graph, you can plot an error bar for all of your measurements.

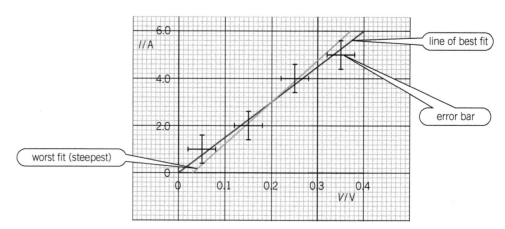

▲ **Figure 4** *Error bars are useful when you draw the line of best fit and the worst line for your measurements*

Your straight best fit line must pass through all the error bars (Figure 4). You would use this line to determine the value of the gradient. How can you determine an approximate value for the uncertainty in the gradient? You would draw the **line of worst fit** – the least acceptable straight line through the data points – this can either be the steepest or the shallowest line.

The absolute uncertainty in the gradient is the positive difference between the gradient of the line of best fit and the gradient of the line of worst fit.

The percentage uncertainty in the gradient can be calculated as

$$\% \text{ uncertainty in gradient} = \frac{\text{absolute uncertainty}}{\text{gradient best fit line}} \times 100\%$$

5d. Physics A Data Sheet

Data, Formulae and Relationships

The data, formulae and relationships in this datasheet will be printed for distribution with the examination papers.

Data

Values are given to three significant figures, except where more – or fewer – are useful.

Physical constants

acceleration of free fall	g	9.81 m s^{-2}
elementary charge	e	$1.60 \times 10^{-19} \text{ C}$
speed of light in a vacuum	c	$3.00 \times 10^{8} \text{ m s}^{-1}$
Planck constant	h	$6.63 \times 10^{-34} \text{ J s}$
Avogadro constant	N_A	$6.02 \times 10^{23} \text{ mol}^{-1}$
molar gas constant	R	$8.31 \text{ J mol}^{-1} \text{K}^{-1}$
Boltzmann constant	k	$1.38 \times 10^{-23} \text{ J K}^{-1}$
gravitational constant	G	$6.67 \times 10^{-11} \text{ N m}^2 \text{kg}^{-2}$
permittivity of free space	ε_0	$8.85 \times 10^{-12} \text{ C}^2 \text{N}^{-1} \text{m}^{-2} \, (\text{F m}^{-1})$
electron rest mass	m_e	$9.11 \times 10^{-31} \text{ kg}$
proton rest mass	m_p	$1.673 \times 10^{-27} \text{ kg}$
neutron rest mass	m_n	$1.675 \times 10^{-27} \text{ kg}$
alpha particle rest mass	m_α	$6.646 \times 10^{-27} \text{ kg}$
Stefan constant	σ	$5.67 \times 10^{-8} \text{ W m}^{-2} \text{K}^{-4}$

Quarks

up quark	$\text{charge} = +\dfrac{2}{3} e$
down quark	$\text{charge} = -\dfrac{1}{3} e$
strange quark	$\text{charge} = -\dfrac{1}{3} e$

Conversion factors

unified atomic mass unit	$1 \text{ u} = 1.661 \times 10^{-27} \text{ kg}$
electronvolt	$1 \text{ eV} = 1.60 \times 10^{-19} \text{ J}$
day	$1 \text{ day} = 8.64 \times 10^{4} \text{ s}$
year	$1 \text{ year} \approx 3.16 \times 10^{7} \text{ s}$
light year	$1 \text{ light year} \approx 9.5 \times 10^{15} \text{ m}$
parsec	$1 \text{ parsec} \approx 3.1 \times 10^{16} \text{ m}$

Mathematical equations

arc length $= r\theta$

circumference of circle $= 2\pi r$

area of circle $= \pi r^2$

curved surface area of cylinder $= 2\pi rh$

surface area of sphere $= 4\pi r^2$

area of trapezium $= \dfrac{1}{2}(a + b)\,h$

volume of cylinder $= \pi r^2 h$

volume of sphere $= \dfrac{4}{3}\pi r^3$

Pythagoras' theorem: $a^2 = b^2 + c^2$

cosine rule: $a^2 = b^2 + c^2 - 2bc \cos A$

sine rule: $\dfrac{a}{\sin A} = \dfrac{b}{\sin B} = \dfrac{c}{\sin C}$

$\sin \theta \approx \tan \theta \approx \theta$ and $\cos \theta \approx 1$ for small angles

$\log(AB) = \log(A) + \log(B)$

(Note: $\lg = \log_{10}$ and $\ln = \log_e$)

$\log\left(\dfrac{A}{B}\right) = \log(A) - \log(B)$

$\log(x^n) = n \log(x)$

$\ln(e^{kx}) = kx$

Formulae and relationships
Module 2 – Foundations of physics
vectors

$$F_x = F \cos \theta$$
$$F_y = F \sin \theta$$

Module 3 – Forces and motion
uniformly accelerated motion

$$v = u + at$$
$$s = \dfrac{1}{2}(u + v)t$$
$$s = ut + \dfrac{1}{2}at^2$$
$$v^2 = u^2 + 2as$$

force

$$F = \dfrac{\Delta p}{\Delta t}$$
$$p = mv$$

turning effects

$$moment = Fx$$
$$torque = Fd$$

density

$$\rho = \dfrac{m}{V}$$

pressure

$$p = \dfrac{F}{A}$$

$$p = h\rho g$$

work, energy and power

$$W = Fx \cos \theta$$

$$efficiency = \dfrac{useful\ energy\ output}{total\ energy\ input} \times 100\%$$

$$P = \dfrac{W}{t}$$

$$P = Fv$$

springs and materials

$$F = kx$$

$$E = \dfrac{1}{2}Fx;\ E = \dfrac{1}{2}kx^2$$

$$\sigma = \dfrac{F}{A}$$

$$\varepsilon = \dfrac{x}{L}$$

$$E = \dfrac{\sigma}{\varepsilon}$$

Module 4 – Electrons, waves and photons

charge $\quad \Delta Q = I\Delta t$

current $\quad I = Anev$

work done $\quad W = VQ;\ W = \varepsilon Q;\ W = VIt$

resistance and resistors

$$R = \dfrac{\rho L}{A}$$
$$R = R_1 + R_2 + \ldots$$
$$\dfrac{1}{R} = \dfrac{1}{R_1} + \dfrac{1}{R_2} + \ldots$$

power $\quad P = VI,\ P = I^2R$ and $P = \dfrac{V^2}{R}$

internal resistance $\quad \varepsilon = I(R + r);\ \varepsilon = V + Ir$

potential divider $\quad V_{out} = \dfrac{R_2}{R_1 + R_2} \times V_{in}$

$$\dfrac{V_1}{V_2} = \dfrac{R_1}{R_2}$$

waves $\quad v = f\lambda$

$$f = \frac{1}{T}$$

$$I = \frac{P}{A}$$

$$\lambda = \frac{ax}{D}$$

refraction $\quad n = \dfrac{c}{v}$

$$n \sin\theta = \text{constant}$$

$$\sin C = \frac{1}{n}$$

quantum physics $\quad E = hf$

$$E = \frac{hc}{\lambda}$$

$$hf = \phi + KE_{max}$$

$$\lambda = \frac{h}{p}$$

Module 5 – Newtonian world and astrophysics

thermal physics $\quad E = mc\Delta\theta$

$$E = mL$$

ideal gases $\quad pV = NkT; \quad pV = nRT$

$$pV = \frac{1}{3}Nm\overline{c^2}$$

$$\frac{1}{2}m\overline{c^2} = \frac{3}{2}kT$$

$$E = \frac{3}{2}kT$$

circular motion

$$\omega = \frac{2\pi}{T}; \quad \omega = 2\pi f$$

$$v = \omega r$$

$$a = \frac{v^2}{r}; \, a = \omega^2 r$$

$$F = \frac{mv^2}{r}; \, F = m\omega^2 r$$

oscillations

$$\omega = \frac{2\pi}{T}; \quad \omega = 2\pi f$$

$$a = -\omega^2 x$$

$$x = A\cos\omega t; \, x = A\sin\omega t$$

$$v = \pm w\sqrt{A^2 - x^2}$$

gravitational field

$$g = \frac{F}{m}$$

$$F = -\frac{GMm}{r^2}$$

$$g = -\frac{GM}{r^2}$$

$$T^2 = \left(\frac{4\pi^2}{GM}\right)r^3$$

$$V_g = -\frac{GM}{r}$$

$$\text{energy} = -\frac{GMm}{r}$$

astrophysics

$$hf = \Delta E; \frac{hc}{\lambda} = \Delta E$$

$$d\sin\theta = n\lambda$$

$$\lambda_{max} \propto \frac{1}{T}$$

$$L = 4\pi r^2 \sigma T^4$$

cosmology

$$\frac{\Delta\lambda}{\lambda} \approx \frac{\Delta f}{f} \approx \frac{v}{c}$$

$$p = \frac{1}{d}$$

$$v = H_0 d$$

$$t = H_0^{-1}$$

Module 6 – Particles and medical physics
capacitance and capacitors

$$C = \frac{Q}{V}$$

$$C = \frac{\varepsilon_0 A}{d}$$

$$C = 4\pi\varepsilon_0 R$$

$$C = C_1 + C_2 + \ldots$$

$$\frac{1}{C} = \frac{1}{C_1} + \frac{1}{C_2} + \ldots$$

$$W = \frac{1}{2}QV; \, W = \frac{1}{2}\frac{Q^2}{C}; \, W = \frac{1}{2}V^2 C$$

$$\tau = CR$$

$$x = x_0 e^{\frac{-t}{CR}}$$

$$x = x_0\left(1 - e^{\frac{-t}{CR}}\right)$$

electric field $\quad E = \dfrac{F}{Q}$

$$F = \frac{Qq}{4\pi\varepsilon_0 r^2}$$

$$E = \frac{Q}{4\pi\varepsilon_0 r^2}$$

$$E = \frac{v}{d}$$

$$V = \frac{Q}{4\pi\varepsilon_0 r}$$

$$\text{energy} = \frac{Qq}{4\pi\varepsilon_0 r}$$

magnetic field $\quad F = BIL\sin\theta$

$$F = BQv$$

electromagnetism

$$\phi = BA\cos\theta$$

$$\varepsilon = -\frac{\Delta(N\phi)}{\Delta t}$$

$$\frac{n_s}{n_p} = \frac{V_s}{V_p} = \frac{I_p}{I_s}$$

radius of nucleus $\quad R = r_0 A^{\frac{1}{3}}$

radioactivity $\quad A = \lambda N; \frac{\Delta N}{\Delta t} = -\lambda N$

$$\lambda t_{\frac{1}{2}} = \ln(2)$$

$$A = A_0 e^{-\lambda t}$$

$$N = N_0 e^{-\lambda t}$$

Einstein's mass-energy
equation $\quad \Delta E = \Delta mc^2$

attenuation of
X-rays $\quad I = I_0 e^{-\mu x}$

ultrasound $\quad Z = \rho c$

$$\frac{I_r}{I_0} = \frac{(Z_2 - Z_1)^2}{(Z_2 + Z_1)^2}$$

$$\frac{\Delta f}{f} = \frac{2v\cos\theta}{c}$$

Glossary

acceleration The rate of change of velocity, a vector quantity

acceleration of free fall The rate of change of velocity of an object falling in a gravitational field, symbol g

air resistance The drag or resistive force experienced by objects moving through air

ammeter A device used to measure electric current — it must be placed in series and ideally have zero resistance

ampère The base SI unit of electric current, symbol A, defined as the current flowing in two parallel wires in a vacuum 1 m apart such that there is an attractive force of 2.0×10^{-7} N per metre length of wire between them

amplitude *(waves)* The maximum displacement from the equilibrium position (can be positive or negative)

angle of incidence The angle between the direction of travel of an incident wave and the normal at a boundary between two media

angle of reflection The angle between the direction of travel of a reflected wave and the normal at a boundary between two media

anion A negatively charged ion, one which is attracted to an anode

anode A positively charged electrode

antiparallel *(vectors)* In the same line but opposite directions

antiphase Particles oscillating completely out of step with each other (one reaches its maximum positive displacement as the other reaches its maximum negative displacement) are in antiphase

Archimedes' principle The upthrust on an object in a fluid is equal to the weight of fluid it displaces

average speed The rate of change in distance calculated over a complete journey

average velocity The change in displacement Δs for a journey divided by the time taken Δt; $\Delta s/\Delta t$

base unit One of seven units that form the building blocks of the SI measurement system

battery A collection of cells that transfers chemical energy into electrical energy

braking distance Distance travelled by a vehicle from the time the brakes are applied until the vehicle stops

breaking strength The stress value at the point of fracture, calculated by dividing the breaking force by the cross-sectional area

brittle Property of a material that does not show plastic deformation and deforms very little (if at all) under high stress

capacitor A component that stores charge, consists of two plates separated by an insulator (dielectric)

cathode A negatively charged electrode

cation A positively charged ion, one which is attracted to a cathode

cell A device that transfers chemical energy into electrical energy

centre of gravity An imaginary point at which the entire weight of an object appears to act

centre of mass A point through which any externally applied force produces straight-line motion but no rotation

charge carrier A particle with charge that moves through a material to form an electric current — for example, an electron in a metal wire

closed system An isolated system that has no interaction with its surroundings

coherence Two waves sources, or waves, that are coherent have a constant phase difference

component One of the two perpendicular vectors obtained by resolving a vector

compression The decrease in length of an object when a compressive force is exerted on it

compression *(waves)* A moving region in which the medium is denser or has higher pressure than the surrounding medium

compressive deformation A change in the shape of an object due to compressive forces

compressive force Two or more forces together that reduce the length or volume of an object

conservation of charge A conservation law which states that electric charge can neither be created nor destroyed — the total charge in any interaction must be the same before and after the interaction

constant speed Motion in which the distance travelled per unit time stays the same

constant velocity Motion in which the change in displacement per unit time stays the same

constructive interference Superposition of two waves in phase so that the resultant wave has greater amplitude than the original waves

conventional current A model used to describe electric current in a circuit — conventional current travels from positive to negative — it is the direction in which positive charges would travel

coulomb The derived SI unit of electrical charge, symbol C — 1 coulomb of electric charge passes a point in one second when there is an electric current of one ampere, $1\,C = 1\,As$

couple A pair of equal and opposite forces acting on a body but not in the same straight line

critical angle The angle of incidence at the boundary between two media that will produce an angle of refraction of 90 °

crystallography A method for determining the structure of a substance by studying the interference patterns produced by waves passing through a crystal of the substance

de Broglie equation An equation relating the wavelength and the momentum of a particle:

$$\lambda = \frac{h}{p}$$

density The mass per unit volume of a substance

derived quantity A quantity that comes from a combination of base units

derived unit A unit used to represent a derived quantity, such as N for force

destructive interference Superposition of two waves in antiphase so that the waves cancel each other out and the resultant wave has smaller amplitude than the original waves

diffraction The phenomenon in which waves passing through a gap or around an obstacle spread out

diode A semiconductor component that allows current only in one particular direction

displacement The distance travelled in a particular direction — it is a vector with magnitude and a direction

displacement *(waves)* The distance from the equilibrium position in a particular direction — displacement is a vector, so it has a positive or a negative value

drag force The resistive force exerted by a fluid on an object moving through it

ductile Property of a material that has a large plastic region in a stress–strain graph, so can be drawn into wires

efficiency The ratio of useful output energy to total input energy, often expressed as a percentage

elastic deformation A reversible change in the shape of an object due to a compressive or tensile force — removal of stress or force will return the object to its original shape and size (no permanent strain)

elastic limit The value of stress or force beyond which elastic deformation becomes plastic deformation, and the material or object will no longer return to its original shape and size when the stress or force is removed

elastic potential energy The energy stored in an object because of its deformation

electric charge A physical property, symbol q or Q, either positive or negative, measured in coulombs, C, or as a relative charge

electric current The rate of flow of charge, symbol I, measured in ampères, A; normally a flow of electrons in metals or a flow of ions in electrolytes

electricity meter A device that measures the electrical energy supplied in kWh to a house from the grid

electrolyte A liquid containing ions that are free to move and so to conduct electricity

electromagnetic spectrum The full range of frequencies of electromagnetic waves, from gamma rays to radio waves

electromagnetic wave Transverse waves with oscillating electric and magnetic field components, such as light and X-rays, that do not need a medium to propagate — they travel at a speed of 3.0×10^8 m s^{-1} in a vacuum

electromotive force (e.m.f.) The work done on the charge carriers per unit charge, symbol V, unit volt, V, measured across a cell, battery or power supply

electron gun A device that uses a large accelerating potential difference to produce a narrow beam of electrons

electronvolt A derived unit of energy used for subatomic particles and photons, defined as the energy transferred to or from an electron when it passes through a potential difference of 1 volt; 1 eV is equivalent to 1.60×10^{-19} J

elementary charge The electric charge equivalent to the charge on a proton, 1.60×10^{-19} C; symbol e

energy The capacity for doing work, measured in joules, J

equilibrium A body is in equilibrium when the net force and net moment acting on it are zero

equilibrium position *(waves)* The resting position for particles in the medium

extension The increase in length of an object when a tensile force is exerted on it

filament lamp An electrical component containing a narrow filament of wire that transfers electrical energy into heat and light

fluid A substance that can flow, including liquids and gases

force A push or pull on an object, measured in newtons, N

force constant A quantity determined by dividing force by extension (or compression) for an object obeying Hooke's law — called constant of proportionality k in Hooke's law, measured in Nm^{-1}

force–extension graph A graph of force against extension (or compression), with the area under the graph equal to the work done on the material

force–time graph A graph of net force against time, with the area under the graph equal to the impulse

free electron An electron in a metal that is not bound to an atom and is free to move — sometimes called a delocalised electron

free fall The motion of an object accelerating under gravity with no other force acting on it

free-body diagram A diagram that represents the forces acting on a single object

frequency *(waves)* The number of wavelengths passing a given point per unit time

fundamental frequency The lowest frequency at which an object (e.g., an air column in a pipe or a string fixed at both ends) can vibrate

fundamental mode of vibration A vibration at the fundamental frequency

gamma rays Short-wavelength electromagnetic waves, with wavelengths from 10^{-10} m to 10^{-16} m

gold-leaf electroscope A device with a metallic stem and a gold leaf that can be used to identify and measure electric charge — a device that was historically used as a voltmeter for measuring large voltages

gradient In a graph, the change in the vertical axis quantity divided by the corresponding change in the horizontal axis quantity

gravitational potential energy The capacity for doing work as a result of an object's position in a gravitational field

harmonic A whole-number multiple of the fundamental frequency

Hooke's law The force applied is directly proportional to the extension of the spring unless the limit of proportionality is exceeded

hysteresis loop A loop-shaped plot obtained when, for example, loading and unloading a material produce different deformations

impulse The area under a force–time graph — the product of force and the time for which the force acts

in phase Particles oscillating perfectly in time with each other (reaching their maximum positive displacement at the same time) are in phase

inelastic collision A collision in which kinetic energy is lost

infrared waves Electromagnetic waves, with wavelengths from 10^{-3} m to 7×10^{-7} m

instantaneous speed The speed at the moment it is measured — speed over an infinitesimal interval of time

intensity *(waves)* The radiant power passing through a surface per unit area

interference Superposition of two progressive waves from coherent sources to produce a resultant wave with a displacement equal to the sum of the individual displacements from the two waves

interference pattern A pattern of constructive and destructive interference formed as waves overlap

internal resistance The resistance of a source of e.m.f. (e.g a cell) due to its construction, which causes a loss in energy/voltage as the charge passes through the source, symbol r, SI unit ohm, Ω

ion An atom that has either lost or gained electrons and so has a net charge

ionic solution An ionic compound dissolved in a liquid to form an electrolyte

I–V **characteristic** A description of the relationship between the electric current in a component and the potential difference across it — in most cases this is usually in the form of a simple graph of I against V

kilowatt-hour A derived unit of energy, most often associated with paying for electrical energy, symbol kWh (1 kWh = 3.6 MJ). Energy in kWh can be calculated by multiplying the power in kW by the time in hours

kinetic energy The energy associated with an object as a result of its motion

Kirchhoff's first law At any point in an electrical circuit, the sum of currents into that point is equal to the sum of currents out of that point, electrical charge is conserved

Kirchhoff's second law In a closed loop of an electrical circuit, the sum of the e.m.f.s is equal the sum of the p.d.s

law of reflection The angle of incidence is equal to the angle of reflection

light-dependent resistor An electrical component with a resistance that decreases as the light intensity incident on it increases

light-emitting diode A type of diode that emits light when it conducts electricity

limit of proportionality The value of stress or force beyond which stress is no longer directly proportional to strain

linear momentum A property of an object travelling in a straight line, the product of its mass and velocity, measured in kg m s^{-1} or N s

loading *(electrical circuits)* Connecting a component or a device across the teminals of a source of e.m.f. or across another component

loading curve A force–extension graph

longitudinal wave A wave in which the medium is displaced in the same line as the direction of energy transfer — oscillations of the medium particles are parallel to the direction of the wave travel

lost volts The potential difference across the internal resistor of a source of e.m.f.

mass Amount of matter, a base quantity measured in kilograms, kg

maximum *(waves)* The point of greatest amplitude in an interference pattern, produced by constructive interference

mean drift velocity The average velocity of electrons as they move through a wire, symbol v, unit ms^{-1}

microwaves Long-wavelength electromagnetic waves, with wavelengths from 10^{-1} m to 10^{-3} m

minimum *(waves)* The point of least amplitude in an interference pattern, produced by destructive interference

moment The product of force and perpendicular distance from a pivot or stated point

monochromatic light Light of a single frequency

negative *(charge)* One type of electric charge; negatively charged objects attract positively charged ones, and repel other negative charges

negative temperature coefficient (NTC) A relationship in which a variable decreases as temperature increases, for example the resistance of NTC thermistors

Newton's first law of motion A body will remain at rest or continue to move with constant velocity unless acted upon by a resultant force

Newton's second law of motion The rate of change of momentum of an object is directly proportional to the resultant force and takes place in the direction of the force

Newton's third law of motion When two objects interact, each exerts an equal but opposite force on the other during the interaction

node For a stationary wave, a point where the amplitude is always zero

non-ohmic component A component that does not obey Ohm's law, e.g filamant lamp and diode

normal An imaginary line perpendicular to a surface such as the boundary between one medium and another (e.g., air and glass)

normal contact force The force exerted by a surface on an object, which acts perpendicularly to the surface

number density The number of free electrons per cubic metre of a material, symbol n, unit m^{-3}

ohm The derived SI unit of resistance, symbol Ω — defined as the resistance of a component that has a potential difference of 1 V per unit ampere

Ohm's law The potential difference across a conductor is directly proportional to the current in the component as long as its temperaure remains constant

ohmic conductor A conductor that obeys Ohm's law

optical fibre A fibre made of glass designed with a varying refractive index in order to totally internally reflect pulses of visible or infrared light travelling through it

oscilloscope An instrument that displays an electrical signal as a voltage against time trace on a screen

out of phase Particles that are neither in phase, nor in antiphase, are out of phase

parallel *(vectors)* In the same line and direction

parallel circuit A type of branching electrical circuit in which there is more than one path for the current — components in parallel have the same potential difference

partially polarised Description of a transverse wave in which there are more oscillations in one particular plane, but the wave is not completely plane polarised — occurs when transverse waves reflect off a surface

path difference The difference in the distance travelled by two waves from the source to a specific point

peak The maximum positive amplitude of a transverse wave

perfectly elastic collision A collision in which no kinetic energy is lost

period *(waves)* The time taken for one complete wavelength to pass a given point

phase difference The difference between the displacements of particles along a wave, or the difference between the displacements of particles on different waves, measured in degrees or radians, with each complete cycle or a difference of one wavelength representing 360° or 2π radians

photoelectric effect The emission of photoelectrons from a metal surface when electromagnetic radiation above a threshold frequency is incident on the metal

photoelectric effect equation Einstein's equation relating the energy of a photon, the work function of a metal, and the maximum kinetic energy of any emitted photoelectrons: $hf = \phi + KE_{max}$

photoelectrons Electrons emitted from the surface of a metal by the photoelectric effect

photon A quantum of electromagnetic energy — photon energy E is given by $E = hf$, where h is the Planck constant and f is the frequency of the electromagnetic radiation

pivot A point about which a body can rotate

Planck constant Symbol h, an important constant in quantum mechanics, 6.63×10^{-34} J s

plane polarised Description of a transverse wave in which the oscillations are limited to only one plane

plastic deformation An irreversible change in the shape of an object due to a compressive or tensile force — removal of the stress or force produces permanent deformation

plumb-line A string with a weight used to provide a vertical reference line

polarisation The phenomenon in which oscillations of a transverse wave are limited to only one plane

polarity The type of charge (positive or negative) or the orientation of a cell relative to a component

polycrystalline graphite Thin layers of graphite with regularly arranged carbon atoms in different orientations

polymeric Description of a material comprising of long-chain molecules, such as rubber, which may show large strains

positive *(charge)* One type of electric charge — positively charged objects attract negatively charged ones, and repel other positive charges

potential difference (pd) Defined as the energy transferred from electrical energy to other forms (heat, light, etc.) per unit charge.

potential divider An electrical circuit designed to divide the potential difference across two or more components (often two resistors) in order to produce a specific output

potential divider equation An equation relating the output potential difference from a simple potential divider containing a pair of resistors:

$$V_{out} = \frac{R_2}{(R_1 + R_2)} \times V_{in}$$

potentiometer An electrical component with three terminals and some form of sliding contact that can be adjusted to vary the potential difference between two of the terminals

power The rate of work done, measured in watts, W

prefix A word or letter placed before another one, for example, 5.0 km is 5.0×10^3 m

pressure The force exerted per unit cross-sectional area, measured in pascals, Pa

principle of conservation of energy The total energy of a closed system remains constant — energy cannot be created nor can it be destroyed

principle of conservation of momentum Total momentum of a system remains the same before and after a collision

principle of moments For a body in rotational equilibrium, the sum of the anticlockwise moments about a point is equal to the sum of the clockwise moments about the same point

principle of superposition of waves When two waves meet at a point the resultant displacement at that point is equal to the sum of the displacements of the individual waves

progressive wave A wave in which the peaks and troughs, or compressions and rarefactions, move through the medium as energy is transferred

projectile An object that is thrown or propelled on the surface of the Earth

P-waves Primary waves — longitudinal waves that travel through the Earth from an earthquake

Pythagoras' theorem The square of the length of the hypotenuse of a right-angled triangle equals the sum of the squares of the lengths of the other two sides

quantisation The availability of some quantities, such as energy or charge, only in certain discrete values

quantity A property of an object, substance, or phenomenon that can be measured

quantum mechanics The branch of physics dealing with phenomena on the very small scale, often less than the size of an atom

radio waves Long-wavelength electromagnetic waves, with wavelengths greater than 10^{-1} m

rarefaction *(waves)* A moving region in which the medium is less dense or has less pressure than the surrounding medium

ray A line representing the direction of energy transfer of a wave, perpendicular to the wavefronts

reflection The change in direction of a wave at a boundary between two different media, so that the wave remains in the original medium

refraction The change in direction of a wave as it changes speed when it passes from one medium to another

refractive index The refractive index of a material $n = \frac{c}{v}$, where c is the speed of light through a vacuum and v is the speed of light through the material

relative charge A simplified measurement of the electric charge of a particle or object, measured as multiples of the elementary charge

resistance A property of a component calculated by dividing the potential difference across it by the current in it, symbol R, unit ohm, Ω

resistivity A property of a material, measured in Ω m, defined as the product of the resistance of a component made of the material and its cross-sectional area divided by its length

resistor An electrical component that obeys Ohm's law, transferring electrical energy to thermal energy

resistor circuit Two or more resistors arranged to provide a specific resistance

resolving a vector Splitting a vector into two component vectors perpendicular to each other

restoring force A force that tries to return a system to its equilibrium position

resultant vector A single vector that has the same effect as two or more vectors added together

scalar quantity A quantity with magnitude (size) but no direction

semiconductor A material with a lower number density than a typical conductor, for example silicon

series An arrangement of electrical components connected end-to-end that means that the current is the same in each component

series circuit A type of electrical circuit where the components are connected end-to-end

SI Système International d'Unités (International System of Units)

standard form Mathematical notation in which a number is shown with the decimal point placed after the first digit, followed by ×10 raised to an appropriate power

standing wave A wave that remains in a constant position with no net transfer of energy and is characterised by its nodes and antinodes — also called a stationary wave

stationary wave A wave that remains in a constant position with no net transfer of energy and is characterised by its nodes and antinodes — also called a standing wave

stiffness The ability of an object to resist deformation

stopping distance The total distance travelled from the time when a driver first sees a reason to stop to the time when the vehicle stops, the sum of the thinking distance and the braking distance

strain see 'tensile strain'

stress see 'tensile stress'

strong material A material with a large value for the ultimate tensile strength

superconductivity A phenomenon in which the resistivity of a material falls to almost zero when the material is cooled below a certain temperature

superposition *(waves)* Overlap of two waves at a point in space

S-waves Secondary waves: transverse waves that travel through the Earth from an earthquake

tensile deformation A change in the shape of an object due to tensile forces

tensile force Equal and opposite forces acting on a material to stretch it

tensile strain The extension per unit length, a dimensionless quantity

tensile stress The force per unit cross-sectional area, measured in Pa

tension The pulling force exerted by a string, cable, or chain on an object

terminal p.d. The potential difference across an electrical power source — when there is no current this is equal to the e.m.f. of the source, but if there is a current in the source this is equal to the e.m.f. minus the lost volts

terminal velocity The constant speed reached by an object when the drag force (and upthrust) is equal and opposite to the weight of the object

thermionic emission The emission of electrons from the surface of a hot metal wire

thermistor An electrical component that has a resistance that decreases as the temperature increases (a negative temperature coefficient)

thinking distance The distance travelled by a vehicle from when the driver first perceives a need to stop to when the brakes are applied

threshold frequency The minimum frequency of the electromagnetic radiation that will cause the emission of an electron from the surface of a particular metal — symbol f_0, measured in Hz

threshold voltage The minimum potential difference at which a diode begins to conduct

time of flight The time taken for an object to complete its motion

timebase The time interval represented by one horizontal square on an oscilloscope screen

torque (of a couple) The product of one of the forces of a couple and the perpendicular distance between the forces

total internal reflection The reflection of all light hitting a boundary between two media back into the original medium when the light is travelling through the medium with the higher refractive index and the incidence angle at the boundary is greater than the critical angle

transverse wave A wave in which the medium is displaced perpendicular to the direction of energy transfer — the oscillations of medium particles are perpendicular to the direction of travel of the wave

triangle of forces Three forces acting at a point in equilbrium, represented by the sides of a triangle

trough The maximum negative amplitude of a transverse wave

ultimate tensile strength The maximum stress that a material can withstand before it breaks

ultraviolet Electromagnetic waves, with wavelengths from 4×10^{-7} m to 10^{-8} m

uniform gravitational field A gravitational field in which the field lines are parallel and the value for g remains constant

unpolarised Description of a transverse wave in which the oscillations occur in many planes

upthrust The upward buoyant force exerted on a body immersed in a fluid

vector quantity A quantity with magnitude (size) and direction

vector triangle A triangle constructed to scale to determine the resultant of two vectors

velocity A vector quantity equal to the rate of change of displacement

visible light Electromagnetic waves, with wavelengths from 4×10^{-7} m to 7×10^{-7} m

volt The derived SI unit of potential difference and electromotive force, symbol V, defined as the energy transferred per unit charge, whether energy is either transferred to or from the charges — 1 V is the p.d. across a component when 1 J of energy is transferred per 1 C passing through the component

voltage See 'potential difference'

voltmeter A device used to measure potential difference — it must be placed in parallel across components and ideally have an infinite resistance

wave equation An equation that relates the frequency f in hertz, the wavelength λ in metres, and the wave speed v in $\mathrm{m\,s^{-1}}$: $v = f\lambda$

wave profile A graph showing the displacement of the particles in the wave against the distance along the wave

wave speed The distance travelled by the wave per unit time

wavefront A line of points in phase with each other in a wave, perpendicular to the direction of energy transfer

wavelength The minimum distance between two points oscillating in phase, for example the distance from one peak to the next or from one compression to the next

wave–particle duality A theory that states that matter has both particle and wave properties and also electromagnetic radiation has wave and particulate (photon) nature

weight The gravitational force on an object, measured in newtons, N

work The product of force and the distance moved in the direction of the force, measured in J

work function The minimum energy needed to remove a single electron from the surface of a particular metal; symbol φ, measured in J

X-rays Short-wavelength electromagnetic waves, with wavelengths from 10^{-8} m to 10^{-13} m, which can be used in medical imaging

yield point A point on a stress–strain graph beyond which the deformation is no longer entirely elastic

Young modulus The ratio of tensile stress to tensile strain when these quantities are directly proportional to each other, measured in Pa

Answers

2.1

$4.5 \times 10^{-8}\,\text{s}$

1 a Distance and time, respectively. [2]

   b $0.60\,\text{m}$ and $0.040\,\text{s}$ (or $4.0 \times 10^{-2}\,\text{s}$) [2]

2 a $0.000\,000\,000\,1\,\text{s}$ (or $1.0 \times 10^{-10}\,\text{s}$) [1]

   b $0.000\,000\,000\,15\,\text{m}$ (or $1.5 \times 10^{-10}\,\text{m}$) [1]

   c $16\,000\,000\,\text{K}$ (or $1.6 \times 10^{7}\,\text{K}$) [1]

3 a $2.0 \times 10^{-10}\,\text{m}$ [1]

   b $4.0 \times 10^{5}\,\text{m}$ [1]

   c $3.5 \times 10^{-5}\,\text{s}$ [1]

   d $2.5 \times 10^{-4}\,\text{A}$ [1]

   e $7.56 \times 10^{-7}\,\text{s}$ [1]

4 a $534\,\text{km}$ [1]

   b $12.74\,\text{Mm}$ [1]

   c $75\,\mu\text{m}$ [1]

   d $140\,\text{nA}$ [1]

2.2

$25°\text{C}$

$-40°\text{F} = -40°\text{C}$

1 base unit for force: base unit for mass × base unit for acceleration

   base unit: $\text{kg} \times \text{m\,s}^{-2}$ [1]

   base unit: $\text{kg\,m\,s}^{-2}$ [1]

2 a base unit for force constant : base unit for force × base unit for extension

   base unit: $\text{kg\,m\,s}^{-2} \div \text{m}$ [1]

   base unit: $\text{kg\,s}^{-2}$ [1]

   b base unit for work done: base unit for force × base unit for distance

   base unit: $\text{kg\,m\,s}^{-2} \times \text{m}$ [1]

   base unit: $\text{kg\,m}^{2}\,\text{s}^{-2}$ [1]

   c base unit for pressure: base unit for force × base unit for area

   base unit: $\text{kg\,m\,s}^{-2} \div \text{m}^{2}$ [1]

   base unit: $\text{kg\,m}^{-1}\,\text{s}^{-2}$ [1]

3 In the last three units, the first letter is a prefix for the factor – nano (10^{-9}), milli (10^{-3}), and mega M (10^{3}). [1]

   There is a space in the first unit showing that it is a derived unit, newton metres. [1]

   The units are: 1 newton metre (energy or work), 1 nanometre (length), one millinewton (force), and one meganewton (force). [1]

4 base unit for number density: inverse of base unit for volume [1]

   base unit: m^{-3} [1]

2.3

1 Both masses should either be in grams or in kilograms.
The answer should be either $0.650\,\text{kg}$ or $650\,\text{g}$. [1]

2 Distance and displacement have the same unit, metres (m). [1]
However, distance is a scalar (magnitude only) and displacement is a vector (magnitude and direction). [1]

3 Both time and energy are scalars. [1]
Dividing these two scalar quantities produces a scalar quantity (power). [1]

4 a distance = $12.0\,\text{cm}$ [1]

   b displacement = $6.0\,\text{cm}$ [1]

   c average speed = $\dfrac{\text{distance}}{\text{time}} = \dfrac{0.120}{20}$ [1]

   average speed = $6.0 \times 10^{-2}\,\text{m\,s}^{-1}$ [1]

5 Displacement is the direct distance between two points. Therefore, it must always be either equal to or less than the distance. [1]

2.4

1 a resultant velocity = $0.5\,\text{m\,s}^{-1}$ [1]

   b resultant velocity = $0.5 + 2.0 = 2.5\,\text{m\,s}^{-1}$ [1]

   c resultant velocity = $0.5 - 1.0 = -0.5\,\text{m\,s}^{-1}$ [1]

2 a A correct triangle drawn, with the sides correctly labelled. [1]
The arrows all 'follow' and are in a clockwise direction. [1]

   $F^{2} = 2.0^{2} + 3.0^{2}$ [1]

   $F = \sqrt{13} = 3.6\,\text{N}$ [1]

   $\theta = \tan^{-1}\left(\dfrac{2.0}{3.0}\right) = 33.7...° = 34°$ (2 s.f.) [1]

   b A correct triangle drawn, with the sides correctly labelled. [1]
The arrows all 'follow' and are in a clockwise direction. [1]

   $F^{2} = 13.5^{2} + 6.0^{2}$ [1]

   $F = \sqrt{218.25} = 14.77... = 14.8\,\text{N}$ (2 s.f.) [1]

   $\theta = \tan^{-1}\left(\dfrac{6.0}{13.5}\right) = 24°$ (2 s.f.) [1]

3 a $v = 0.90 + 0.30 = 1.2\,\text{m\,s}^{-1}$ [1]
Direction: due north [1]

   b $v = 0.90 - 0.30 = 0.60\,\text{m\,s}^{-1}$ [1]
Direction: due north [1]

   c $v^{2} = 0.30^{2} + 0.90^{2}$ [1]

   $v = \sqrt{0.90} = 0.949... = 0.95\,\text{m\,s}^{-1}$ (2 s.f.) [1]

   $\theta = \tan^{-1}\left(\dfrac{0.30}{0.90}\right) = 18°$ (2 s.f.)

   Direction: bearing of 18° from the north (18° east of north) [1]

2.5

1 $\theta = 0°$: $F_x = F\cos\theta = 10\cos 0° = 10\,\text{N}$ [1]

  $\theta = 45°$: $F_x = F\cos\theta = 10\cos 45° = 7.1\,\text{N}$ [1]

  $\theta = 90°$: $F_x = F\cos\theta = 10\cos 90° = 0\,\text{N}$ [1]

  The horizontal component of the force decreases as the angle θ is increased. [1]

2 Horizontal: $F_x = F\cos\theta = 1650\cos 35° = 1352\,\text{N}$ [1]

  Vertical: $F_y = F\sin\theta = 1650\sin 35° = 946\,\text{N}$ [1]

3 a Northwards: $F_x = F\cos\theta = 350\cos 40° = 268\,\text{N}$ [1]

  b Eastwards: $F_y = F\sin\theta = 350\sin 40° = 224.98... = 220\,\text{N}$ (2 s.f.) [1]

  (Note: this is the same as $350\cos 50°$)

4 Vertical force: $F_x = F\cos\theta = 6.5\cos 20° = 6.1\,\text{kN}$ [1]

  Horizontal force: $F_y = F\sin\theta = 6.5\sin 20° = 2.2\,\text{kN}$ [1]

2.6

1 The forces are equivalent to 1.0 N to the right and 1.0 N downward. [1]

  $F^2 = 1.0^2 + 1.0^2$ [1]

  $F = 1.4\,\text{N}$ [1]

  Direction: $\theta = \tan^{-1}\left(\dfrac{1.0}{1.0}\right) = 45°$ [1]

2 Total force in x direction $= 8.0\cos 20° + 8.0\cos 20°$
  $= 15.0\,\text{kN}$ [1]

  Total force in y direction $= 8.0\sin 20° - 8.0\sin 20° = 0$ [1]

  Resultant force $F = 15.0\,\text{kN}$ [1]

  The resultant force is in the x direction (from left to right). [1]

3 Total force in x direction $=$
  $12.5 + 11.8\cos 25° + 8.7\cos 35° = 30.3\,\text{kN}$ [3]

  Total force in y direction $=$
  $11.8\sin 25° + 12.5\sin 0° - 8.7\sin 35° \approx 0\,\text{kN}$ [1]

  Resultant force $F = 30.3\,\text{kN}$ [1]

  The resultant force is in the x direction (from left to right). [1]

4 $F^2 = 4.0^2 + 3.0^2 - 2 \times 4.0 \times 3.0 \times \cos 40°$ (cosine rule) [1]

  $F = 2.57...$ [1]

  $\dfrac{3.0}{\sin\theta} = \dfrac{2.57...}{\sin 40}$ (sine rule) [1]

  $\theta = 48.6... = 49°$ (2 s.f.) [1]

3.1

1 a $v = \dfrac{x}{t} = \dfrac{180}{9.0}$ [1]

  $v = 20\,\text{ms}^{-1}$ [1]

  b $v = \dfrac{x}{t} = \dfrac{2000}{6.5 \times 60}$ [1]

  $v = 5.12... = 5.1\,\text{ms}^{-1}$ (2 s.f.) [1]

2 $v = \dfrac{x}{t} = \dfrac{19.2}{24 \times 3600}$ [1]

  $v = 2.22... \times 10^{-4} = 2.2 \times 10^{-4}\,\text{ms}^{-1}$ (2 s.f.) [1]

3 $x = vt = 31 \times 19$ [1]

  $x = 589 = 590\,\text{m}$ (2 s.f.) [1]

4 $t = \dfrac{x}{v} = \dfrac{12000 \times 10^3}{240}$ [1]

  $t = 5.0 \times 10^4\,\text{s}$ [1]

  $t = \dfrac{5.0 \times 10^4}{3600} = 13.8... = 14\,\text{h}$ (2 s.f.) [1]

5 a $x = (25 \times 2.0 \times 60) + 800$ [1]

  $x = 3800\,\text{m}$ [1]

  b $v = \dfrac{x}{t} = \dfrac{3800}{120 + 50}$ [1]

  $v = 22.35... = 22\,\text{ms}^{-1}$ (2 s.f.) [1]

6 $v = \text{gradient}$ [1]

  $v = \dfrac{\Delta x}{\Delta t} = \dfrac{1600 - 400}{90 - 60}$ [1]

  $v = 40\,\text{ms}^{-1}$ [1]

3.2

1 From $t = 0$ to $t = 4.0\,\text{s}$: constant velocity (moving away). [1]

  From $t = 4.0\,\text{s}$ to $t = 6.0\,\text{s}$: constant velocity (coming back). [1]

  The magnitude of the velocity between $t = 4.0\,\text{s}$ to $t = 6.0\,\text{s}$ is greater than the magnitude of the velocity between $t = 0$ to $t = 4.0\,\text{s}$. [1]

2 $v = \text{gradient}$ [1]

  velocity v at $t = 2.0\,\text{s}$: $v = \dfrac{\Delta s}{\Delta t} = \dfrac{0.40}{4.0}$ [1]

  $v = 0.10\,\text{ms}^{-1}$ [1]

  velocity v at $t = 5.0\,\text{s}$: $v = \dfrac{\Delta s}{\Delta t} = \dfrac{0 - 0.40}{6.0 - 4.0}$ [1]

  $v = -0.20\,\text{ms}^{-1}$ (in opposite direction from earlier velocity) [1]

3 total distance $\times$ travelled $= 0.40 \times 2 = 0.80\,\text{m}$ and total time $t = 6.0\,\text{s}$ [1]

  $v = \dfrac{x}{t} = \dfrac{0.80}{6.0}$ [1]

  $v = 0.133... \approx 0.13\,\text{ms}^{-1}$ (2 s.f.) [1]

4 a $v = \dfrac{x}{t} = \dfrac{2\pi r}{t}$ [1]

  $v = \dfrac{2\pi \times 0.80}{8.0}$ [1]

  $v = 0.63\,\text{ms}^{-1}$ (2 s.f.) [1]

  b change in displacement $= \sqrt{0.80^2 + 0.80^2}$
  $= 1.131...\,\text{m}$ [1]

  $v = \dfrac{\Delta s}{\Delta t} = \dfrac{1.131...}{\frac{1}{4} \times 8.0}$ [1]

  $v = 0.57\,\text{ms}^{-1}$ (2 s.f.) [1]

3.3

1 $a = \dfrac{\Delta v}{\Delta t} = \dfrac{8.0 - 0}{12}$ [1]

  $a = 0.67\,\text{ms}^{-2}$ (2 s.f.) [1]

2 $a = \dfrac{\Delta v}{\Delta t} = \dfrac{10 - 40}{60}$ [1]

  $a = -0.50\,\text{ms}^{-2}$ (magnitude of $0.50\,\text{ms}^{-2}$) [1]

3 From $t = 0$ to $t = 1.0$ s: constant acceleration. [1]

From $t = 1.0$ s to $t = 3.0$ s: constant acceleration of smaller magnitude. [1]

4 a Maximum acceleration is when gradient of v–t is maximum; this occurs at $t < 1.0$ s. [1]

$a = \dfrac{\Delta v}{\Delta t} = \dfrac{2.0 - 0}{1.0}$ [1]

$a = 2.0\,\mathrm{m\,s^{-2}}$ [1]

b Graph showing constant acceleration between $t = 0$ and $t = 1.0$ s and a different constant acceleration between $t = 1.0$ s and $t = 3.0$ s. [1]

Constant values of accelerations $2.0\,\mathrm{m\,s^{-2}}$ and $0.5\,\mathrm{m\,s^{-2}}$ marked on the graph. [1]

5 a $a = \dfrac{\Delta v}{\Delta t} = \dfrac{7.5 - 0}{0.75}$ [1]

$a = 10\,\mathrm{m\,s^{-2}}$ [1]

b a = gradient [1]

$a = \dfrac{\Delta v}{\Delta t} = -\dfrac{0 - 10}{1.50 - 0.25}$ [1]

$a = -8.0\,\mathrm{m\,s^{-2}}$ (The negative sign implies deceleration.) [1]

3.4

1 From $t = 0$ to $t = 75$ s: constant acceleration. [1]

From $t > 75$ s: constant velocity (of $15\,\mathrm{m\,s^{-1}}$). [1]

2 a (displacement or distance = area under the graph)

distance $= \left(\dfrac{1}{2} \times 75 \times 15\right) + (15 \times 25)$ (area of triangle and rectangle) [2]

distance $= 937.5 = 940$ m (2 s.f.) [1]

b $v = \dfrac{\Delta s}{\Delta t} = \dfrac{937.5}{100}$ [1]

$v = 9.4\,\mathrm{m\,s^{-1}}$ (2 s.f.) [1]

3 a (displacement or distance = area of triangle + area of rectangle + area of trapezium)

distance $= \left(\dfrac{1}{2} \times 10 \times 4.0\right) + (10 \times 5.0) +$

$\left[\dfrac{1}{2}(10 + 5.0) \times 4.0\right]$ [3]

distance $= 100$ m [1]

b $v = \dfrac{\Delta s}{\Delta t} = \dfrac{100}{13.0}$ [1]

$v = 7.69\ldots = 7.7\,\mathrm{m\,s^{-1}}$ (2 s.f.) [1]

4 a Constant deceleration (allow 'acceleration'). [1]

Stops momentarily at $t = 4.0$ s. [1]

After 4.0 s, it starts to return. [1]

b total distance = area enclosed by line and t-axis [1]

total distance $= \left(\dfrac{1}{2} \times 4.0 \times 4.0\right) \times 2 = 16$ m [1]

c average speed $= \dfrac{\text{distance}}{\text{time}} = \dfrac{16}{8.0}$ [1]

average speed $= 2.0\,\mathrm{m\,s^{-1}}$ [1]

average velocity $= \dfrac{\text{change in displacement}}{\text{time}}$;

change in displacement $= 0$ [1]

Therefore, average velocity $= 0$ [1]

3.5

 Look at the derivation of $s = ut + \dfrac{1}{2}at^2$ and follow a similar process.

1 $s = ?,\ u = 13.4\,\mathrm{m\,s^{-1}},\ v = 22.3\,\mathrm{m\,s^{-1}},\ t = 8.7$ s;

$s = \dfrac{1}{2}(u + v)t$ [1]

$s = \dfrac{1}{2} \times (13.4 + 22.3) \times 8.7$ [1]

$s = 155$ m (3 s.f.) [1]

2 $a = ?,\ s = 200$ m, $v = 4.2\,\mathrm{m\,s^{-1}},\ u = 3.2\,\mathrm{m\,s^{-1}}$;

$v^2 = u^2 + 2as$ [1]

$a = \dfrac{v^2 - u^2}{2s} = \dfrac{4.2^2 - 3.2^2}{2 \times 200} = 0.019\ \mathrm{m\,s^{-2}}$ (2 s.f.) [1]

3 $s = 100$ m, $a = ?,\ t = 9.58$ s; [1]

$a = \dfrac{2s}{t^2} = \dfrac{2 \times 100}{9.58^2}$ [1]

$a = 2.17\ldots = 2.2\ldots\,\mathrm{m\,s^{-2}}$ (2 s.f.) [1]

4 a $s = 400$ m, $u = 0,\ a = ?,\ t = 4.6$ s; [1]

$a = \dfrac{2s}{t^2} = \dfrac{2 \times 400}{4.6^2}$ [1]

$a = 37.8\ldots = 38\,\mathrm{m\,s^{-2}}$ (2 s.f.) [1]

b $u = 0,\ a = 37.8\ldots\,\mathrm{m\,s^{-2}},\ t = 4.6$ s; $v = u + at$ [1]

$v = 0 + (37.8\ldots \times 4.6)$ [1]

$v = 173.9\ldots = 170\,\mathrm{m\,s^{-1}}$ (2 s.f.) [1]

5 $s = ?,\ u = 0,\ a = 9.81\,\mathrm{m\,s^{-2}},\ t_1 = 3.0$ s and $t_2 = 5.0$ s [1]

distance $= \left[\dfrac{1}{2} \times 9.81 \times 5.0^2\right] - \left[\dfrac{1}{2} \times 9.81 \times 3.0^2\right]$ [1]

distance $= 78.48 = 78$ m (2 s.f.) [1]

6 $s = 30$ m, $u = 28\,\mathrm{m\,s^{-1}},\ v = 0,\ a = ?$; $v^2 = u^2 + 2as$ [1]

$a = \dfrac{0 - 28^2}{2 \times 30}$ [1]

$a = -13.06\ldots = -13\,\mathrm{m\,s^{-2}}$ (magnitude is $13\,\mathrm{m\,s^{-2}}$ to 2 s.f) [1]

3.6

1 stopping distance = thinking distance + braking distance [1]

stopping distance $= (22 \times 1.5) + 38$ [1]

stopping distance $= 71$ m [1]

2 thinking distance = speed of car × reaction time [1]

Assuming the reaction time is constant, thinking distance $\propto$ speed of car. [1]

3 a braking distance $s = 75$ m (from Table 1) [1]

$s = 75$ m, $u = 31.1\,\mathrm{m\,s^{-1}},\ v = 0,\ a = ?$; $v^2 = u^2 + 2as$ [1]

$a = \dfrac{0 - 31.1^2}{2 \times 75}$ [1]

$a = -6.448\ldots = -6.4\,\mathrm{m\,s^{-2}}$ (magnitude is $6.4\,\mathrm{m\,s^{-2}}$ to 2 s.f.) [1]

b $u = 31.1\,\mathrm{ms^{-1}}$, $v = 0$, $a = -6.448\,\mathrm{ms^{-2}}$, $t = ?$;

$v = u + at$ [1]

$t = \dfrac{0 - 31.1}{6.448}$ [1]

$t = 4.8\,\mathrm{s}$ (2 s.f.) [1]

4 thinking distance = area under graph up to
$0.50\,\mathrm{s} = 20 \times 0.50 = 10\,\mathrm{m}$ [1]

braking distance = area under graph from

$t = 0.50\,\mathrm{s}$ to $3.5\,\mathrm{s} = \dfrac{1}{2} \times 3.0 \times 20 = 30\,\mathrm{m}$ [1]

stopping distance = $10 + 30 = 40\,\mathrm{m}$ [1]

5 $v^2 = u^2 + 2as$ [1]

$v = 0$, therefore, $0 = u^2 + 2as$ or $s = \dfrac{u^2}{2a}$ [1]

For a constant deceleration of magnitude a, $s \propto u^2$. [1]

3.7

1 Gradient calculated using a large triangle.
gradient = $4.89\,\mathrm{m\,s^{-2}}$ for the results;
allow $\pm 0.03\,\mathrm{m\,s^{-2}}$

2 $g = 2 \times$ gradient = 2×4.89
$g = 9.78\,\mathrm{m\,s^{-2}}$

3 % difference = $\dfrac{9.78 - 9.81}{9.81} \times 100$

% difference = $(-)0.31\%$

1 The acceleration of free fall is the same for both and
equal to g ($9.81\,\mathrm{ms^{-2}}$). [1]

Assumption: There is negligible air resistance acting
on the falling objects. [1]

2 $s = H = ?$, $u = 0$, $a = 9.81\,\mathrm{ms^{-2}}$, $t = 2.3\,\mathrm{s}$; $s = ut + \dfrac{1}{2}at^2$ [1]

$H = \dfrac{1}{2} \times 9.81 \times 2.3^2$ [1]

$H = 25.9... = 26\,\mathrm{m}$ (2 s.f.) [1]

3 Object dropped from a given height. [1]

Time of fall measured and measurements repeated. [1]

Explanation of how g is determined using

$s = \dfrac{1}{2}gt^2$ [1]

Correct account of how experiment is made precise,
e.g: Drop a heavy object from a larger distance, so
percentage uncertainty in the distance is smaller. [1]

4 a $s = 9.5\,\mathrm{m}$, $u = 0\,\mathrm{ms^{-1}}$, $t = 1.5\,\mathrm{s}$, $a = ?$ [1]

$a = \dfrac{2s}{t^2} = \dfrac{2 \cdot 9.5}{1.5^2}$ [1]

$a = g = 8.44... \approx 8.4\,\mathrm{ms^{-2}}$ (2 s.f.) [1]

b The presence of drag (air resistance) would give a
lower experimental value for the acceleration of
free fall than the true value. [1]

5 a $s = 0.125\,\mathrm{m}$ ($\pm 0.001\,\mathrm{m}$), $u = 0\,\mathrm{ms^{-1}}$,
$t = 4 \times 0.04 = 0.16\,\mathrm{s}$, $a = ?$ [1]

$a = \dfrac{2s}{t^2} = \dfrac{2 \cdot 0.125}{0.16^2}$ [1]

$a = g = 9.77... = 9.8\,\mathrm{ms^{-2}}$ (2 s.f.) [1]

b $s = 0.071\,\mathrm{m}$ ($\pm 0.001\,\mathrm{m}$), $u = 0\,\mathrm{ms^{-1}}$,
$t = 3 \times 0.04 = 0.12\,\mathrm{s}$, $a = ?$ [1]

$a = \dfrac{2s}{t^2} = \dfrac{2 \cdot 0.071}{0.12^2}$ [1]

$a = g = 9.86...\,\mathrm{ms^{-2}}$ [1]

average acceleration = $\dfrac{(9.77 + 9.86...)}{2} =$
$9.8\,\mathrm{ms^{-2}}$ (2 s.f.) [1]

3.8

1 $v = \sqrt{7.0^2 + 9.0^2}$ [1]

$v = 11\,\mathrm{ms^{-1}}$ (2 s.f.) [1]

2 a Vertically:
$s = 29\,\mathrm{m}$, $u = 0$, $a = 9.81\,\mathrm{ms^{-2}}$, $t = ?$; [1]

$t = \sqrt{\dfrac{2s}{a}} = \sqrt{\dfrac{2 \times 29}{9.81}}$ [1]

$t = 2.4\,\mathrm{s}$ (2 s.f.) [1]

b Horizontally: [1]
distance = $vt = 320 \times 2.4315$ [1]
distance = $780\,\mathrm{m}$ (2 s.f.) [1]

c vertical velocity = $9.81 \times 2.4315 = 23.8...\,\mathrm{ms^{-1}}$ [1]
$v^2 = 320^2 + 23.8...^2$ [1]
$v = 321\,\mathrm{ms^{-1}}$ (3 s.f.) [1]

3 a Vertically:
$s = ?$, $u = 22.0\sin 35°$, $v = 0$, $a = -9.81\,\mathrm{ms^{-2}}$;
$v^2 = u^2 + 2as$ [2]

$s = \dfrac{0 - (22\sin 35)^2}{2 \cdot 9.81}$ [1]

$s = 8.1\,\mathrm{m}$ (2 s.f.) [1]

b Vertically:
$v = -22.0\sin 35°$, $u = 22.0\sin 35°$, $a = -9.81\,\mathrm{ms^{-2}}$,
$t = ?$; $v = u + at$ [1]

$t = \dfrac{-22.0\sin 35 - 22.0\sin 35}{9.81}$ [1]

$t = 2.57...\,\mathrm{s}$ [1]
Horizontally:
distance = $vt = 22\cos 35° \times 2.57...$ [1]
distance = 46 (2 s.f.) [1]

4 v–t graph shows straight line of negative gradient. [1]
The initial vertical velocity is $+13\,\mathrm{ms^{-1}}$ ($12.6\,\mathrm{ms^{-1}}$)
at $t = 0$. [1]
The final velocity is $-13\,\mathrm{ms^{-1}}$ at about $2.6\,\mathrm{s}$. [1]

4.1

1 Graph showing $m = m_0$ when v is much smaller than c. The curve is asymptotic to $v = c$ line.

2 a $m = \dfrac{m_0}{\sqrt{1-(v/c)^2}} = \dfrac{9.1 \times 10^{-31}}{\sqrt{1-0.10^2}}$

$m = 9.146 \times 10^{-31} \approx 9.1 \times 10^{-31}$ kg

b $m = \dfrac{m_0}{\sqrt{1-(v/c)^2}} = \dfrac{9.1 \times 10^{-31}}{\sqrt{1-0.999^2}}$

$m = 2.04 \times 10^{-29}$ kg (an increase in the mass by a factor of about 22)

3 The mass would be infinite when $v = c$.
It would require an infinite force or an infinite amount of energy to move the electron.

1 a $F = ma;\ a = \dfrac{F}{m} = \dfrac{500}{1200}$ [1]

$a = 0.42\,\text{ms}^{-2}$ (2 s.f.) [1]

b $v = u + at = 0 + (0.417 \times 6.0)$ [1]

$v = 2.5\,\text{ms}^{-1}$ (2 s.f.) [1]

2 $W = mg;\ m = \dfrac{W}{g} = \dfrac{1.1}{9.81}$ [1]

$m = 0.112...$ kg, therefore $m = 110$ g (2 s.f.) [1]

3 $F = ma;\ a = \dfrac{F}{m} = \dfrac{5800}{0.046}$ [1]

$a = 1.3 \times 10^5\,\text{ms}^{-2}$ (2 s.f.) [1]

4 $a = \dfrac{(v-u)}{t} = \dfrac{28}{9.6}$ [1]

$a = 2.92\,\text{ms}^{-2}$ (2 s.f.) [1]

5 net force $= 10^{-16}\sqrt{2.0^2 + 1.5^2}$ [1]

net force $= 2.5 \times 10^{-16}\,\text{N}$ [1]

$a = \dfrac{F}{m} = \dfrac{2.5 \times 10^{-16}}{1.7 \times 10^{-27}}$ [1]

$a = 1.5 \times 10^{11}\,\text{ms}^{-2}$ (2 s.f.) [1]

direction: $\theta = \tan^{-1}\left(\dfrac{2.0}{1.5}\right) = 53°$ to the 1.5×10^{-16} N force. [2]

6 $v^2 = u^2 + 2as;\ a = \dfrac{(0 - 420^2)}{2 \times 0.098}$ [1]

$a = 9.0 \times 10^5\,\text{ms}^{-2}$ [1]

$F = ma = 0.0080 \times 9.0 \times 10^5,\ F = 7.2 \times 10^3\,\text{N}$ [1]

4.2

1 The centre of mass will be at the 50 cm mark (the middle) only if the ruler is uniform in both shape and density of the material. This may not have been the case [1]

2 a [1]

centre of mass

b [1]

centre of mass

c [1]

centre of mass

3 By trial and error, horizontally balance the card on the edge of the ruler. [1]

Mark points on the card and then draw a straight line to show the line along which the card was balanced. The centre of mass will lie on this line. [1]

Repeat the procedure above for a different orientation of the card and draw another straight line. [1]

The centre of mass is located at the point of intersection of these two lines. [1]

4 The weight of the ball is due to the plastic wall. [1]

The ball is a symmetrical object so the centre of mass will be in the centre. [1]

5 Push the object with the point of a pencil (or equivalent) and by trial and error determine the point on the object where the object does not rotate when pushed. [1]

The centre of mass will lie along the line of action of the force applied by the pencil when there is no rotation. [1]

Repeat this procedure from a different orientation. [1]

The centre of mass is located at the point of intersection of these two (imaginary) lines of action. [1]

4.3

1 a [1]

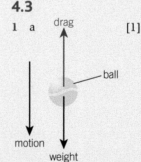

b [1]

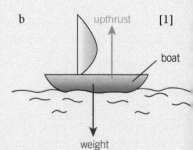

2 net force = ma; net force = 8.0×1.5 [1]

net force = N – weight; N = weight + net force [1]

$N = (8.0 \times 9.81) + (8.0 \times 1.5)$ [1]

$N = 90\,N$ (2 s.f.) [1]

3 $a = g\sin 30°$ [1]

$a = 9.81 \times \sin 30°$ [1]

$a = 4.9\,ms^{-2}$ (2 s.f.) [1]

4 net force = $mg\sin 30° - R$, where R = friction [1]

net force = $(0.020 \times 9.81 \times 0.5) - 0.10 = -1.9 \times 10^{-3}$ [1]

$a = \dfrac{F}{m}$; $a = \dfrac{-1.9 \times 10^{-3}}{0.020}$ [1]

$a = -0.095\,ms^{-2}$ [1]

4.4

> ⚛ **1** The motion sensor may just track the position of the cylinder and, therefore, cannot be used to investigate the motion of the ball. [1]
>
> **2** Mark the positions of two points along the length of the cylinder (towards its bottom end). [1]
> Measure the separation between these points with a metre rule. [1]
> Gently drop the ball into the fluid. Start a timer when it passes the upper mark and then stop the timer as it passes the lower mark. Record this time of fall. [1]
> Repeat the step above and record another time of fall. [1]
> Calculate the average time of fall. [1]
> The terminal velocity is determined by dividing the distance between the two marks by the average time of fall. [1]

1 Change her area by changing shape (e.g., extending arms or opening a parachute). [1]

2 The weight is equal to the drag force at terminal velocity. [1]

drag = $mg = 0.120 \times 9.81$ [1]

drag = $1.2\,N$ (2 s.f.) [1]

3 drag $\propto$ speed^2 [1]

The drag will increase by a factor of $3^2 = 9$. [1]

drag = $1.0 \times 9 = 9.0\,kN$ [1]

4 Skydiver:

$W = mg$; $m = \dfrac{W}{g} = \dfrac{800}{9.81}$ [1]

$m = 81.5\ldots\,kg$ [1]

net force = ma; $800 - 300 = 81.5\ldots a$ [1]

$a = \dfrac{500}{81.5\ldots}$ [1]

$a = 6.1\,ms^{-2}$ (2 s.f.) [1]

Ball:

$W = mg$; $m = \dfrac{W}{g} = \dfrac{2.0}{9.81}$ [1]

$m = 0.203\ldots\,kg$ [1]

net force = ma; $2.0 - 2.5 = 0.203\ldots a$ [1]

$a = \dfrac{-0.5}{0.203\ldots}$ [1]

$a = -2.5\,ms^{-2}$ (2 s.f.) (deceleration) [1]

5 a $D = 0.20 \times 1.5^2 = 0.45\,N$ [1]

$W = mg = 0.30 \times 9.81 = 2.943\,N$ [1]

net force = ma; $2.943 - 0.45 = 0.30a$ [1]

$a = \dfrac{2.493}{0.30}$ [1]

$a = 8.3\,ms^{-2}$ (2 s.f.) [1]

5 b At terminal velocity, drag is equal to weight. Therefore, $D = 2.94\,N$. [1]

$D = 2.94 = 0.20v^2$ [1]

$v = \sqrt{\dfrac{2.94}{0.20}}$ [1]

$v = 3.8\,ms^{-1}$ (2 s.f.) [1]

4.5

1 2.0 N force: moment = $Fx = 2.0 \times 0.10 = 0.2\,Nm$ (anticlockwise) [1]

4.0 N force: moment = $Fx = 4.0 \times 0 = 0$ [1]

6.0 N force: moment = $Fx = 6.0 \times 0.18 = 1.08\,Nm$ (clockwise) [1]

2 a clockwise moment = $(18 \times 0.150) + (35 \times 0.360)$ [2]

clockwise moment = $15\,Nm$ (2 s.f.) [1]

b sum of clockwise moments = sum of anticlockwise moments [1]

$15.3 = 0.032 \times F$ [1]

$F = 480\,N$ (2 s.f.) [1]

3 The line of action of the weight must fall beyond the base. [1]

$\tan\theta = \dfrac{1.5}{5.0}$ [1]

$\theta = 17°$ (2 s.f.) [1]

4 sum of clockwise moments = sum of anticlockwise moments [1]

$(18 \times 0.40) + (20 \times 0.60) = F\sin 60° \times 0.75$ [3]

$F = \dfrac{19.2}{(0.75\sin 60°)}$ [1]

$F = 30\,N$ (2 s.f.) [1]

4.6

1 It will move forward. [1]

It will also spin about its centre (of mass) as it moves. [1]

2 force $\approx 2\,N$ (allow a force in the range 0.5–10 N) [1]

torque = $Fd = 2 \times 0.04$ [1]

torque = $0.08\,Nm$ [1]

3 a It will move to the right. [1]

It will also spin in a clockwise direction. [1]

b It will spin in an anticlockwise direction. [1]

This is because there is a net couple in the anti-clockwise direction. [1]

4 a moment = $(F \times [d + x]) - F \times x$ [1]

moment = $Fd + Fx - Fx = Fd$ [1]

b The moment about **A** is the same as the torque of the couple. [1]

4.7

1 The net force is zero – the resultant of the three forces is zero. [1]

2 **a** A correct triangle drawn, with the sides correctly labelled. [1]

   The arrows all 'follow' and are in a clockwise direction. [1]

   b $T^2 = 5.0^2 + 12^2$ [1]

   $T = \sqrt{169} = 13\,\text{N}$ [1]

   c The resultant in any direction must be zero. [1]

   The resultant of the 5.0 N and 12 N forces is equal to T but in the opposite direction. [1]

3 $\sin 60° = \dfrac{T_1}{3.8 \times 10^3}$ [1]

   $T_1 = 3.8 \times 10^3 \times \sin 60° = 3.3 \times 10^3\,\text{N (2 s.f.)}$ [1]

   $\sin 30° = \dfrac{T_2}{3.8 \times 10^3}$ [1]

   $T_2 = 3.8 \times 10^3 \times \sin 30° = 1.9 \times 10^3\,\text{N (2 s.f.)}$ [1]

4 **a** $\dfrac{T_1}{\sin 40} = \dfrac{0.500 \times 9.81}{\sin 110}$ (sine rule) [1]

   $T_1 = 3.4\,\text{N (2 s.f.)}$ [1]

   $\dfrac{T_2}{\sin 30} = \dfrac{0.500 \times 9.81}{\sin 110}$ (sine rule) [1]

   $T_2 = 2.6\,\text{N (2 s.f.)}$ [1]

   b If the strings are horizontal, with the angles zero, there can be no upward vertical component of forces from either T_1 or T_2 to balance the downward weight mg. [1]

   It would be impossible for the slotted masses to be in equilibrium, therefore, the strings cannot be horizontal. [1]

4.8

1 $m = \rho V = 1.3 \times 140$ [1]

   $m = 180\,\text{kg (2 s.f.)}$ [1]

2 $p = \dfrac{F}{A} = \dfrac{8.0}{\pi \times (3.75 \times 10^{-3})^2}$ [1]

   $p = 2.4 \times 10^5\,\text{Pa (2 s.f.)}$ [1]

3 $\rho = \dfrac{m}{V} = \dfrac{0.080}{(85-70) \times 10^{-6}}$ (1 cm^3 = 10^{-6} m^2) [1]

   $\rho = 5.3 \times 10^3\,\text{kg m}^{-3}\,\text{(2 s.f.)}$ [1]

4 area $A = (0.15 \times 0.075)\,\text{m}^2$ [1]

   $F = pA = 1.0 \times 10^5 \times (0.15 \times 0.075)$ [1]

   $F = 1.1 \times 10^3\,\text{N (2 s.f.)}$ [1]

5 volume $V = \frac{4}{3}\pi r^3 = \frac{4}{3} \times \pi \times (12 \times 10^3)^3$ [1]

   $\rho = \dfrac{m}{V} = \dfrac{3.0 \times 10^{24}}{\frac{4}{3} \times \pi \times (12 \times 10^3)^3}$ [1]

   $\rho = 4.1 \times 10^{11}\,\text{kg m}^{-3}\,\text{(2 s.f.)}$ [1]

6 mass of gold $= 0.582 \times 3.34 \times 10^{-6} \times 1.93 \times 10^4$
   $= 3.75... \times 10^{-2}\,\text{kg}$ [1]

   mass of copper $= 0.418 \times 3.34 \times 10^{-6}$
   $\times 8.96 \times 10^3 = 1.25... \times 10^{-2}\,\text{kg}$ [1]

   $\rho = \dfrac{(3.75...+1.25...) \times 10^{-2}}{3.34 \times 10^{-6}}$ [1]

   $\rho = 1.50 \times 10^4\,\text{kg m}^{-3}\,\text{(3 s.f.)}$ [1]

   Assumption: gold and copper do not react together and their atoms do not become more or less densely packed when they are mixed. [1]

4.9

1 pressure from water, $p = \rho gh = 1.03 \times 10^3 \times 9.81 \times 10 = 1.01... \times 10^5\,\text{Pa}$

   atmospheric pressure $= 1.01... \times 10^5\,\text{Pa}$

   Therefore, total pressure is $2.02 \times 10^5\,\text{Pa (2 s.f.)}$

2 force $F = pA = (\rho gh + \text{atmospheric pressure})A$

   $F = [(1.03 \times 10^3 \times 9.81 \times 330) + 1.01... \times 10^5] \times 1.2... \times 10^{-2}$

   $F = 4.12 \times 10^4$ (2 s.f.)

3 Pressure in the lungs at a depth of 10 m is twice the (atmospheric) pressure at the water surface.

   If the diver does not exhale, this air will expand without a way to escape because of the pressure difference of about 10^5 Pa between the air in the lungs of the diver and the surrounding air.

1 $p = \rho gh = 1.35 \times 10^4 \times 9.81 \times 0.765$ [1]

   $p = 1.01 \times 10^5\,\text{Pa (2 s.f.)}$ (the same as atmospheric pressure on Earth's surface) [1]

2 $p = \rho gh = 1.0 \times 10^3 \times 9.81 \times 610$ [1]

   $p = 6 \times 10^6\,\text{Pa (2 s.f.)}$ [1]

3 Immediately after release, the ball experiences two forces – upthrust and weight. [1]

   Upthrust is greater than the weight of the ball, so the ball accelerates upwards. [1]

   As it travels vertically upwards it also experiences drag due to water. [1]

   As it pops out of the water, the only force is weight, so it decelerates and falls back onto the surface of the water. [1]

   Eventually it remains still on the surface of the water, with upthrust equal to weight. [1]

4 **a** upthrust $= 1.54 - 1.34 = 0.20\,\text{N}$ [1]

   b upthrust = weight of water displaced; mass of water $= \dfrac{0.20}{9.81} = 2.03... \times 10^{-2}\,\text{kg}$ [1]

   (volume of water displaced = volume V of bar)

   $V = \dfrac{2.03... \times 10^{-2}}{1.0 \times 10^3} = 2.03... \times 10^{-5}\,\text{m}^3$ [1]

   $\rho = \dfrac{m}{V} = \dfrac{(1.54 / 9.81)}{2.03... \times 10^{-5}}$ [1]

   $\rho = 7.70 \times 10^3\,\text{kg m}^{-3}\,\text{(3 s.f.)}$ [1]

5 weight of block = $(\rho_s \times x^3)g$, where x = length of each side of cube [1]

weight of water displaced = $(\rho \times f x^3)g$, where f is the fraction of volume submerged [1]

weight of block = weight of water displaced [1]

$(\rho_s \times x^3)g = (\rho \times f x^3)g$

Therefore, $f = \dfrac{\rho_s}{\rho}$, as required. [1]

5.1

1 $W = Fx = 24 \times 0.50$ [1]

$W = 12\,\text{J}$ [1]

2 $W = Fx = 430 \times 1000$ [1]

$W = 4.3 \times 10^5\,\text{J}$ [1]

3 $W = Fx = mg \times x = 60 \times 9.81 \times 5.8$ [1]

$W = 3.4 \times 10^3\,\text{J}$ (2 s.f.) [1]

4 $x = 3.1 \times \sin 10 = 0.538...\,\text{m}$ [1]

$W = Fx = mg \times x = 38 \times 9.81 \times 0.538...$ [1]

$W = 200\,\text{J}$ (2 s.f.) [1]

5 $W = Fx \cos \theta = 65 \times 5.0 \times \cos 52$ [1]

$W = 200\,\text{J}$ [1]

All the work done is transformed to thermal energy. [1]

6 work done = change in KE [1]

$F \times 0.030 = 1.4 \times 10^3$ [1]

$F = \dfrac{1.4 \times 10^3}{0.030}$ [1]

$F = 4.7 \times 10^4\,\text{N}$ (2 s.f.) [1]

5.2

1 a Potential means 'stored'. [1]

b Kinetic energy to thermal energy. [1]

2 Heat [1]

$20 - 5 = 15\,\text{J}$ [1]

3 a Electrical energy to light. [1]

b Electrical energy to sound. [1]

4 a The car is travelling at constant velocity, so KE of car does not change. [1]

b Thermal energy = $100 - 25 = 75\%$ [1]

5.3

1 $E_k = \dfrac{1}{2}mv^2 = \dfrac{1}{2} \times 1500 \times 10^2$ [1]

$E_k = 7.5 \times 10^4\,\text{J}$ [1]

2 $E_p = mgh = 9.4 \times 10^4 \times 9.81 \times 1500$ [1]

$E_p = 1.4 \times 10^9\,\text{J}$ (2 s.f.) [1]

3 loss in energy = change in GPE [1]

loss in energy = $mg\Delta h = 0.120 \times 9.81 \times (0.90 - 0.70)$ [1]

loss in energy = $0.24\,\text{J}$ (2 s.f.) [1]

4 a $E_p = mgh = 400 \times 9.81 \times 55$ [1]

$E_p = 2.15... \times 10^5\,\text{J} = 2.2 \times 10^5\,\text{J}$ (2 s.f.) [1]

b Energy is conserved, therefore kinetic energy = $2.2 \times 10^5\,\text{J}$ (2 s.f.) [1]

c $E_k = \dfrac{1}{2}mv^2; v = \sqrt{\dfrac{2E_k}{m}}$ [1]

$v = \sqrt{\dfrac{2 \times 2.15... \times 10^5}{400}}$ [1]

$v = 33\,\text{m s}^{-1}$ (2 s.f.) [1]

5 $v = \sqrt{2gh} = \sqrt{2 \times 9.81 \times 110}$ [1]

$v = 46\,\text{m s}^{-1}$ (2 s.f.) [1]

Assumption: All GPE transferred to KE – no losses. [1]

6 $E_p = mgh = 0.80 \times 9.81 \times 1.2 = 9.4176\,\text{J}$ [1]

$E_k = \dfrac{1}{2}mv^2 = \dfrac{1}{2} \times 0.80 \times 3.2^2 = 4.096\,\text{J}$ [1]

work done against drag = $9.4176 - 4.096$
$= 5.3\,\text{J}$ (2 s.f.) [1]

7 $E_k = \dfrac{1}{2}mv^2 = \dfrac{1}{2} \times 0.030 \times 240^2 = 864\,\text{J}$ [1]

work done = change in KE; $864 = F \times 0.085$ [1]

$F = \dfrac{864}{0.085} = 1.0 \times 10^4\,\text{N}$ (2 s.f.) [1]

5.4

1 $P = \dfrac{W}{t} = \dfrac{240}{30}$ [1]

$P = 8.0\,\text{W}$ [1]

2 energy = $Pt = 2000 \times 60$ [1]

energy = $1.2 \times 10^5\,\text{J}$ [1]

3 electrical energy = $Pt = 60 \times 3600$ [1]

electrical energy = $2.16 \times 10^5\,\text{J}$ [1]

light energy = $0.05 \times 2.16 \times 10^5$
$= 1.1 \times 10^4\,\text{J}$ (2 s.f.) [1]

4 $E_k = \dfrac{1}{2}mv^2 = \dfrac{1}{2} \times 1200 \times 18^2$ [1]

$E_k = 1.944 \times 10^5\,\text{J}$ [1]

rate of work done = $\dfrac{1.944 \times 10^5}{20} = 9.7 \times 10^3\,\text{J s}^{-1}$ or W (2 s.f.) [1]

5 $P = Fv = (4 \times 210 \times 10^3) \times 250$ [1]

$P = 2.1 \times 10^8\,\text{W}$ [1]

6 work done = $Fx = 15 \times 1.4 = 21\,\text{J}$ [1]

input energy = $3.5 \times 30 = 105\,\text{J}$ [1]

efficiency = $\dfrac{21}{105} \times 100 = 20\%$ [1]

7 rate of loss of GPE = $\dfrac{600 \times 10^6}{0.40} = 1.5 \times 10^9\,\text{J s}^{-1}$ or W [1]

(mass per second) $\times g \times h = 1.5 \times 10^9$ [1]

mass per second = $\dfrac{1.5 \times 10^9}{9.81 \times 50} = 3.05... \times 10^6\,\text{kg s}^{-1}$ [1]

volume per second = $\dfrac{3.05... \times 10^6}{\rho} = \dfrac{3.05... \times 10^6}{1000}$ [1]

volume per second = $3.1 \times 10^3\,\text{m}^3\text{s}^{-1}$ (2 s.f.) [1]

6.1

1 Extension is proportional to load, so new extension is $\dfrac{6.0 \times 12}{4.0} = 18\,\text{mm}$

Assumption: the spring obeys Hooke's law.

2 $k = \dfrac{F}{x} = \dfrac{4.0 \cdot 9.81}{0.012} = 3270\,\text{N}\,\text{m}^{-1}$

1 Both obey Hooke's law and elastic behaviour. [1]

A has a greater value of force constant because of the larger gradient. [1]

2 Straight-line graph up to 5.0 N. [1]

Correct value of extension (8 mm) shown on the graph. [1]

Curved section beyond the elastic limit. [1]

3 a $k = \dfrac{F}{x} = \dfrac{4.0}{5 \cdot 10^{\,3}}$ [1]

$k = 800\ \text{N}\,\text{m}^{-1}$ [1]

b $F = \dfrac{32}{5} \cdot 4.0$ [1]

$F = 26\ \text{N}$ (2 s.f.) [1]

Assumption: Hooke's law is obeyed. [1]

4 a

F / N
0.98
1.96
2.94
3.92
4.91
5.89
6.87

Correct values of F in the final column. [1]

b Correct labelling of axes, including the units. [1]

Correct plotting of all points. [1]

Correct straight line through first 5 points followed by a smooth curve. [1]

c The graph shows a linear relationship between force and length. [1]

The graph crosses the L-axis at the original length of the spring; therefore, the graph shows Hooke's law is obeyed. [1]

Force at the elastic limit is about 4.9 N. [1]

d force constant = gradient of the linear section of the graph [1]

force constant $\approx 23\,\text{N}\,\text{m}^{-1}$ [1]

5 a force = $mg = (0.280 \times 9.81)$ N and $x = 0.294 - 0.200 = 0.094$ m [1]

$k = \dfrac{F}{x} = \dfrac{0.280 \cdot 9.81}{0.094}$ [1]

$k = 29.2\,\text{N}\,\text{m}^{-1}$ (3 s.f.) [1]

b (i) Force is shared equally by each spring. The extension of each spring is halved. [1]

Since $k = \dfrac{F}{x}$, this implies that $\dfrac{k \propto 1}{x}$. [1]

The combined force constant is doubled; force constant = $29.2 \times 2 = 58.4\,\text{N}\,\text{m}^{-1}$ [1]

b (ii) Force is the same in each spring. The extension of combined springs is doubled. [1]

Since $k = \dfrac{F}{x}$, this implies that $k \propto \dfrac{1}{x}$. [1]

The combined force constant is halved; force constant = $\dfrac{29.2}{2} = 14.6\,\text{N}\,\text{m}^{-1}$ [1]

6.2

1 $E = \dfrac{1}{2}Fx = \dfrac{1}{2} \cdot 18 \cdot 0.20$ [1]

$E = 1.8\,\text{J}$ [1]

2 $E = \dfrac{1}{2}kx^2$, therefore: $1.5 = \dfrac{1}{2}k \cdot (2.0 \cdot 10^{\,3})^2$ [1]

$k = \dfrac{1.5 \cdot 2}{(2.0 \cdot 10^{\,3})^2}$ [1]

$k = 7.5 \times 10^5\,\text{N}\,\text{m}^{-1}$ [1]

3 work done = area under graph (area of trapezium) [1]

work done = $\dfrac{1}{2} \times 0.10 \times (5 + 15)$ [1]

work done = $1.0\,\text{J}$ [1]

4 elastic potential energy = $E = \dfrac{1}{2}kx^2 = \dfrac{1}{2} \cdot 120 \cdot 0.04^2$ [1] [1]

elastic potential energy = $9.6 \times 10^{-2}\,\text{J}$ [1]

Assume all elastic potential energy is transferred to gravitational potential energy. [1]

$mgh = 9.6 \times 10^{-2}\,\text{J}$ [1]

$h = \dfrac{9.6 \cdot 10^{\,2}}{0.008 \cdot 9.81} = 1.2\,\text{m}$ (2 s.f.) [1]

6.3

1 The molecular chains are easier to untangle for smaller forces, hence the F–x graph has a smaller gradient.

The molecular chains are difficult to extend further when they are fully extended, hence the $F - x$ graph has a steeper gradient.

2 Rubber returns to its original length when the force is completely removed, and so shows elastic behaviour. The hysteresis loop shows that all the energy stored is not returned – some energy is transferred to thermal energy. So rubber is poor at storing energy.

3 a Not all the energy stored in the material is returned – so the landings are less bumpy.

b The work done on the material is greater than the energy returned. The area of the hysteresis loop is the energy transferred to thermal energy, so the tyres warm up.

1   Rubber will return to its original length when the forces are removed. [1]

2   The force–extension graph for rubber is not a straight line through the origin. [1]

It does not obey Hooke's law: force is not proportional to extension. Therefore, a rubber band cannot have a force constant. [1]

3   The metal wire shows elastic behaviour up to its elastic limit and polythene does not show any elastic behaviour. [1]

Polythene shows plastic deformation, as does the metal wire beyond its elastic limit. [1]

4   thermal energy = area of the hysteresis loop [1]

area of loop ≈ 4.0 squares [1]

Therefore, thermal energy ≈ 4.0 × (0.1 × 10) = 4 J [1]

6.4

1   area under graph = $\frac{1}{2} \times \sigma \times \varepsilon$

area under graph = $\frac{1}{2} \times \frac{F}{\rho} \times \frac{x}{L}$

area under graph = $\frac{\frac{1}{2}Fx}{V}$, $V = AL$ – the volume of the material.

elastic potential energy = $\frac{1}{2}Fx$

Therefore, the area under the graph is energy stored per unit volume.

2   Energy is equivalent to work done, and work done = force × distance.

Force is given by the equation $F = ma$. It has base unit $kg\,m\,s^{-2}$.

(Volume has base units m^3.)

Therefore, energy per unit volume has base units:

$\frac{kg\,m\,s^{-2} \times m}{m^3} = kg\,m^{-1}\,s^{-2}$

stress = force/area

(Area has base units m^2.)

Therefore, stress has base units: $\frac{kg\,m\,s^{-2}}{m^2} = kg\,m^{-1}\,s^{-2}$

Both base units are the same.

3   energy per unit volume = $\frac{1}{2} \cdot$ stress $\cdot$ strain

$120 \times 10^6 = \frac{1}{2} \times$ stress $\times 0.40$

stress = 6.0×10^8 Pa

Assumption: the silk obeys Hooke's law.

1   It is the maximum stress that can be applied to a material before it breaks. [1]

2   Cast iron is a brittle material.

3   $\sigma = \frac{6.3}{\pi \times (0.10 \times 10^{-3})^2} = 2.00... \times 10^8$ Pa [1]

$\varepsilon = \frac{1.048 - 1.035}{1.035} = 1.25... \times 10^{-2}$ [1]

Young modulus = $\frac{2.00... \times 10^8}{1.25... \times 10^{-2}} = 1.6 \times 10^{10}$ Pa (2 s.f.) [1]

4   (ultimate tensile strength = $\frac{F}{A}$) [1]

$220 \times 10^6 = \frac{F}{\pi \times (0.60 \times 10^{-3})^2}$ [1]

$F = 220 \times 10^6 \times \pi \times (0.60 \times 10^{-3})^2$ [1]

$F = 250$ N (2 s.f.) [1]

5   a $E = \sigma \div \varepsilon = \frac{F}{A} \div \frac{x}{L}$ [1]

$E = \frac{F}{A} \times \frac{L}{x} = \frac{FL}{Ax}$ [1]

b $E = \frac{FL}{Ax} = \frac{300 \times 2.500}{\pi \times (0.42 \times 10^{-3})^2 \times 1.4 \times 10^{-2}}$ [1]

$E = 9.7 \times 10^{10}$ Pa (2 s.f.) [1]

c   The elastic limit of the material is not exceeded, so stress ∝ strain. [1]

7.1

1   The resultant force is zero. [1]

2   Force on the Earth = 600 N [1]

The force is the same because the person and the Earth interact with each other. According to Newton's third law, the magnitude of the force experienced by each is the same. [1]

3   The force experienced by each is the same but in opposite directions. [1]

Hence, the resultant force = 0. [1]

4   The runner exerts a backward force on the Earth. [1]

According to Newton's third law, the Earth exerts an equal forward force on the person. [1]

5   a   The acceleration of free fall would be $9.81\,m\,s^{-2}$, assuming there is no drag. [1]

b   The force on the Earth is the same as the weight of the bird (Newton's third law). [1]

$F = ma$

$a = \frac{F}{m} = \frac{150}{6.0 \times 10^{24}}$ [1]

$a = 2.5 \times 10^{-23}\,m\,s^{-1}$ [1]

7.2

1   Total energy and linear momentum are both conserved. [1]

2   The loss of momentum is numerically equal to the gain in momentum (principle of conservation of momentum); therefore, $\Delta p = 120\,kg\,m\,s^{-1}$. [1]

$\Delta v = \frac{\Delta p}{m} = \frac{120}{2.0}$ [1]

$\Delta v = 60\,m\,s^{-1}$ [1]

3   The initial momentum is zero; therefore, the final momentum must also be zero.

momentum of cannon = momentum of shell [1]

$1200 \times v = 20 \times 300$ [1]

$v = 5.0\,m\,s^{-1}$ [1]

4 a total initial momentum = total final momentum

$(300 \times 2.5) + (400 \times -4.0) = 300v + 0$ [1]

$v = \dfrac{-850}{300}$ [1]

$v = -2.8\,\text{m s}^{-1}$ (2 s.f.) [1]

b initial KE $= \dfrac{1}{2} \times 300 \times 2.5^2 + \dfrac{1}{2} \times 400 \times 4.0^2$

$= 4.13... \times 10^3\,\text{J}$ [1]

final KE $= \dfrac{1}{2} \times 300 \times 2.8^2 = 1.1176 \times 10^3\,\text{J}$ (2 s.f.) [1]

loss in KE $= 4.13... \times 10^3 - 1.1176 \times 10^3$

$= 3.0 \times 10^3\,\text{J}$ (2 s.f.) [1]

c The kinetic energy is not conserved. [1]

Therefore, the collision is inelastic. [1]

7.3

1 $F = \dfrac{\Delta p}{\Delta t} = \dfrac{1.2 \times 10^4}{5.0}$ [1]

$F = 2.4 \times 10^3\,\text{N}$ [1]

2 $\Delta p = F \times \Delta t = 150 \times 0.025$ [1]

$\Delta p = 3.75 = 3.8\,\text{kg m s}^{-1}$ (2 s.f.) [1]

3 $F = \dfrac{\Delta p}{\Delta t} = \dfrac{1.0 \times 10^6 \times 3.4 \times 10^3 - 2.3 \times 10^6 \times 1.2 \times 10^3}{200}$ [1]

$F = 3.2 \times 10^6\,\text{N}$ [1]

4 $\Delta p = (0.150 \times -15) - (0.150 \times 15) = -4.5\,\text{kg m s}^{-1}$ [1]

$F = \dfrac{\Delta p}{\Delta t} = \dfrac{4.5}{0.025}$ (magnitude only) [1]

$F = 180\,\text{N}$ [1]

5 change in momentum *per second* $= 2.5 \times 4.0$

$= 10\,\text{kg m s}^{-2}$ [1]

Therefore, force $= 10\,\text{N}$ [1]

7.4

1 Since $p = \dfrac{h}{\lambda}$, momentum p is inversely proportional to the wavelength λ.

2 $p = \dfrac{h}{\lambda} = \dfrac{6.63 \times 10^{-34}}{500 \times 10^{-9}}$

$p = 1.3 \times 10^{-27}\,\text{kg m s}^{-1}$ (2 s.f.)

3 area $= \dfrac{1}{9.1 \times 10^{-6}}$

area $= 1.1 \times 10^5\,\text{m}^2$ (2 s.f.)

1 Ns and kg m s^{-1} [2]

2 impulse $= F \times \Delta t = 200 \times 5.0$ [1]

impulse $= 1000\,\text{Ns}$ [1]

3 $\Delta p = $ impulse $= 1.1\,\text{Ns}$

$\Delta v = \dfrac{\Delta p}{m} = \dfrac{1.1}{0.050}$ [1]

$\Delta v = 22\,\text{m s}^{-1}$ [1]

4 a change in momentum = area under graph [1]

$\Delta p = \dfrac{1}{2} \times 1.5 \times 10^{-17} \times 6.0 \times 10^{-6}$

$= 4.5 \times 10^{-23}\,\text{kg m s}^{-1}$ [1]

b $4.5 \times 10^{-23} = 1.7 \times 10^{-27}v - (1.7 \times 10^{-27} \times 5.0 \times 10^4)$ [1]

$v = \dfrac{1.3 \times 10^{-22}}{1.7 \times 10^{-27}}$ [1]

$v = 7.6 \times 10^4\,\text{m s}^{-1}$ (2 s.f.) [1]

7.5

1 The angle would be 90°. [1]

2 The diagram is incorrect because:

After the collision, the total momentum in the original direction of travel of X is $5\,\text{kg m s}^{-1}$. This is not the same as the initial momentum in this direction. [1]

Also, there can be no momentum at right angles to the original direction of travel of X because the initial momentum in this direction was zero. [1]

3 The initial momentum must equal the final momentum. The final momentum is the vector sum of the two momentums. [1]

Therefore

initial momentum $= \sqrt{3.0^2 + 2.0^2}$ [1]

initial momentum $= 3.6\,\text{kg m s}^{-1}$ (2 s.f.) [1]

4 final momentum = total initial momentum [1]

final momentum $= \sqrt{(0.250 \times 3.0)^2 \times (0.150 \times 4.0)^2}$ [1]

final momentum $= 0.96...\,\text{kg m s}^{-1}$ [1]

final velocity $v = \dfrac{0.96...}{0.400} = 2.4\,\text{m s}^{-1}$ (2 s.f.) [1]

8.1

1 The charge was due to an excess of electrons, each with charge e. The total charge on the drop must be ne, where n is the number of excess of electrons.

2 Measure diameter to determine the radius, r, and then use the equations for volume and density to determine the mass.

$\rho = \dfrac{m}{V}$; therefore, $m = \rho V$ and $V = \dfrac{4}{3}\pi r^3$;

therefore, $m = \rho \dfrac{4}{3}\pi r^3$

Weight is given by $W = mg$; therefore, the weight of the drop is given by $W = \rho \dfrac{4}{3}\pi r^3 g$

3 To allow the uncertainty and so the accuracy to be determined. To ensure the data does not appear to be more accurate than it actually is.

1 a $2.0 \times 1.60 \times 10^{-19}\,\text{C} = 3.2 \times 10^{-19}\,\text{C}$ [1]

b $-5.0 \times 1.60 \times 10^{-19}\,\text{C} = -8.0 \times 10^{-19}\,\text{C}$ [1]

c $-12 \times 1.60 \times 10^{-19}\,\text{C} = -1.9 \times 10^{-18}\,\text{C}$ (2 s.f.) [1]

d $41 \times 1.60 \times 10^{-19}\,\text{C} = 6.6 \times 10^{-18}\,\text{C}$ (2 s.f.) [1]

2 **a** $-10\,e$ must be 10 electrons, each with charge e. [1]

b $e = 1.60 \times 10^{-19}\,C$ number of electrons [1]

$$= \frac{\text{total charge}}{\text{charge on each electron}}$$ [1]

$$\frac{15}{1.60 \times 10^{-19}} = \text{number of electrons}$$ [1]

$$\frac{15}{1.60 \times 10^{-19}} = 9.4 \times 10^{19}\ \text{electrons (2 s.f.)}$$ [1]

3 $I = \dfrac{Q}{t}$ therefore $Q = I\ t$ [1]

$t = 4.0$ hours $= 14\,400$ seconds [1]

$\Delta Q = 500 \times 10^{-3} \times 14\,400 = 7200\,C$ [1]

[1 mark for 2.0 C]

4 $Q = ne$ therefore

$Q = 5.0 \times 10^{14} \times 1.60 \times 10^{-19} = 8.0 \times 10^{-5}\,C$ [1]

$I = \dfrac{Q}{t} = \dfrac{8.0 \times 10^{-5}}{1.0} = 8.0 \times 10^{-5}\ A$ [1]

5 number of electrons

$$= \frac{\text{total charge}}{\text{charge on each electron}}$$

$$= \frac{9000}{1.60 \times 10^{-19}} = 5.625 \times 10^{22}\,\text{electrons}$$ [1]

$\Delta t = 2.0$ hours $= 7200\,s$ [1]

average number of electrons per second

$$= \frac{\text{number of electrons}}{\text{time taken}} = \frac{5.625 \times 10^{22}}{7200}$$

$= 7.8 \times 10^{18}$ electrons per second (2 s.f.) [1]

6 Two weeks $= 1.2 \times 10^{6}\,s$.

$\Delta Q = I\Delta t = 6.2 \times 1.2 \times 10^{6} = 7.44 \times 10^{6}\,C$ [1]

number of electrons $= \dfrac{\text{total charge}}{\text{charge on each electron}}$

$$= \frac{7.44 \times 10^{6}}{1.60 \times 10^{-19}}$$

$$= 4.7 \times 10^{25}\,\text{electrons (2 s.f.)}$$ [1]

7 1 As is equivalent to 1 C [1]

$1.0\,mAh = 1.0 \times 10^{-3}\,Ah = 60 \times 10^{-3}\,A\,min = 3.6\,As$;
therefore, $1.0\,mA\,h = 3.6\,C$ [1]

$5000\,mAh = 5000 \times 3.6 = 18000\,C$ [1]

8.2

1 Conventional current is from the positive terminal to the negative terminal. [1]

Electrons flow from the negative terminal to the positive terminal. [1]

2 Any valid comparisons, examples include:
Similarities:
Both are examples of flows of charge. [1]

Differences:

In metals the charge carriers are electrons. In ionic solutions the charge carriers are ions. [1]

In metals the charge carriers are negative. In ionic solutions the charge carriers can be positive or negative. [1]

In metals electrons flow in the opposite direction to the conventional current. In ionic solutions the charge carriers can either flow in the same direction or in the opposite direction to the conventional current. [1]

Must contain at least one similarity and one difference for 4 marks.

3 As Figure 5 with labelled electrodes [1]

Cu^{2+} moving towards negative cathode [1]

SO_4^{2-} moving towards positive anode [1]

4 Cations move from positive to negative (towards the cathode). The same direction as the conventional current. [1]

Anions move from negative to positive (towards the anode). The opposite direction to the conventional current. [1]

5 $I = \dfrac{Q}{t}$ and $\Delta t = 3.0$ minutes $= 180\,s$ [1]

Each cation has a relative charge of +2e.

$\Delta Q = 2 \times 1.60 \times 10^{-19} \times 6.0 \times 10^{14} = 1.9 \times 10^{-4}\,C$ [1]

$I = \dfrac{1.9 \times 10^{-4}}{180} = 1.1 \times 10^{-6}\,A = 1.1\,\mu A$ (2 s.f.) [1]

8.3

1 The total/net charge in any interaction must be the same before and after the interaction. [2]

[A simple, 'The charge in any interaction must be the same before and after the interaction' gains 1 mark]

2 As Figure 2 [1]

The sum of the current into a point must equal the sum of the current out of the point. [1]

3 **a** **i** 7 A towards the 2 A

ii 5 A away from the junction [both required for 1]

b **iii** 4 A towards the junction [1]

c **iv** 2 A to the left [1]

v 5 A to the left [1]

vi 7 A towards the junction [1]

4 Current in wire A =

$I = \dfrac{1.9 \times 10^{21} \times 1.60 \times 10^{-19}}{60} = 5.06...\,A$ [2]

Current in wire B = 15 A − 5.06... A = 9.9 A (2 s.f.) [1]

5 Discussion should include:

Charge must be conserved

Charge is due to electrons/ions

Therefore, the total number of electrons/ions must be conserved

Current is a flow of charge

Rate of flow of charge into a point must be equal to the rate of flow of charge from that point

[1 mark for each valid point, with up to three total marks]

6 Two protons have a net charge of $+2e$ (3.20×10^{-19} C)   [1]

Any particles created in the collision must give rise to the same net charge. For example, if the positive charges are measured after the collision and found to be $+5e$, this suggests a particle (or several particles) with a charge of $-3e$ must have been created, ensuring the net charge remains at $+2e$.   [1]

8.4

> **1** Show that the equation is homogenous with respect to base units.
>
> $I = A\, n\, e\, v$
>
> $I\,[\text{A}], A\,[\text{m}^2], n\,[\text{m}^{-3}], e\,[\text{C}], v\,[\text{m s}^{-1}]$
>
> Therefore
>
> $[\text{A}] = [\text{m}^2] \times [\text{m}^{-3}] \times [\text{C}] \times [\text{m s}^{-1}]$
>
> $[\text{A}] = [\text{C s}^{-1}]$
>
> $[\text{A}] = [\text{A}]$
>
> **2 a** mean drift velocity increases
>
> **b** mean drift velocity increases
>
> **c** mean drift velocity halves

1 Conductors have the greatest number density   [1]
followed by semiconductors, and then insulators   [1]

2 $I = A\, n\, e\, v$ and n for copper $= 8.5 \times 10^{28}\,\text{m}^{-3}$ [1]

$I = 5.50 \times 10^{-8} \times 8.5 \times 10^{28} \times$

$1.60 \times 10^{-19} \times 2.0 \times 10^{-3}$ [1]

$I = 1.5\,\text{A}$ (2 s.f.)   [1]

3 $I = A\, n\, e\, v$ therefore $v = \dfrac{I}{A\, n\, e}$ [1]

$v = \dfrac{500 \times 10^{-3}}{7.10 \times 10^{-6} \times 5.86 \times 10^{28} \times 1.60 \times 10^{-19}}$ [1]

$v = 7.51 \times 10^{-6}\,\text{ms}^{-1}$ (3 s.f.)   [1]

4 a From $v = \dfrac{I}{A\, n\, e}$ it follows that $v \propto \dfrac{1}{A}$ [1]

Therefore, if the cross-sectional area increases, the mean drift velocity decreases.   [1]

b From $v = \dfrac{I}{A\, n\, e}$ it follows that $v \propto \dfrac{1}{n}$.   [1]

As copper has a higher n than zinc, n increases, so the mean drift velocity decreases.   [1]

c From $v = \dfrac{I}{A\, n\, e}$ it follows that $v \propto \dfrac{1}{A}$.   [1].

As $A = \pi r^2$, if r decreases by a factor of 3, A will decrease by a factor of 9 (3^2).   [1]

As A decreases by a factor of 9, the mean drift velocity must increase by a factor of 9.   [1]

5 $I = A\, n\, e\, v$; therefore, $v = \dfrac{I}{A\, n\, e}$ [1]

$A = \pi r^2$; therefore, $A = \pi \times \left(\dfrac{1.0 \times 10^{-3}}{2} \right)^2$ [1]

$= 7.85... \times 10^{-7}\,\text{m}^2$

$v = \dfrac{I}{A\, n\, e} = \dfrac{3.0 \times 10^{-3}}{7.85... \times 10^{-7} \times 6.6 \times 10^{28} \times 1.60 \times 10^{-19}}$

$v = 3.6 \times 10^{-7}\,\text{ms}^{-1}$ (2 s.f.)   [1]

6 $I = A\, n\, e\, v$; therefore, $n = \dfrac{I}{A\, v\, e}$ [1]

$n = \dfrac{I}{A\, v\, e} = \dfrac{12 \times 10^{-3}}{8.2 \times 10^{-6} \times 72 \times 1.60 \times 10^{-19}}$ [1]

$n = 1.27... \times 10^{20}$ [1]

Giving a ratio of $\dfrac{1.27... \times 10^{20}}{8.5 \times 10^{28}}$ or 1.5×10^{-9} :

1 (2 s.f.)   [1]

9.1

1 a Diode   [1]

b Thermistor   [1]

c Capacitor   [1]

2 a Cell and lamp   [1]

Correctly drawn in series   [1]

b Battery   [1]

Resistor and ammeter   [1]

Correctly drawn in series   [1]

c Power supply and two resistors   [1]

Correctly drawn in series   [1]

3 Cell, open switch, lamp in series   [1]

Voltmeter correctly drawn in parallel with lamp   [1]

4 Any two from   [2]

Circuit not complete

Negative terminal labelled incorrectly

Positive terminal labelled incorrectly

5 Power supply, ammeter and lamp   [1]

Correctly drawn in series   [1]

Voltmeter in parallel across lamp   [1]

9.2

1 Power supply/cell/battery, voltmeter and lamp   [1]

Voltmeter in parallel across lamp   [1]

2 Potential difference is used when work is done by the charge carriers. A transfer of energy from the charge carriers to the component, transferring electrical energy into other forms.   [1]

Electromotive force is used when work is done on the charge carriers. A transfer of energy to the charge carriers from the cell/battery/power supply, transferring other forms of energy (chemical, light, etc.) into electrical energy.   [1]

3 $V = \dfrac{W}{Q}$; therefore, $W = VQ$ [1]

$W = 80 \times 4.0 = 320\,J$ [1]

4 Calculate the p.d. across a filament lamp when $168\,J$ of energy is transferred to the lamp by $14\,C$ of charge. [2]

$V = \dfrac{W}{Q}$ [1]

$V = \dfrac{168}{14} = 12\,V$ [1]

5 One volt is the potential difference across a component when $1\,J$ of energy is transferred per unit charge passing through the component. $1\,V$ is equal to $1\,J$ of energy transferred per coulomb of charge. [1]

$V = \dfrac{W}{Q}$, V [V], W [J], Q [C]

From $\Delta Q = I\Delta t$ [C] = [A s]

Therefore

$[V] = [J] \times [A^{-1}] \times [s^{-1}]$ [1]

From $W = Fx$ [J] = [N m]

Therefore:

$[V] = [N] \times [m] \times [A^{-1}] \times [s^{-1}]$

From $F = ma$ [N] = [kg m s^{-2}] [1]

Therefore:

$[V] = [kg] \times [m] \times [s^{-2}] \times [m] \times [A^{-1}] \times [s^{-1}]$

Which becomes:

$[V] = [kg] \times [m^2] \times [s^{-3}] \times [A^{-1}]$

Therefore, in base units the volt is equal to

kg m^2 s^{-3} A^{-1} [1]

6. Time $= 6.0 \times 60 \times 60 = 21600\,s$ [1]

$\Delta Q = I\Delta t = 500 \times 10^{-3} \times 21600 = 10800\,C$ [1]

$V = \dfrac{W}{Q}$; therefore, $W = VQ$ [1]

$W = 1000 \times 10800 = 11\,MJ$ (2 s.f.) [1]

9.3

1 Total $V = 50 \times 100 = 5000\,V$

$eV = \dfrac{1}{2}mv^2$ therefore $v = \sqrt{\dfrac{2eV}{m}}$

$v = \sqrt{\dfrac{2 \times 1.60 \times 10^{-19} \times 5000}{9.11 \times 10^{-31}}}$

$v = 4.2 \times 10^7\,ms^{-1}$ (2 s.f.)

2 Electrons must spend the same time in each electrode as they move along the accelerator (in order to ensure they leave the electrode at the same time as the potential difference changes and accelerates them towards the next one). The electrons are moving faster as they move along the accelerator. Therefore, electrodes must increase in length.

1 Electrons are emitted from the hot wire/filament at the rear of the electron gun. [1]

There is a large p.d. between the filament and an anode. [1]

Electrons are accelerated towards the anode. [1]

They pass through a hole/gap in the anode. [1]

2 $eV = \dfrac{1}{2}mv^2$ therefore kinetic energy $= eV$ [1]

kinetic energy $= 1.60 \times 10^{-19} \times 12000$ [1]

kinetic energy $= 1.9 \times 10^{-15}\,J$ (2 s.f.) [1]

3 kinetic energy $= \dfrac{1}{2}mv^2$ therefore [1]

$v = \sqrt{\dfrac{2 \times 1.8 \times 10^{-15}}{9.11 \times 10^{-31}}} = 6.0 \times 10^7\,ms^{-1}$ (2 s.f.) [1]

4 $v = 0.09 \times 3.00 \times 10^8 = 2.7 \times 10^7\,ms^{-1}$ [1]

$eV = \dfrac{1}{2}mv^2$ therefore $V = \dfrac{\frac{1}{2}mv^2}{e}$ [1]

$V = \dfrac{\frac{1}{2} \times 9.11 \times 10^{-31} \times \left(2.7 \times 10^7\right)^2}{1.60 \times 10^{-19}}$ [1]

$V = 2100\,V$ (2 s.f.) [1]

5 Velocity of electron will be greater than the proton. [1]

The kinetic energy of each will be the same (as they have the same magnitude charge). [1]

Mass of proton is greater, so travels more slowly at the same kinetic energy. [1]

9.4

1 Graph of I against V [1] with a straight line through the origin. [1]

$V \propto I$ [1] for a conductor at constant temperature [1]

2 $R = \dfrac{V}{I} = \dfrac{5.2}{0.50} = 10\,\Omega$ (2 s.f.) [1]

3 $R = \dfrac{V}{I} = \dfrac{2.4}{80 \times 10^{-3}} = 30\,\Omega$ [1]

4 One ohm is the resistance of a component when a p.d. of $1\,V$ is produced per ampere of current. [1]

(See 9.2 question 3 to get the base units of the volt.)

$R = \dfrac{V}{I}$

R [Ω], V [kg m^2 s^{-3} A^{-1}], I [A] [1]

$[\Omega] = [kg\,m^2\,s^{-3}\,A^{-1}] \times [A^{-1}]$

Which becomes:

$[\Omega] = [kg\,m^2\,s^{-3}\,A^{-2}]$

Therefore, in base units the ohm is equal to kg m^2 s^{-3} A^{-2} [1]

5 $I = \dfrac{Q}{t} = \dfrac{54}{180} = 0.30\,A$ [1]

$R = \dfrac{V}{I}$ therefore $V = IR = 0.30 \times 1200 = 360\,V$ [1]

6 $I = \dfrac{Q}{t}$ and $V = \dfrac{W}{Q}$ therefore $V = \dfrac{W}{I\,t}$ [1]

$$V = \frac{500}{1.5 \times 60} = 5.55...\,\text{V}$$ [1]

$$R = \frac{V}{I} = \frac{5.55...}{1.5} = 3.7\,\Omega \ (2\ \text{s.f.})$$ [1]

9.5

> **1** Power supply, ammeter, voltmeter, variable resistor/potentiometer
>
> **2** Increasing the resistance of the variable resistor will reduce the current in the circuit.
>
> **3** Set up either circuit as shown in Figure 2.
> Record values for *I* and *V*.
> Adjust variable resistor/potential divider to produce a range of values.
> Include negative values.
> Repeat and average.
> Plot a graph of *I* against *V*.

1 **a** A [1] $V \propto I$ [1]

    b Component A: $R = \dfrac{V}{I} = \dfrac{4.0}{2.0} = 2.0\,\Omega$ [1]

        Component B: $R = \dfrac{V}{I} = \dfrac{4.0}{1.6} = 2.5\,\Omega$ [1]

2 Graph of *I* against *V* with axes labelled (including units)   [1]

    Points plotted correctly   [1]

    Line of best fit drawn   [1]

3 **a** $1.8\,\Omega +/- 0.2\,\Omega$ [1]

    b $2.2\,\Omega +/- 0.2\,\Omega$ [1]

    c $4.5\,\Omega +/- 0.3\,\Omega$ [1]

4 Filament lamp   [1]

    Any three from: [3]

    As the current increases more electrons/charge carriers pass through the lamp per unit time/second.

    The rate of collisions between electrons and positive ions increases.

    Each collision transfers energy to the ions/ions gain more energy.

    Ions vibrate more.

    Temperature of the wire increases.

    Resistance increases.

5 Graph of *I* against *V*.

    Two straight lines through the origin with different gradients   [1]

    Steeper line labelled room temperature / shallower line labelled higher temperature.   [1]

    Explanation.

    Hotter wire has a greater resistance.   [1]

    This results in a lower current at the same p.d. / shallower line.   [1]

9.6

1 Any two from: [2]

    Voltmeter connected in series

    Voltmeter in the wrong place (i.e., not across diode)

    Missing ammeter

    Diode wrong way round

2 A and B   [1]

    Diode in series with B is the correct orientation to allow current in bulb B.   [1]

    Diode in series with C will not allow current in bulb C.   [1]

3 Graph of I against V with axes labelled   [1]

    Axes include correct units   [1]

    Points plotted correctly   [1]

    Line of best fit drawn   [1]

4 **a** $\infty\,\Omega$ / very large   [1]

    b $650\,\Omega +/- 1000\,\Omega$ [1]

    c $12\,\Omega +/- 5\,\Omega$ [1]

5 **a** 0 values for the current on the negative side as p.d. increase.   [1]

    Current remains 0 until a small increase in positive p.d.   [1]

    Current then increases linearly.   [1]

    b On the negative side the $-I$ is directly proportional to $-V$ (as the current is in the resistor only).   [1]

        On the positive side the graph is the same as a standard diode (as the current passes through the diode when its resistance drops low enough).   [1]

9.7

> **1** Resistance of a normal wire drops as it gets cooler.
> Resistance of a superconductor drops as it gets cooler, but at a critical temperature falls to $0\,\Omega$.
>
> **2** No energy transferred to heat when there is a current in a component.
> Allows very high currents.

1 Resistance applies to a particular component.   [1]

    Resistivity is a property of the material.   [1]

2 $R = \dfrac{\rho L}{A}$ [1]

    $R = \dfrac{1.7 \times 10^{-8} \times 1.0}{3.32 \times 10^{-6}}$ [1]

    $R = 5.1\,\text{m}\Omega$ (2 s.f.)   [1]

3 Increasing the temperature of the metal:

    Positive ions in the metal have more energy.   [1]

    The positive ions vibrate more.   [1]

    Increasing the resistivity.   [1]

4 $R = \dfrac{\rho L}{A}$ therefore $\rho = \dfrac{RA}{L}$ [1]

$A = \pi r^2$ [1]

$A = \pi \times \left(3.0 \times 10^{-3}\right)^2 = 2.82... \times 10^{-5} \, \text{m}^2$ [1]

$\rho = \dfrac{170 \times 2.82... \times 10^{-5}}{12}$ [1]

$\rho = 4.0 \times 10^{-4} \, \Omega\text{m}$ (2 s.f.) [1]

5 a $R = \dfrac{\rho L}{A}$ therefore $R \propto L$. As L doubles R doubles. [1]

b $R = \dfrac{\rho L}{A}$ therefore $R \propto \dfrac{1}{A}$ [1]. As A doubles R halves. [1]

c As $A = \pi r^2$ if r halves A decreases by a factor of 4 (2^2) [1]

$R = \dfrac{\rho L}{A}$ therefore $R \propto \dfrac{1}{A}$. A decreases by a factor of 4, R increases by a factor of 4. [1]

d As $V = L \times A$, if V remains constant, if L doubles then A halves. [1]

From $R \propto L$ as L doubles R doubles and from $R \propto \dfrac{1}{A}$ as A halves R doubles. [1]

Therefore R increases by a factor of 4. [1]

6 a Reduce the impact of random errors [1]

To ensure the wire is of uniform diameter [1]

b i Graph of R against L with axes labelled [1]

Axes include correct units [1]

Points plotted correctly [1]

Line of best fit drawn [1]

ii gradient $= \dfrac{R}{L}$ [1]

From $R = \dfrac{\rho L}{A}$, $\dfrac{R}{L} = \dfrac{\rho}{A}$ therefore gradient $= \dfrac{\rho}{A}$ and $\rho = \text{gradient} \times A$ [1]

$A = \pi \times \left(\dfrac{0.46 \times 10^{-3}}{2}\right)^2 = 1.66... \times 10^{-7} \, \text{m}^2$ [1]

$\rho = \text{gradient} \times A$

$\rho = -6.12 \times 1.66... \times 10^{-7}$

$\rho = 1.0 \times 10^{-6} \, \Omega\text{m}$ (2 s.f.) $+/- \, 0.2 \times 10^{-6} \, \Omega\text{m}$ [1]

c Adjusting the variable resistor keeps the current at 0.50 A / Reducing the resistance of the variable resistor as the length of the wire increases keeps the current at 0.50 A [1]

Changing the current will change the temperature of the wire. [1]

Changing the temperature will affect the resistance of the wire. [1]

7 Any four from: [4]

Set up circuit as Figure 1.

Record values of V and I for different thicknesses of wire of the same length (L).

Adjust variable resistor (if present) to ensure I remains constant.

Repeat readings and average.

Calculate R for different thicknesses using $R = \dfrac{V}{I}$

Calculate A of wire using $A = \pi r^2$

Plot a graph of R against $\dfrac{1}{A}$

Measure gradient.

$gradient = \rho L$

Find resistivity of the wire from $\rho = \dfrac{\text{gradient}}{L}$

9.8

1 a As the temperature drops the resistance increases. [1]

b Use the thermistor as a sensor. At a given temperature the thermistor will have a specific resistance. As the resistance changes this can be used to adjust the temperature of the lorry. If the resistance drops, the temperature should be increased, and so on, keeping the temperature at a set level. [2]

2 Graph of R against T with axes labelled (include correct units), smooth curve of decreasing gradient [1]

With significant change in R over the range of 100–300 °C [1]

Large change in R over the range of temperatures found in ovens [1]

3 Calculate R using $R = \dfrac{V}{I}$ [1]

Graph of R against T with axes labelled [1]

Axes include correct units [1]

Points plotted correctly [1]

Line of best fit drawn [1]

4 a $64 \, \Omega \, +/- \, 5 \, \Omega$ [1]

b $15 \, \Omega \, +/- \, 3 \, \Omega$ [1]

5 $R = \dfrac{V}{I} = \dfrac{1.03}{45.2 \times 10^{-3}} = 22.8 \, \Omega$ [2]

Temperature $= 17 \, ^\circ\text{C} \, +/- \, 2 \, ^\circ\text{C}$ [1]

9.9

1 To reduce the effect of the atmosphere (IR absorbed by water vapour) / to reduce the impact of cloud cover (i.e. it is more likely there is a cloud-free day).

2 When the planet passes in front of the star there is a small but detectable drop in intensity (as some of the light is blocked by the planet). This drop in intensity might be detectable by an LDR, resulting in a slight increase in resistance.

1. LDR resistance increases as it gets darker. [1]
   LDR resistance decreases as it gets lighter. [1]
2. LED should be LDR [1]
   Table missing units [1]
3. Any valid comparisons, examples include: [4]
   Resistance of resistor and thermistor affected by temperature (unaffected by light intensity)
   Resistance of thermistor increases as it gets colder.
   Resistance of resistor decreases as it gets colder.
   Resistance of LDR changes as the light level changes (unaffected by temperature).
   Thermistor and LDR are made from semiconductors.
4. a 20–40 (arbitrary units) [1]
   b 0.06–0.08 (arbitrary units) [1]

9.10

1. Circuit diagram with power supply/cell/battery, filament lamp in series with an ammeter. Voltmeter connected in parallel across lamp. [1]
   Measure V and I and calculate power using $P = IV$. [1]
2. $P = IV$ [1]
   $P = IV = 5.0 \times 8.0 = 40\,W$ [1]
3. a $P = IV$ therefore $I = \dfrac{P}{V}$ [1]
   $I = \dfrac{1200}{20} = 60\ A$ [1]
   b $P = \dfrac{W}{t}$ therefore $W = Pt$ [1]
   $W = 1200 \times 3600 = 4.3\,MJ$ (2 s.f.) [1]
4. a $P = I^2R$ therefore at constant current $P \propto R$ [1]
   If R doubles P doubles. [1]
   b $P = I^2R$ therefore if the resistance remains unchanged $P \propto I^2$ [1]
   If I doubles P increases by a factor of 4 (2^2). [1]
5. Find the watt in base units.
   From $P = \dfrac{W}{t} \rightarrow [W] = [J\ s^{-1}]$
   From $W = Fx \rightarrow [J] = [N] \times [m]$
   Therefore $[W] = [N\ m\ s^{-1}]$
   From $F = ma \rightarrow [N] = [kg\ m\ s^{-2}]$
   Therefore the watt in base units is $kg\,m^2\,s^{-3}$. [2]
   Express V A in base units.
   From $V = \dfrac{W}{Q} \rightarrow [V] = [J\ C^{-1}]$
   From $W = Fx \rightarrow [J] = [N] \times [m]$
   Therefore $[V] = [N\ m\ C^{-1}]$
   From $F = ma \rightarrow [N] = [kg\,m\,s^{-2}]$
   Therefore $[V] = [kg\,m^2\,s^{-2}\,C^{-1}]$
   From $\Delta Q = I\Delta t \rightarrow [C] = [A\ s]$
   Therefore $[V] = [kg\,m^2\,s^{-3}\,A^{-1}]$
   Therefore $[V\ A] = [kg\,m^2\,s^{-3}\,A^{-1}\ A] = [kg\,m^2\,s^{-3}]$ [2]

9.11

1. $P = \dfrac{W}{t}$ therefore $W = Pt$ [1]
   $W = 60 \times 7200 = 430\,kJ$ [1]
2. number of units $= 0035387 - 0034512 = 875$ units [1]
   Cost $= 875 \times 0.12 = £105$ [1]
3. From definition 1 kW h is the energy transferred by a 1 kW device in 1 hour. [1]
   $P = \dfrac{W}{t}$ therefore $W = Pt$ [1]
   $W = 1000 \times 3600 = 3.6\,MJ$ [1]
4. a $P = \dfrac{W}{t}$ therefore $W = Pt$ [1]
   In SI units, $P = 9000\,W$ and $t = 900$ s [1]
   $W = 9000 \times 900 = 8.1\,MJ$ [1]
   b $P = \dfrac{W}{t}$ therefore $W = Pt$ [1]
   In kW h units, $P = 9.0\,kW$ and $t = 0.25$ hours [1]
   $W = 9 \times 0.25 = 2.3\,kW\ h$ [1]
5. $P = \dfrac{W}{t}$ therefore $W = Pt$
   In kW h units, $P = 0.060\,kW$ and $t = 840$ hours [1]
   $W = 0.060 \times 840 = 50.4$ kW h [1]
   Cost $= 50 \times 0.112 = £5.60$ [1]
6. Filament lamps: in kW h units, $P = 18 \times 0.100 = 1.8\,kW$ and LEDs: in kW h units, $P = 18 \times 0.015 = 0.27\,kW$ [1]
   $t = 60 \times 60 \times 2 \times 365 = 2.62 \times 10^6\,s$ [1]
   Energy transferred by lamps $=$
   $W = 1.8 \times 10^3 \times 2.62 \times 10^6 = 4.716 \times 10^9\,J$ [1]
   Energy transferred by LEDs $=$
   $W = 0.27 \times 10^3 \times 2.62 \times 10^6 = 707.4 \times 10^6\,J$ [1]
   Energy saved $= 4.716 \times 10^9 - 707.4 \times 10^6 = 4.0$ GJ [1]

10.1

1. A: 2.0 A [1]
   B: 2.0 A [1]
   C: 0.6 A [1]
   D: 3.2 A [1]
   E: 1.2 A [1]
   F: 1.8 A [1]
2. A: 4.0 V [1]
   B: 12 V [1]
   C: 9.0 V [1]
   D: 8.0 V [1]
   E: 4.0 V [1]
   F: 4.0 V [1]
   G: 3.0 V [1]
3. a Two resistors in series with power supply. [1]
   5.0 V across each resistor. [1]
   b Two resistors in parallel with power supply. [1]
   10 V across each resistor. [1]

4 a Two resistors in series with power supply. 3.3 V across one resistor (the one with lower resistance) [1] and 6.7 V across the other resistor (the one with higher resistance). [1]

b Two resistors in parallel with power supply. 10 V across each resistor. [2]

5 Current in the added branch. [1]

Current in the previous branches remains unaffected. [1]

More current drawn from supply. [1]

6 a Correctly drawn circuit with two lamps on the first branch [1] and one lamp on the second. [1]

b First branch has twice the resistance as the second. [1]

Therefore, the lamps on this branch have half the current of the second branch through them. [1]

First branch 2.0 A through each lamp [1]

Second branch 4.0 A [1]

c $P = IV$

First branch: Current in each bulb = 2.0 A and p.d. = 6.0 V [1] therefore $P = IV = 2.0 \times 6.0 = 12$ W [1]

Second branch: Current in bulb = 4.0 A and p.d. = 12 V [1] therefore $P = IV = 4.0 \times 2 = 48$ W [1]

d As before, the first branch has twice the resistance as the second therefore the lamps on this branch have half the current of the second branch through them. [1]

First branch $\left(1\frac{2}{3}\right)$ 1.7 A through each lamp [1]

Second branch has twice the current. $\left(3\frac{1}{3}\right)$ 3.3 A [1]

10.2

1 a resistance increases. [1]

b resistance decreases. [1]

2 a $R = 5.0 + 9.0 = 14\,\Omega$ [1]

b $\dfrac{1}{R} = \dfrac{1}{6.0} + \dfrac{1}{4.0} = \dfrac{5}{12}$ [1]

therefore $R = \dfrac{12}{5} = 2.4\ \Omega$ [1]

c $\dfrac{1}{R} = \dfrac{1}{10} + \dfrac{1}{15} + \dfrac{1}{20} = \dfrac{13}{60}$ [1]

therefore $R = \dfrac{60}{13} = 4.6\ \Omega$ [1]

3 a $\dfrac{1}{R} = \dfrac{1}{(150+100)} + \dfrac{1}{200} = \dfrac{9}{1000}$ [1]

therefore $R = \dfrac{1000}{9} = 111\,\Omega$ (3 s.f.) [1]

b For the parallel part of the circuit

$\dfrac{1}{R} = \dfrac{1}{(20+10)} + \dfrac{1}{20} = \dfrac{1}{12}$ [1]

therefore $R = \dfrac{12}{1} + 60 = 72\,\Omega$ [1]

c For each branch $\dfrac{1}{R} = \dfrac{1}{10} + \dfrac{1}{10} + \dfrac{1}{10} = \dfrac{3}{10}$ [1]

therefore $R = \dfrac{10}{3} \times 3 = 10\ \Omega$ [1]

4. Single $50\,\Omega$

Single $100\,\Omega$

Single $200\,\Omega$ [1 mark in total for all three singles correct]

$50\,\Omega$ in series $100\,\Omega$ [1]

$50\,\Omega$ in series $200\,\Omega$ [1]

$200\,\Omega$ in series $100\,\Omega$ [1]

All three in series [1]

$50\,\Omega$ in parallel $100\,\Omega$ [1]

$50\,\Omega$ in parallel $200\,\Omega$ [1]

$200\,\Omega$ in parallel $100\,\Omega$ [1]

All three in parallel [1]

($50\,\Omega$ in parallel $100\,\Omega$) in series with $200\,\Omega$ [1]

($50\,\Omega$ in parallel $200\,\Omega$) in series with $100\,\Omega$ [1]

($200\,\Omega$ in parallel $100\,\Omega$) in series with $50\,\Omega$ [1]

($50\,\Omega$ in series $100\,\Omega$) in parallel with $200\,\Omega$ [1]

($50\,\Omega$ in series $200\,\Omega$) in parallel with $100\,\Omega$ [1]

($200\,\Omega$ in series $100\,\Omega$) in parallel with $50\,\Omega$ [1]

5 $\dfrac{1}{R} = \dfrac{1}{R_1} + \dfrac{1}{R_2} + \dfrac{1}{R_3}$ therefore $\dfrac{1}{R} - \left(\dfrac{1}{R_1} + \dfrac{1}{R_2}\right) = \dfrac{1}{R_3}$ [1]

$\dfrac{1}{1030} - \left(\dfrac{1}{2200} + \dfrac{1}{4700}\right) = \dfrac{1}{R_3} = 303... \times 10^{-6}\,\Omega^{-1}$ [1]

$R_3 = \dfrac{1}{303... \times 10^{-6}} = 3290\,\Omega$ (3 s.f.) [1]

10.3

1 Four marks for all 7, three marks for 6 correct, two marks for 5 correct, one mark for 4 correct, zero marks for less than 4 correct.

I: current

V: potential difference

W: work done/energy transferred

P: power

R: resistance

Q and ΔQ: charge

t and Δt: time

2 a $\Delta Q = I\Delta t$ [1]

$\Delta Q = 0.30 \times 45 = 14$ C (2 s.f.) [1]

b $W = Pt$ [1]

$W = 0.20 \times 120 = 24$ J [1]

3 Current through resistor B = 0.50 − 0.40 = 0.10 A [1]

b Resistor A: $V = IR$ [1]

therefore $V = 0.50 \times 5.0 = 2.5$ V [1]

Resistor D: $= IV$, $V = \dfrac{P}{I}$ [1]

therefore $V = \dfrac{0.50}{0.50} = 1.0$ V [1]

c $R = \dfrac{V}{I} = \dfrac{1.0}{0.50} = 2.0\,\Omega$ [2]

4 Resistor A: Using $P = I^2R$, $P = 0.50^2 \times 5.0$
= 1.3 W (2 s.f.) [1]

Resistor B: Using $P = I^2R$, $P = 0.10^2 \times 8.0 = 80$ mW [1]

Resistor C: The p.d across B,
$V = IR = 0.10 \times 8.0 = 0.80$ V = p.d across C [1]
$P = 0.40 \times 0.80 = 320$ mW [1]

Resistor D: 0.50 W
Total power = 1.3 + 0.080 + 0.32 + 0.50 = 2.2 W [1]

therefore $V = \dfrac{P}{I} = \dfrac{2.2}{0.50} = 4.4$ V [1]

10.4

1 **a** $\varepsilon = V + Ir$ therefore $V = \varepsilon - Ir$

Total e.m.f = 1.5 + 1.5 = 3.0 V

Total internal resistance = 0.75 + 0.75 = 1.5 Ω

$V = \varepsilon - Ir = 3.0 - (0.80 \times 1.5) = 1.8$ V

b Find the current in the cell. $\varepsilon = I(R + r)$ therefore
$\dfrac{\varepsilon}{(R + r)} = I = \dfrac{3.0}{(11.5)} = 0.26...$ A

$V = \varepsilon - Ir = 3.0 - (0.26... \times 1.5) = 2.6$ V (2 s.f.)

2 **a** Total e.m.f = 1.5 V

Total internal resistance = $\dfrac{1}{R} = \dfrac{1}{R_1} + \dfrac{1}{R_2}$,
$R = 0.38$ Ω

$V = \varepsilon - Ir = 1.5 - (0.80 \times 0.38) = 1.2$ V

b Find the current in the cell. $\varepsilon = I(R + r)$ therefore
$\dfrac{\varepsilon}{(R + r)} = I = \dfrac{1.5}{(10.38)} = 0.14$ A

$V = \varepsilon - Ir = 1.5 - (0.14 \times 0.38) = 1.4$ V

---

1 Graph of terminal p.d. against I with axes labelled
Axes include correct units
Points plotted correctly
Line of best fit drawn

2 Gradient is constant.

3 Intercept = e.m.f = 1.6 V
Gradient = internal resistance = 0.50 Ω

4 To ensure it does not heat up, changing its internal resistance.

---

1 A low internal resistance is needed to provide a large current [1], for example, in a car battery.

A high internal resistance is needed to ensure a high current cannot be produced (for safety reasons) [1]
For example in a school high voltage power supply. [1]

2 As the current increases the terminal p.d decreases. [2]

3 **a** Lost volts = $Ir = 1.5 \times 2.0 = 3.0$ V [1]

Terminal p.d. = e.m.f – lost volts = 9.0 – 3.0
= 6.0 V [1]

Using $P = I^2R$, $P = 1.5^2 \times 2.0 = 4.5$ W [1]

Using $V = IR$, $R = \dfrac{6.0}{1.5} = 4.0$ Ω [1]

---

4 **a** For the parallel part of the circuit
$\dfrac{1}{R} = \dfrac{1}{90} + \dfrac{1}{45} = \dfrac{1}{30}$ therefore $R = \dfrac{30}{1} + 50 = 80$ Ω [2]

b Terminal p.d. = $IR = 0.10 \times 80 = 8.0$ V [1]

Lost volts = 12 – 8.0 = 4.0 V [1]

c Lost volts = Ir therefore
$r = \dfrac{\text{lost volts}}{I} = \dfrac{4.0}{0.1} = 40$ Ω [1]

5 Graph of terminal p.d against I with axes labelled (include correct units), showing a straight line with a negative gradient [1]

Y-axis intercept labelled e.m.f [1]

Gradient labelled -r [1]

Second line with double the e.m.f and double the gradient [1]

10.5

1 The resistance of this part of the potential divider drops. Lowering V_{out}.

2 **a** $V_{out} = \dfrac{R_2}{(R_1 + R_2)} \times V_{in}$

$= \dfrac{4700}{(2200 + 4700)} \times 12 = 8.2$ V (2 s.f.)

b Resistance of the loaded part of the circuit
$= \dfrac{1}{R} = \dfrac{1}{4700} + \dfrac{1}{10000} = \dfrac{147}{470000}$ therefore

$R = \dfrac{470000}{147} = 3197.2...$ Ω

$V_{out} = \dfrac{R_2}{(R_1 + R_2)} \times V_{in}$

$= \dfrac{3197.2...}{(2200 + 3200)} \times 12 = 7.1$ V (2 s.f.)

c Resistance of the loaded part of the circuit =
$\dfrac{1}{R} = \dfrac{1}{4700} + \dfrac{1}{100} = \dfrac{12}{1175}$ therefore

$R = \dfrac{1175}{12} = 97.9...$ Ω

$V_{out} = \dfrac{R_2}{(R_1 + R_2)} \times V_{in}$

$= \dfrac{97.9...}{(2200 + 98)} \times 12 = 0.51$ V (2 s.f.)

---

1 Potential divider circuit drawn with two resistors in series. [1]

Diagram labelled with 20 V V_{in} and V_{out} across one of the resistors. [1]

Both resistors must have the same resistance. [1]

Therefore, the p.d. will be shared equally between them, each one receiving 10 V. [1]

2 **a** $V_{out} = \dfrac{R_2}{(R_1 + R_2)} \times V_{in} = \dfrac{90}{(270 + 90)} \times 12 = 3.0$ V [2]

$V_{out} = \dfrac{R_2}{(R_1 + R_2)} \times V_{in} = \dfrac{120}{(30 + 120)} \times 60 = 48$ V [2]

3 **a** Potential divider with two resistors drawn correctly. V_{out} connected across $30\,\Omega$ resistor [1]

b Potential divider with two resistors drawn correctly. V_{out} connected across $90\,\Omega$ resistor [1]

4 $\dfrac{V_1}{V_2} = \dfrac{R_1}{R_2}$ therefore $R_2 = \dfrac{R_1 \times V_2}{V_1}$ [1]

$V_1 = 360 - 3.0 = 357\,V$ [1]

$R_2 = \dfrac{110 \times 3.0}{357} = 0.92\,\Omega$ (2 s.f.) [1]

10.6

> **1** Potential divider containing a resistor and thermistor in series.
> V_{out} connected across the resistor.
>
> **2** Maximum:
> $V_{out} = \dfrac{R_2}{(R_1 + R_2)} \times V_{in} = \dfrac{50 \cdot 10^6}{(1000 + 50 \cdot 10^6)} \times 9.0$
>
> $= 9.0\,V$ $(8.9998\; V)$
>
> Minimum: $V_{out} = \dfrac{R_2}{(R_1 + R_2)} \times V_{in}$
>
> $= \dfrac{500}{(1000 + 500)} \times 9.0 = 3.0\,V$

1 Any valid two: [2]

Can be made very compact

Uses fewer components

Can easily be made into a rotary dial.

Allows the full range of output potential difference from 0 V to V_{in}.

2 Potential divider containing a resistor and an LDR in series. [2]

V_{out} connected across the LDR. [1]

3 **a** $V_{2200} = 12 - 6.0 = 6.0\,V$ [1]

$R_2 = \dfrac{2200 \times 6.0}{6.0} = 2200\,\Omega$ [1]

(As 6.0 V is half of V_{in}, the resistance of the thermistor must equal the resistance of the resistor)

b $V_{2200} = 12 - 10 = 2.0\,V$ [1]

$R_2 = \dfrac{2200 \times 10}{2.0} = 11000\,\Omega$ [1]

c $V_{2200} = 12 - 1.0 = 11\,V$ [1]

$R_2 = \dfrac{2200 \times 1.0}{11} = 200\,\Omega$ [1]

4 Increasing the temperature would increase the value of V_{out}. [2]

5 **a** Graph of R against T with axis labelled (include correct units) and a smooth curve of decreasing gradient [1]

Values at 0°C and at 100°C drawn correctly. [1]

b Find V_1 when $V_{out} = 4.0\,V$. $V_1 = 12 - 4.0 = 8.0\,V$ [1]

$\dfrac{V_1}{V_2} = \dfrac{R_1}{R_2}$ therefore $R_2 = \dfrac{R_1 \times V_2}{V_1}$.

$R_2 = \dfrac{220 \times 4.0}{8.0} = 110\,\Omega$ [1]

Read value of temperature at $110\,\Omega$. [1]
(around 30 °C – depending on the sketch graph)

6 Allows V_{out} to be varied [1] at a specific temperature. [1]

11.1

1 Any valid three, for example (answers must contain at least one similarity and one difference) [3]

Similarities:

Progressive waves

Transfer energy

Differences:

Transverse wave – oscillations are perpendicular to the direction of the wave's movement.

Longitudinal wave – oscillations are parallel to the direction of the wave's movement.

Transverse wave – contains peaks and troughs.

Longitudinal wave – contains compressions and rarefactions.

2 Transverse wave – Fix one end of the slinky, hold the other end, move this end perpendicular to the body of the slinky. [1]

Longitudinal wave – Fix one end of the slinky, hold the other end, move this end parallel to the body of the slinky. [1]

3 A: Vertically downwards [1]

B: Vertically upwards [1]

C: Vertically downwards [1]

4 Particles are closer together [1]

Stronger restoring force/vibrations are passed more rapidly from one particle to the next. [1]

5 Diagram should include: Compressions [1], Rarefactions [1], particles vibrating parallel to the direction of energy transfer [1]

11.2

> **1** **a** $\varphi = \dfrac{x}{\lambda} \times 360^\circ = \dfrac{20}{40} \times 360^\circ = 180^\circ$
>
> **b** $\varphi = \dfrac{x}{\lambda} \times 360^\circ = \dfrac{40}{40} \times 360^\circ = 360^\circ$
>
> **c** $\varphi = \dfrac{x}{\lambda} \times 360^\circ = \dfrac{80}{40} \times 360^\circ = 720^\circ$
>
> **2** **a** $x = \dfrac{\varphi \times \lambda}{360^\circ} = \dfrac{90^\circ \times 1.60}{360^\circ} = 40\,cm$
>
> **b** $x = \dfrac{\varphi \times \lambda}{360^\circ} = \dfrac{540^\circ \times 1.60}{360^\circ} = 2.4\,m$
>
> **c** $5\pi\,rad = 900^\circ$ $x = \dfrac{\varphi \times \lambda}{360^\circ} = \dfrac{900^\circ \times 1.60}{360^\circ}$
> $= 4.0\,m$

1 Increasing the timebase results in a smaller
   time period for each square on the screen.
   The wave trace will appear more compressed.

2 Period of oscillation = 0.02 s, therefore one complete
   cycle will be completed in each square.

1 $v = f\lambda$ [1]

   $v = 2.0 \times 0.50 = 1.0 \text{ m s}^{-1}$ [1]

2 A: down [1]

   D: up [1]

3 Connect a microphone to an oscilloscope. [1]
   Blow the whistle and record the number of divisions
   n between successive peaks of the signal displayed on
   the oscilloscope. [1]

   Find the period T by multiplying n by the time base. [1]

   Frequency f of the sound calculated using $f = 1/T$. [1]

4 $f = \dfrac{1}{T}$ [1]

   $f = \dfrac{1}{2.0 \times 10^{-3}} = 500 \text{ Hz}$ [1]

   $v = f\lambda$ therefore $\lambda = \dfrac{v}{f}$ [1]

   $\lambda = \dfrac{340}{500} = 0.68 \text{ m}$ [1]

5 a i Same shaped wave profile as in question. [1]

         Wave profile shifted a quarter-cycle to
         the right. [1]

      ii Same shaped wave profile as in question. [1]

         Wave profile shifted a half-cycle to the right. [1]

   b i 0.3 m [1]

      ii 0.0 m [1]

      iii 0.3 m [1]

6 a 90° or $\dfrac{\pi}{2}$ rad [1]

   b 180° or π rad [1]

   c 270° or $\dfrac{3}{2}\pi$ rad [1]

11.3

1 Any valid three, for example (answers must contain
   at least one similarity and one difference) [3]

   Similarities:

   Property of all waves

   Frequency does not change

   Differences:

   Reflection – speed and wavelength do not change

   Refraction – speed and wavelength change

   Reflection – wave does not change medium

   Refraction – wave changes from one medium to
   another

2 Normal drawn in each example (at 90° to the
   surface). [1]

In each case the angle of incidence = angle of
reflection. [3]

3 Normal drawn correctly and partial reflection
   shown in diagram. [1]

   Ray of light bends away from the normal. [1]

4 $v = f\lambda$ therefore if f is constant [1]
   $v \propto \lambda$ [1]

5 General shape, including a different wavelength
   at each end of the pool [1]

   λ increases moving towards the deep end [1]

   λ reduces moving towards the shallow end [1]

11.4

1 Only transverse waves can be plane polarised. [1]

   Sound is a longitudinal wave. [1]

   Therefore, sound waves cannot be plane polarised

2 Any two examples of transverse waves: [2]

   e.g.

   (Visible) Light; Microwaves; Radio waves; Infrared
   waves; Ultraviolet; S waves

3 Diffraction effects are most significant when the
   wavelength is a similar size to the gap/obstacle. [1]

   Wavelength of light is much smaller than most
   gaps / sound waves have a larger wavelength. [1]

4 3.0 m wave diffracts more. [1]

   3.0 cm wave does not really diffract as the wavelength
   is much smaller than the size of the gap. [1]

   3.0 m wave diffracts significantly as the
   wavelength is the same size of the gap. [1]

5 Higher frequency means a smaller wavelength. [1]

   Radio waves have a longer wavelength; therefore,
   they diffract over the hill reaching the bottom of
   the valley. [1]

   The TV signal has a shorter wavelength; therefore,
   does not diffract as significantly, failing to reach
   the bottom of the valley. [1]

6 a $\lambda = \dfrac{v}{f} = \dfrac{340}{1200} = 0.28 \text{ m}$ (2 s.f.) [1]

      Very similar to the gap therefore significant
      diffraction. [1]

   b $\lambda = \dfrac{v}{f} = \dfrac{340}{1.0 \times 10^6} = 340 \mu\text{m}$ [1]

      Much smaller than the gap therefore no
      diffraction. [1]

11.5

1 Graph of intensity against distance with axis
   labelled (including units), points plotted correctly
   and line of best fit drawn.

2 Use $I = \dfrac{k}{distance^2}$ to check if the data follows an
   inverse square relationship.

1 **a** Intensity increases by a factor of 9 (3^2). [1]

   b Intensity decreases by a factor of 16 (4^2). [1]

2 $I = \dfrac{P}{A}$ [1]

   $I = \dfrac{P}{A} = \dfrac{400}{20} = 20\,\text{W m}^{-2}$ [1]

3 $I = \dfrac{P}{4\pi r^2}$ [1]

   $I = \dfrac{60}{4 \times \pi \times (20)^2}$ [1]

   $I = 12\,\text{mW m}^{-2}$ (2 s.f.) [1]

4 From $I = \dfrac{P}{A}$ if the power is constant [1]

   then: $I \propto \dfrac{1}{A}$ [1]

   As the area reduces [1]

   Intensity increases. [1]

5 Intensity = $1.4\,\text{kW m}^{-2}$ therefore power received by each $8.0\,\text{m}^2$ panel is: $1400 \times 8.0 = 11\,200\,\text{W}$ [1]

   Total power received = $11\,200 \times 2 = 22\,400\,\text{W}$ [1]

   $P = \dfrac{W}{t}$ therefore $W = Pt$ [1]

   $W = 22\,400 \times 7200 = 160$ MJ [1]

6 **a** $I = \dfrac{P}{4\pi r^2}$ therefore $P = I \times 4\pi r^2$ [1]

   $P = 1.0 \times 10^{-4} \times 4 \times \pi \times (15)^2 = 0.28$ W (2 s.f.) [1]

   $I = \dfrac{P}{4\pi r^2}$ therefore the intensity at 120 m

   $= I = \dfrac{0.28}{4 \times \pi \times 120^2} = 1.6 \times 10^{-6}\,\text{W m}^{-2}$ (2 s.f.) [1]

   b Intensity has fallen by a factor of 65. [1]

   As intensity $\propto$ (amplitude)^2 the amplitude will have decreased by a factor of 8.1 ($\sqrt{65}$) (2 s.f.) [1]

1 Lower chance of cloud cover, intensity of electromagnetic waves is greater (less energy absorbed by the atmosphere)

2 All EM waves can be detected, no atmospheric distortion, no weather/cloud cover
Drawbacks: Cost, difficult to repair

3 **a** **i** Gamma rays, X-rays, Ultraviolet, longer λ of IR, radio waves longer than around 10 m

   ii most wavelengths of visible light, microwaves, radio waves up to 10 m

   b $v = f\lambda$ therefore if $f = \dfrac{v}{\lambda}$

   Highest frequency corresponds to a λ of around 200 nm:

   $f = \dfrac{v}{\lambda} = \dfrac{3.00 \times 10^8}{200 \times 10^{-9}} = 1.5 \times 10^{15}$ Hz

   Lowest frequency corresponds to a λ of around

   10 m: $f = \dfrac{v}{\lambda} = \dfrac{3.00 \times 10^8}{10} = 30\,\text{MHz}$

11.6

1 Gamma rays, X-rays, ultraviolet, visible light, infrared, microwaves, radio waves [2]

   (1 mark if one incorrect, 0 marks if more than one incorrect)

2 Polarisation [1]

3 **a** $\lambda = \dfrac{v}{f} = \dfrac{3.00 \times 10^8}{88 \times 10^6} = 3.4$ m (2 s.f.) [1]

   b $\lambda = \dfrac{v}{f} = \dfrac{3.00 \times 10^8}{2.4 \times 10^9} = 0.13$ m (2 s.f.) [1]

   c $\lambda = \dfrac{v}{f} = \dfrac{3.00 \times 10^8}{9.0 \times 10^{16}} = 3.3$ nm (2 s.f.) [1]

4 $v = f\lambda$ therefore if $f = \dfrac{v}{\lambda}$

   Highest frequency:

   $f = \dfrac{v}{\lambda} = \dfrac{3.00 \times 10^8}{400 \times 10^{-9}} = 7.5 \times 10^{14}$ Hz [1]

   Lowest frequency:

   $f = \dfrac{v}{\lambda} = \dfrac{3.00 \times 10^8}{700 \times 10^{-9}} = 4.3 \times 10^{14}$ Hz (2 s.f.) [1]

5 $t = \dfrac{distance\ travelled}{speed} = \dfrac{150 \times 10^9}{3.00 \times 10^8}$ [1]

   $t = 500\,\text{s}$ [1]

6 distance travelled = speed × time taken [1]

   Time taken for pulse to reach aircraft = 0.28 μs [1]

   distance travelled = $3.00 \times 10^8 \times 0.28 \times 10^{-6}$ = 84 m [1]

1 When the Polaroids are aligned the intensity is at a maximum value.
As one is rotated the intensity is reduced.
When the second Polaroid is aligned at 90° to the first, the intensity is zero.

2 As the grille is rotated the intensity falls.
It falls to zero when the gaps in the grille are in the opposite plane to the microwaves (after 90°).
As the grille is rotated further the intensity increases again.
It reaches the maximum value again when the gaps in the grille are in the same plane as the microwaves (after 180°).

3 Holes will not allow any orientation of plane polarised microwaves through.

11.7

1 Only transverse waves can be plane polarised. [1]

2 At 0°, 180°, and 360° the Polaroids are aligned, so the maximum intensity is received. [1]

   At 90° and 270° the Polaroids are aligned in opposite planes, so the intensity falls to zero. [1]

3 **a** It must be plane polarised. [1]

    b Rotate the Polaroid further [1] until it is at 90° from the minimum intensity. [1]

       In this orientation the Polaroid is aligned in the same plane as the light emitted from the screen. [1]

4 **a** $20 \times 9.0 \times 10^{-4} = 1.8 \times 10^{-3}$ W [1]

    b $\dfrac{100}{28} = 3.6$ (2 s.f.) [1]

11.8

1 As the speed of light through the material decreases the refractive index increases. [1]

    From $n = \dfrac{c}{v}$ [1] the refractive index is inversely proportional to the speed of light through the material $n \propto \dfrac{1}{v}$ [1]

2 $n = \dfrac{c}{v}$ [1]

    $n = \dfrac{3.00 \times 10^8}{220 \times 10^6} = 1.4$ (2 s.f.) [1]

3 $n = \dfrac{c}{v}$ therefore $v = \dfrac{c}{n}$ [1]

    $v = \dfrac{3.00 \times 10^8}{1.33}$ [1]

    $v = 2.3 \times 10^8 \, \mathrm{m\,s^{-1}}$ (2 s.f.) [1]

4 $n_1 \sin\theta_1 = n_2 \sin\theta_2$ therefore $n_2 = \dfrac{n_1 \sin\theta_1}{\sin\theta_2}$ [1]

    $n_2 = \dfrac{1.10 \times \sin 51}{\sin 36} = 1.5$ (2 s.f.) [1]

5 Normal and partial reflection shown [1]

    Angle θ_2 is less than θ_1 [1]

    Angle $\theta_2 = 43°$ [2]

    (from workings below)

    $n_1 \sin\theta_1 = n_2 \sin\theta_2$ therefore $\sin\theta_2 = \dfrac{n_1 \sin\theta_1}{n_2}$

    $= \dfrac{1.47 \times \sin 45}{1.52} = 0.638...$

    Angle $\theta_2 = \sin^{-1} 0.638.. = 43°$

6 $n_1 \sin\theta_1 = n_2 \sin\theta_2$ therefore $\sin\theta_2 = \dfrac{n_1 \sin\theta_1}{n_2}$ [1]

    $\sin\theta_2 = \dfrac{2.42 \times \sin 20}{1.33} = 0.622..$ [1]

    $\theta_2 = \sin^{-1} 0.622... = 38°$ (2 s.f.) [1]

11.9

    1 In order for TIR the light must be travelling from a material of higher refractive index to one of a lower refractive index.

    2 The pulse reflected multiple times arrives after the pulse transmitted through the centre of the fibre. The pulse reflected multiple times has a lower amplitude when it leaves the fibre than the pulse transmitted through the centre of the fibre.

    3 Light follows a curved path (sinusoidal).

1 From $\sin C = \dfrac{1}{n}$ [1]

    If the refractive index decreases the critical angle increases. [1]

2 $\sin C = \dfrac{1}{n}$ therefore $C = \sin^{-1}\left(\dfrac{1}{n}\right)$ [1]

    $C = \sin^{-1}\left(\dfrac{1}{2.42}\right) = 24°$ (2 s.f.) [1]

3 $\sin C = \dfrac{1}{n}$ therefore $n = \dfrac{1}{\sin C}$ [1]

    $n = \dfrac{1}{\sin 42.8} = 1.47$ (3 s.f.) [1]

4 $C = 36°$ [1]

    $\sin C = \dfrac{1}{n}$ therefore $n = \dfrac{1}{\sin C}$

    $n = \dfrac{1}{\sin 36}$ [1]

    $n = 1.7$ (2 s.f.) [1]

5 **a** At the critical angle therefore light travels along boundary. [2]

    b Below the critical angle therefore light is partially reflected and refracted. [2]

    c Above the critical angle therefore light is totally internally reflected. [2]

6 $n = \dfrac{c}{v} = \dfrac{3.00 \times 10^8}{185 \times 10^6} = 1.62...$ [1]

    $\sin C = \dfrac{1}{n}$ therefore $C = \sin^{-1}\left(\dfrac{1}{n}\right)$ [1]

    $C = \sin^{-1}\left(\dfrac{1}{1.62...}\right) = 38.0°$ (3 s.f.) [1]

12.1

1 Destructive interference [1]

    In order to cancel out the sound waves from the surrounding environment [1]

2 As figure 3. [2]

3 intensity $\propto$ (amplitude)^2 [1]

    As the amplitude doubles the intensity will increase by a factor of 4 (2^2) [1]

    [1 mark only if mentioned only increasing]

4 Two clearly distinct waves with the correct wavelength [1] and amplitude. [1]

    Waves have a clear sinusoidal shape. [1]

    Constructive interference where the waves both have positive displacement [1] and where the waves both have negative displacement. [1]

    Destructive interference where one wave has positive displacement and the other has negative displacement. [1]

5 See Figure 1. [6]

12.2

1 Path difference at the 1st order maxima = 1λ.

$$f = \frac{v}{\lambda} = \frac{340}{0.28} = 1200\,\text{Hz} \ (2\ \text{s.f.})$$

2 If frequency is halved, the wavelength is doubled. Maxima and minima would be further apart (the path difference at the 1st order maxima would be doubled).

1 Use a ruler to carefully measure the path from the centre of the first slit to the 1st order maxima. Use a ruler to carefully measure the path from the centre of the second slit to the 1st order maxima. Calculate the difference.

Path difference at the 1st order maxima = 1λ.

Repeat for different maxima and minima (being careful to relate the path difference to the wavelength (e.g., path difference at the 2nd order maxima = 2λ, etc)). Average the values for the wavelength.

2 Most significant diffraction when the gap size = wavelength.

$$\lambda = \frac{v}{f} = \frac{3.00 \times 10^8}{24 \times 10^9} = 1.3\,\text{cm} \ (2\ \text{s.f.})$$

3 Use the metal sheets to create a single gap. Place in front of the microwave source and rotate the source through $180°$. If the intensity received by the receiver drops then rises again the source is plane polarised.

1 Diagram showing light reflecting off the top surface (obeying the law of reflection) and light reflecting off the bottom surface after travelling through the oil (refracting on entry and exit).

2 The light travelling through the oil travels a greater distance (approx. 2 x the thickness of the oil).

3 Wavelength of red light is greater than blue light.

The path difference must be greater in order to produce destructive interference; therefore, the oil needs to be thicker.

1 a Constructive [1]
  b Constructive [1]
  c Destructive [1]
2 Increases [1]
  Zero in the centre (at the central maxima) [1]
  Moving through $180°$ or π rad at the first-order minima, $360°$ or 2π rad at the first-order maxima, etc [1]
  Reaching $1080°$ or 6π rad at the third-order maxima. [1]
3 At the second-order maxima the path difference = 2λ [1]
  Therefore, the wavelength = 4.5 cm [1]

4 a the wavelength of the sound;
    $$\lambda = \frac{v}{f} = \frac{340}{2000} = 0.17\,\text{m}$$ [1]
  b At a phase difference of 5π radians the path difference = 2.5λ [1]
    Path difference = $2.5 \times 0.17 = 0.43$ m [1]
  c At the second-order minima the phase difference = 3π radians [1]
    At a phase difference of 3π radians the path difference = 1.5λ [1]
    Path difference = $1.5 \times 0.17 = 0.26\,\text{m}$ [1]

12.3

1 There would be green light in place of red. The separation between fringes would be smaller due to the shorter wavelength.

2 This would reduce percentage uncertainty in measurements as the distance will be greater.

3 $\lambda = \frac{ax}{D}$ therefore $x = \frac{\lambda D}{a}$

   $$x = \frac{632.8 \times 10^{-9} \times 10}{0.50 \times 10^{-3}} = 13\,\text{mm}$$

1 Waves must be coherent to form a stable interference pattern [1]

   Using a monochromatic source, a single slit and double slit results in two sources of coherent light [1]

2 $\lambda = \frac{ax}{D}$ [1]

   $$\lambda = \frac{0.6 \times 10^{-3} \times 1.4 \times 10^{-3}}{1.6} = 530\,\text{nm}$$ [1]

3 $x = 8.3$ mm [1]

   x measured across several fringes [1]

   $$\lambda = \frac{1.0 \times 10^{-3} \times 8.3 \times 10^{-3}}{15} = 550\,\text{nm} \ +/-\ 50\,\text{nm}$$ [1]

4 $\lambda = \frac{ax}{D}$ therefore $D = \frac{ax}{\lambda}$ [1]

   $$D = \frac{0.40 \times 10^{-3} \times 1.8 \times 10^{-3}}{610 \times 10^{-9}}$$ [1]

   $D = 1.2$ m (2 s.f.) [1]

5 a As λ increases x increases. [1]
  b $x = \frac{\lambda D}{a}$ therefore if other factors remain constant $x \propto \frac{1}{a}$ [1]
    As a doubles x halves. [1]
  c $x = \frac{\lambda D}{a}$ therefore if other factors remain constant $x \propto D$ [1]
    As D increases by a factor of 3 x increases by a factor of 3. [1]
  d Double the frequency and the wavelength halves. [1]
    $x = \frac{\lambda D}{a}$ therefore if other factors remain constant $x \propto \lambda$ [1]
    As λ halves x halves. [1]

12.4

1 Measure the distance between adjacent nodes (or antinodes) – this is equal to $\frac{\lambda}{2}$.
Multiply the average distance by 2.

2 $\lambda = \frac{v}{f} = \frac{3.00 \times 10^8}{5.0 \times 10^9} = 0.060\,\text{m}$

Therefore the distance between nodes = 0.030 m

3 Close to the metal sheet both waves which form the stationary wave have similar amplitudes (they have travelled similar distances from the source).
At other nodes there is a greater difference between the amplitudes of the two waves (as they have travelled different distances). Resulting in non-perfect cancellation.

1 Comprised of oscillations/vibrations. [1]

Particles in the waves have frequency, period of oscillation and amplitude. [1]

2 a They are in antiphase [1]
b They are in phase [1]

3 The distance between adjacent nodes is equal to $\frac{\lambda}{2}$. [1]
The wavelength of the parent waves = 0.30 × 2 = 0.60 m [1]

4 At the node the amplitude = 0 [1]

Moving away from the node the amplitude increases (reaching a maximum at the antinode). [1]

Moving past the antinode, the amplitude reduces back to zero at the node. [1]

5 a (i) 180° or π rad. [1]
(ii) 360° or 2π rad. [1]
b (i) Flat line [1]
(ii) Simple sinusoidal wave (two waves on top of each other) [2]
(iii) As first diagram [1]

12.5

1 107 Hz

2 See Table 1
a At fundamental frequency
b At 2nd harmonic
c No pattern forms (not an integer multiple of the fundamental frequency)

3 As T increases v increases. From $v = f\lambda$ as λ is constant (as the length of the string is constant), f increases. As T increases by a factor of 2, f increases by a factor of $\sqrt{2}$.

1 A stretched string fixed at one end [1] with a vibration generator (connected to a signal generator) at the other end. [1]
Adjust the frequency of the signal generator. [1]

2 If L increases f_o decreases. [1]
If L doubles f_o halves. [1]

3 a 3rd harmonic, therefore the fundamental frequency must be $\frac{120}{3} = 40$ Hz. [1]
b Length of string = 0.36 m, therefore the distance between nodes = 0.12 cm [1]
Wavelength = 2 × 0.12 = 0.24 cm [1]
(Using $\lambda = \frac{2}{3}L$ results in both marks)
c (i) 4th harmonic, see Table 1 [1]
(ii) no pattern formed (not an integer multiple of f_o) [1]

4 a 0.5 Hz +/- 0.05 Hz [1]
b Intensity increases at integer multiple of f_o. [1]
Due to the string also vibrating at these harmonics. [1]

5 a 6th harmonic, node to node distance = $\frac{0.90}{6} = 0.15$ m [1]
Wavelength = 2 × 0.15 = 0.30 cm [1]
b $v = f\lambda$ [1]
$v = 3600 \times 0.30 = 1080$ m s^{-1} [1]

12.6

1 See figure 3 at the fundamental frequency
2 a gradient = Lf
The wavelength of the progressive waves = 4L.
$v = f\lambda$, becomes $v = f \times 4L$ therefore $\frac{v}{4} = fL$.
b Gradient = 85 m s^{-1} +/– 5 m s^{-1}
Therefore speed of sound = 85 × 4 = 340 m s^{-1}
c Tuning fork is slightly above the tube.
Measurements of L contain slight systematic error and are shorter than $\frac{\lambda}{4}$.

1 Connect speaker to signal generator. [1]
Position speaker in front of solid surface, with microphone in between. [1]
Adjust frequency of sound until a number of nodes and antinodes are detected using the microphone. [1]
Measure the distance between nodes = $\frac{\lambda}{2}$. Therefore the wavelength of the sound waves = node-to-node distance × 2 [1]

2 a $L = \frac{\lambda}{4}$ therefore $\lambda = 4L$ [1]
$\lambda = 4 \times 1.2 = 4.8$ m [1]
b $f = \frac{v}{\lambda} = \frac{340}{4.8} = 71$ Hz [1]

3    $L = \dfrac{\lambda}{2}$ therefore $\lambda = 2\,L$ [1]

    $\lambda = 2 \times 1.2 = 2.4$ m   [1]

    $f = \dfrac{v}{\lambda} = \dfrac{340}{2.4} = 142$ Hz (3 s.f.)   [1]

4   Nodes: B and E   [2]

    Antinodes: A, D and G   [3]

    (1 mark for each correctly identified. Deduct 1 mark for each incorrect letter, minimum mark = 0)

5   At $2f_0$ there would need to be an antinode at each end of the tube.   [1]

    In a closed tube there must be a node at the closed end.   [1]

13.1

1   a $E = 6.63 \times 10^{-34} \times 1.02 \times 10^{14}$

       $= 6.76 \times 10^{-20}$ J (3 s.f.)   [1]

   b $E = 6.63 \times 10^{-34} \times 97.0 \times 10^{6}$

       $= 6.43 \times 10^{-26}$ J (3 s.f.)   [1]

   c $E = 6.63 \times 10^{-34} \times 6.00 \times 10^{14}$

       $= 3.98 \times 10^{-19}$ J (3 s.f.)   [1]

2   Violet [1]   Highest frequency. [1]

    From $E = hf$, the higher the frequency the greater the energy.   [1]

3    $E = \dfrac{hc}{\lambda}$ therefore $\lambda = \dfrac{hc}{E}$ [1]

    $\lambda = \dfrac{6.63 \times 10^{-34} \times 3.00 \times 10^{8}}{3.32 \times 10^{-18}}$ [1]

    $\lambda = 59.9$ nm (3 s.f.)   [1]

4   a 6.3×10^{18} eV (2 s.f.)   [1]

   b   206 keV (3 s.f.)   [1]

   c 3.8×10^{12} eV (2 s.f.)   [1]

5   Note, the exact values in this question may vary depending on the method used to calculate the frequency.

    Radio: $f = 9.7$ MHz   [1]

    Infrared: $f = 2.5 \times 10^{13}$ Hz   [1]

    Visible – red: $f = 4.5 \times 10^{14}$ Hz   [1]

    Visible – green: $f = 5.6 \times 10^{14}$ Hz   [1]

    Visible – blue: $f = 6.5 \times 10^{14}$ Hz   [1]

    UV: $f = 1.4 \times 10^{15}$ Hz   [1]

    X-ray: $f = 2.5 \times 10^{17}$ Hz   [1]

    gamma: $f = 3.6 \times 10^{20}$ Hz   [1]

6   a $E = \dfrac{6.63 \times 10^{-34} \times 3.00 \times 10^{8}}{4.50 \times 10^{-10}} = 4.42 \times 10^{-16}$ J   [1]

       $= 2760$ eV   [1]

   b $E = \dfrac{6.63 \times 10^{-34} \times 3.00 \times 10^{8}}{600 \times 10^{-9}} = 3.3 \times 10^{-19}$ J   [1]

       $= 2.1$ eV   [1]

7   a $eV = \dfrac{hc}{\lambda}$ [1]

     $V = \dfrac{hc}{e\lambda} = \dfrac{6.63 \times 10^{-34} \times 3.00 \times 10^{8}}{1.60 \times 10^{-19} \times 620 \times 10^{-9}}$

     $= 2.0$ V (2 s.f.)   [1]

   b   General shape of I–V characteristic for diode (see topic 9.6)   [3]

     Threshold p.d. lower for the red LED   [1]

     Red photons have a lower frequency [1], therefore less energy (lower threshold p.d) is required   [1]

8   Energy of each photon,

    $E = \dfrac{hc}{\lambda} = \dfrac{6.63 \times 10^{-34} \times 3.00 \times 10^{8}}{405 \times 10^{-9}} = 4.91... \times 10^{-19}$ J   [1]

    10 mW $= 10 \times 10^{-3}$ J s^{-1} [1]

    Number of photons $= \dfrac{10 \times 10^{-3}}{4.91... \times 10^{-19}}$

    $= 2.0 \times 10^{-16}$ photons per second (2 s.f.)   [1]

9   Connect the LED to a variable supply.   [1]

    A safety resistor is also connected in series with the LED.   [1]

    Connect a voltmeter across the LED.   [1]

    Slowly increase the p.d. across the LED until it just emits light. Record the p.d. V across the LED.   [1]

    The Planck constant is calculated using $eV = hc/\lambda$ or $h = eV\lambda/c$, where e is the alimentary charge and c is the speed of light in a vacuum.   [1]

    Improvement: Carry out the experiment in a dark room or place a black tube over the LED to judge when the LED just starts to emit light.   [1]

13.2

1   Become (positively) charged   [1]

    Gold leaf would rise/move away from the stem   [1]

2   a   No emission [1] Infrared photons are below threshold frequency (have insufficient energy)   [1]

   b   Emission of photoelectrons as blue photons are above threshold frequency   [1]

3   Energy transferred to each electron comes from a single photon in a one-to-one interaction.   [1]

    Energy of each photon depends on its frequency ($E = hf$).   [1]

    Greater the frequency, the higher the energy of the photon and so the greater the maximum kinetic energy of the electron.   [1]

4   Maximum [1] wavelength that would cause photoelectric emission from the surface of a metal. [1]

5   Increased emission   [1]

    Number of emitted electrons per second would quadruple.   [1]

    Quadrupling the intensity results in four times the number of photons; therefore, four times the number of electrons emitted per second.   [1]

13.3

1. $hf = \phi + KE_{MAX} = 3.77 \times 10^{-19} + 2.68 \times 10^{-19}$ [1]

   $hf = 6.45 \times 10^{-19}$ J [1]

2. a. $KE_{MAX} = hf - \phi = 5.20 - 4.08 = 1.12$ eV [1]

   b. $hf < \phi$ [1] therefore no electrons emitted [1]

3. $hf_o = \phi$ therefore $f_o = \dfrac{\phi}{h}$

   Zinc: $\phi = 4.30$ eV $= 6.88 \times 10^{-19}$ J [1]

   Sodium: $\phi = 2.36$ eV $= 3.776 \times 10^{-19}$ J [1]

   Zinc: $f_o = \dfrac{6.88 \times 10^{-19}}{6.63 \times 10^{-34}} = 1.04 \times 10^{15}$ Hz (3 s.f.) [1]

   Sodium: $f_o = \dfrac{3.776 \times 10^{-19}}{6.63 \times 10^{-34}} = 5.70 \times 10^{14}$ Hz (3 s.f.) [1]

4. With monochromatic radiation all photons have the same frequency and therefore the same energy. [1]

   Each electron requires a specific energy to free it from the surface of the metal (the lowest energy required to free an electron is equal to the work function of the metal). [1]

   The kinetic energy of the emitted electron is a result of the remaining energy after the electron has been freed. [1]

5. $hf = \phi + KE_{MAX}$ therefore $KE_{MAX} = hf - \phi$

   Sodium: $\phi = 2.36$ eV $= 3.78 \times 10^{-19}$ J

   $KE_{MAX} = 6.63 \times 10^{-34} \times 1.48 \times 10^{15} - 3.78 \times 10^{-19}$

   $= 6.03 \times 10^{-19}$ J (3 s.f.) [1]

   $KE_{MAX} = \dfrac{1}{2}mv^2$ therefore $v = \sqrt{\dfrac{2 \times KE_{MAX}}{m}}$ [1]

   $v = \sqrt{\dfrac{2 \times 6.03 \times 10^{-19}}{9.11 \times 10^{-31}}}$ [1]

   $v = 1.15$ Mm s^{-1} (3 s.f.) [1]

6. $hf = \phi + KE_{MAX}$ therefore $\phi = hf - KE_{MAX} = (6.63 \times 10^{-34}) \times (8.0 \times 10^{14}) - (1.36 \times 1.60 \times 10^{-19})$ [1]

   $\phi = 3.128 \times 10^{-19} = 3.1 \times 10^{-19}$ J (2 s.f.) [1]

   $hf_0 = \phi$ therefore $f_0 = \dfrac{\phi}{h} = \dfrac{3.128 \times 10^{-19}}{6.63 \times 10^{-34}}$ [1]

   $f_0 = 4.7 \times 10^{14}$ Hz (2 s.f.) [1]

13.4

1. a. It decreases.

   b. It decreases by a factor of 1 over $\sqrt{3}$.

   c. It increases by a factor of $\sqrt{10}$.

2. a. 2.03×10^{-12} m

   b. 5.01×10^{-11} m

3. 1.23×10^{-10} m

1. Adjust the accelerating p.d. This changes the velocity/momentum of the electron and therefore its wavelength.

2. They are too penetrating. They may damage the material.

3. $E_K = 40$ eV $= 6.4 \times 10^{-18}$ J

   $\varepsilon_K = \dfrac{1}{2}mv^2$ therefore

   $v = \sqrt{\dfrac{2 \times \varepsilon_K}{m}} = \sqrt{\dfrac{2 \times 6.4 \times 10^{-18}}{9.11 \times 10^{-31}}} = 3.7 \times 10^6$ m s^{-1}

   $\lambda = \dfrac{h}{p}$ as momentum, $p = mv$ we can say $\lambda = \dfrac{h}{mv}$

   $\lambda = \dfrac{6.63 \times 10^{-34}}{9.11 \times 10^{-31} \times 3.7 \times 10^6} = 2.0 \times 10^{-10}$ m (2 s.f.)

   Similar size to the space between the atoms in the crystal, resulting in significant diffraction.

1. $\lambda = \dfrac{h}{p}$ [1]

   $\lambda = \dfrac{6.63 \times 10^{-34}}{1.67 \times 10^{-19}} = 3.97 \times 10^{-15}$ m (3 s.f.) [1]

2. The proton has a greater mass [1]

   Therefore, at the same velocity the momentum of the proton is greater than the electron [1]

   As $\lambda = \dfrac{h}{p}$ the wavelength of the proton must be smaller [1]

3. Wavelength of the electron is very small [1]

   In order to observe diffraction the electron must pass through a gap a similar size to its wavelength [1]

   This does not happen in the course of most experiments using electrons [1]

4. a. $v = \dfrac{h}{\lambda m} = \dfrac{6.63 \times 10^{-34}}{3.63 \times 10^{-10} \times 9.11 \times 10^{-31}}$ [1]

   $v = 2.00 \times 10^6$ m s^{-1} [1]

   b. $v = \dfrac{h}{\lambda m} = \dfrac{6.63 \times 10^{-34}}{4.85 \times 10^{-12} \times 9.11 \times 10^{-31}}$ [1]

   $v = 150 \times 10^6$ m s^{-1} [1]

5. a. $\lambda = \dfrac{h}{p}$ as momentum, $p = mv$ therefore $\lambda = \dfrac{h}{mv}$

   $\lambda = \dfrac{6.63 \times 10^{-34}}{9.11 \times 10^{-31} \times 4.20 \times 10^7}$ [1]

   $\lambda = 1.73 \times 10^{-11}$ m [1]

   b. $v = \dfrac{h}{\lambda m} = \dfrac{6.63 \times 10^{-34}}{1.73 \times 10^{-11} \times 1.67 \times 10^{-27}}$ [1]

   $v = 22.9$ km s^{-1} (3 s.f.) [1]

6. $0.25\,c = 0.25 \times 3.00 \times 10^8 = 75 \times 10^6$ m s^{-1} [1]

   $\lambda = \dfrac{h}{mv} = \dfrac{6.63 \times 10^{-34}}{9.11 \times 10^{-31} \times 75 \times 10^6}$

   $= 9.7 \times 10^{-12}$ m (2 s.f.) [1]

Index

Acknowledgements

p2-3: Sarahbean/Shutterstock; **p6-7**: Martyn F. Chillmaid/ Science Photo Library; **p8**: Nasa/Vrs/Science Photo Library; **p10**: Tr3gin/Shutterstock; **p12**: Aleksandar Todorovic/ Shutterstock; **p14**: Piotr Wawrzyniuk/Shutterstock; **p16**: Roger Bamber/Alamy; **p18**: US Coast Guard Photo/Alamy; **p20-21**: OlegDoroshin/Shutterstock; **p22**: Paul White - Transport Infrastructures/Alamy; **p24**: Bikeriderlondon/Shutterstock; **p27** (T): Villiers Steyn/Shutterstock; **p27** (B): Christoff/Shutterstock; **p29**: Pavel Polkovnikov/Shutterstock; **p31**: Denis Tabler/ Shutterstock; **p32**: Peteri/Shutterstock; **p35** (T): Bob Mawby/ Shutterstock; **p35** (B): SP-Photo/Shutterstock; **p37**: Ria Novosti/ Science Photo Library; **p38**: Kenneth Eward/Biografx/Science Photo Library; **p40**: Dick Kenny/Shutterstock; **p46**: Andrew Brookes/National Physical Laboratory/Science Photo Library; **p47**: NASA; **p49** (T): Heidi Coppock-Beard/Getty Images; **p49** (B): Don Hammond/Design Pics/Alamy; **p51**: Greg Epperson/ Shutterstock; **p52**: Greg Epperson/Shutterstock; **p54** (T): Robert L Kothenbeutel/Shutterstock; **p54** (B): Justin Kase z12z/Alamy; **p57**: Awe Inspiring Images/Shutterstock; **p60**: Lassedesignen/ Shutterstock; **p62**: Monkey Business Images/Shutterstock; **p65**: Lawrence Livermore National Laboratory/ University Of California/Science Photo Library; **p66**: Rob Hyrons/ Shutterstock; **p67**: Alexis Rosenfeld/Science Photo Library; **p69** (T): Mevans/iStockphoto; **p69** (B): GARY DOAK/Alamy; **p72**: Baloncici/iStockphoto; **p74** (T): Mar.K/Shutterstock; **p74** (B): Jesus Keller/Shutterstock; **p75**: John Hanley/ Shutterstock; **p76**: Tom Hirtreiter/Shutterstock; **p77**: Germanskydiver/Shutterstock; **p79**: Pearl Bucknall/Alamy; **p80**: Bikeriderlondon/Shutterstock; **p82**: David Parker/Science Photo Library; **p84** (T): Danlsaunders/iStockphoto; **p84** (B): Charistoone-images/Alamy; **p86**: Peter Marshall/Alamy; **p88**: Irina Papoyan/Shutterstock; **p90**: Melis/Shutterstock; **p92**: Hellen Grig/Shutterstock; **p96**: Nitinut380/Shutterstock; **p100**: NASA/Science Photo Library; **p101**: Iliuta Goean/Shutterstock; **p102**: NASA/Science Photo Library; **p103**: Ken Schulze/ Shutterstock; **p104**: RobertCrum/iStockphoto; **p105**: fStop Images - Caspar Benson/Getty Images; **p107**: Loren Winters, Visuals Unlimited/Science Photo Library; **p109**: Science Photo Library; **p115** (T): Weber/Shutterstock; **p115** (B): Sunny Forest/ Shutterstock; **p116-117**: Matteis/Look at Sciences/Science Photo Library; **p118**: Jhaz Photography/Shutterstock; **p122**: Sergey Nivens/Shutterstock; **p123**: Mihai Simonia/Shutterstock; **p124**: Jon Le-Bon/Shutterstock; **p125**: CERN/Science Photo Library; **p127** (T): Nobeastsofierce/Shutterstock; **p127** (B): Oktay Ortakcioglu/iStockphoto; **p130**: Iceninephoto/iStockphoto; **p134**: Science Photo Library; **p135**: Trevor Clifford Photography/Science Photo Library; **p136**: Peter Menzel/Science Photo Library; **p138** (T): Dino Osmic/Shutterstock; **p138** (B): Martyn F. Chillmaid/Science Photo Library; **p139**: Gavran333/ Shutterstock; **p141**: Francisco Javier Gil/Shutterstock; **p144**: Power and Syred/Science Photo Library; **p147** (T): Alexandru Nika/Shutterstock; **p147** (B): Science Source/Science Photo Library; **p149**: Idea for life/Shutterstock; **p152**: David Parker/ Science Photo Library; **p153** (T): Spfotocz/Shutterstock; **p153** (B): Martyn F. Chillmaid/Science Photo Library; **p156**: Martyn F. Chillmaid/Science Photo Library; **p157**: NASA/JPL-Caltech/ Science Photo Library; **p158**: Prill/Shutterstock; **p159**: StockPhotosArt/Shutterstock; **p161** (T): Pi-Lens/Shutterstock; **p161** (B): Martyn F. Chillmaid/Science Photo Library; **p162**: Martyn F. Chillmaid/Science Photo Library; **p166**: Jim Corwin/ Science Photo Library; **p173**: S Corvaja/European Space Agency/ Science Photo Library; **p177**: Public Health England/Science Photo Library; **p182**: Paulthepunk/iStockphoto;

p186: Trevor Clifford Photography/Science Photo Library; **p190** (T): Xieyouding/iStockphoto; **p190** (B): Jag_cz/ Shutterstock; **p193**: Furtseff/Shutterstock; **p201**: Philippe Plailly/Science Photo Library; **p202**: Adrian Davies/Alamy; **p203**: M-gucci/iStockphoto; **p206**: American Spirit/ Shutterstock; **p213**: 123dartist/Shutterstock; **p214**: Asharkyu/ Shutterstock; **p220** (T): Mr Twister/Shutterstock; **p220** (B): Berenice Abbott/Science Photo Library; **p223**: TFoxFoto/ Shutterstock; **p226**: Edward Kinsman/Science Photo Library; **p227**: Russell Shively/Shutterstock; **p230**: Ad_doward/ iStockphoto; **p231**: Andrew Lambert Photography/Science Photo Library; **p233**: Aastock/Shutterstock; **p238**: PR Michel Zanca/ISM/Science Photo Library; **p243**: European Space Agency/P. Carril/Science Photo Library; **p247**: PobladuraFCG/ iStockphoto; **p250**: Evan Oto/Science Photo Library; **p251**: Andrew Lambert Photography/Science Photo Library; **p253**: Science Photo Library; **p257**: General-fmv/Shutterstock; **p266** (T): Mino Surkala/Shutterstock; **p266** (B): Tim Roberts Photography/Shutterstock; **p267**: Gecko753/iStockphoto; **p270** (T): Dotshock/Shutterstock; **p270** (B): Charistoone-stock/Alamy